The Body

ALSO BY BILL BRYSON

The Lost Continent
The Mother Tongue
Neither Here Nor There
Made in America
Notes from a Small Island
A Walk in the Woods
I'm a Stranger Here Myself
In a Sunburned Country
Bryson's Dictionary of Troublesome Words
Bill Bryson's African Diary
A Short History of Nearly Everything
A Short History of Nearly Everything: Special Illustrated Edition
The Life and Times of the Thunderbolt Kid
Shakespeare: The World as Stage
Bryson's Dictionary for Writers and Editors
At Home: A Short History of Private Life
At Home: A Short History of Private Life: Illustrated Edition
One Summer
The Road to Little Dribbling: Adventures of an American in Britain

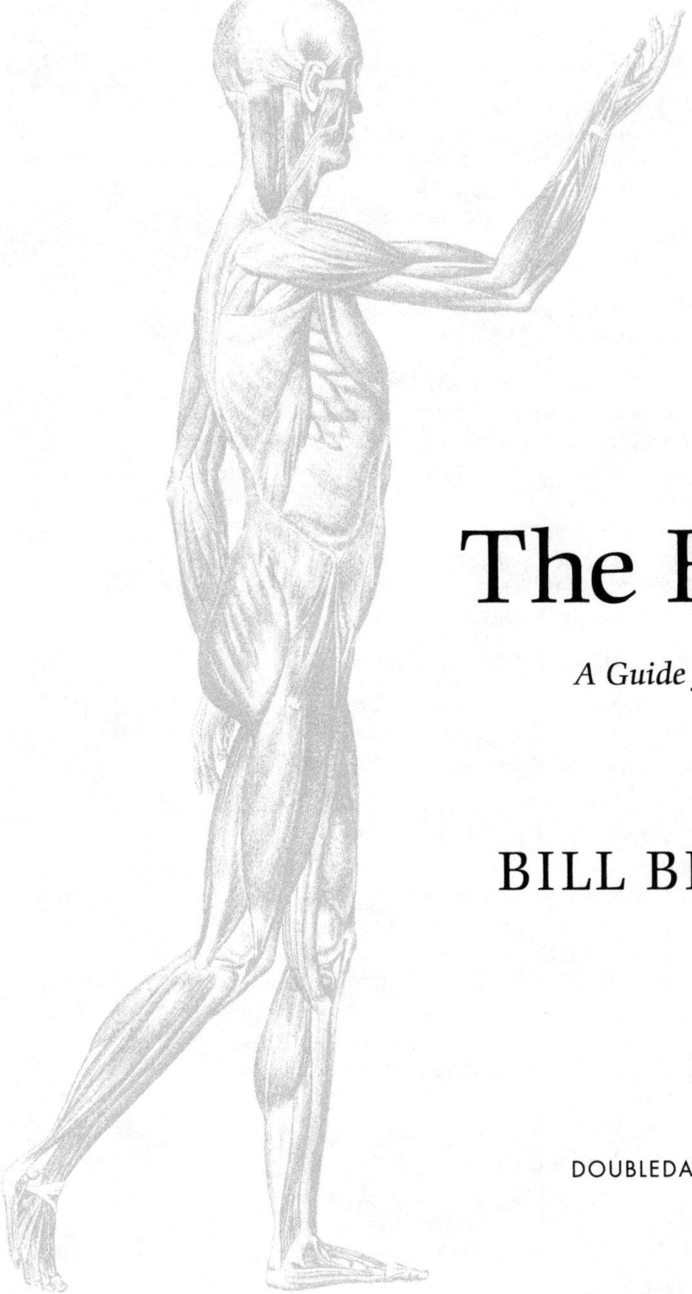

The Body

A Guide for Occupants

BILL BRYSON

DOUBLEDAY / NEW YORK

www.doubleday.com

Grateful acknowledgment is made to Harvard University Press for permission to reprint excerpts from *The Poems of Emily Dickinson: Variorum Edition*, edited by Ralph W. Franklin, Cambridge, Mass.: The Belknap Press of Harvard University Press. Copyright © 1998 by the President and Fellows of Harvard College. Copyright © 1951, 1955 by the President and Fellows of Harvard College, copyright renewed 1979, 1983 by the President and Fellows of Harvard College. Copyright © 1914, 1918, 1919, 1924, 1929, 1930, 1932, 1935, 1937, 1942 by Martha Dickinson Bianchi. Copyright © 1952, 1957, 1958, 1963, 1965 by Mary L. Hampson.

Pages 449–50 constitute an extension of this copyright page.

Book design by Michael Collica
Jacket design by John Gall
Jacket composite image by Aleksandr Andreev and robuart, both Shutterstock

Library of Congress Cataloging-in-Publication Data
Names: Bryson, Bill, author.
Title: The body: a guide for occupants / Bill Bryson.
Description: First edition. | New York: Doubleday, 2019. | Includes bibliographical references and index.
Identifiers: LCCN 2019012407 | ISBN 9780385539302 (hardcover) | ISBN 9780385539319 (ebook)
Subjects: LCSH: Human anatomy. | Human physiology. | BISAC: SCIENCE / Life Sciences / Human Anatomy & Physiology. | MEDICAL / Anatomy. | HUMOR / General.
Classification: LCC QM23.2 .B79 2019 | DDC 612–dc23
LC record available at https://lccn.loc.gov/2019012407

MANUFACTURED IN THE UNITED STATES OF AMERICA

9 8 7 6 5 4 3 2 1

First United States Edition

To Lottie. Welcome to you, too.

CONTENTS

The Body

1 HOW TO BUILD A HUMAN

How like a god!

—WILLIAM SHAKESPEARE

LONG AGO, WHEN I was a junior high school student in Iowa, I remember being taught by a biology teacher that all the chemicals that make up a human body could be bought in a hardware store for $5.00 or something like that. I don't recall the actual sum. It might have been $2.97 or $13.50, but it was certainly very little even in 1960s money, and I remember being astounded at the thought that you could make a slouched and pimply thing such as me for practically nothing.

It was such a spectacularly humbling revelation that it has stayed with me all these years. The question is, was it true? Are we really worth so little?

Many authorities (for which possibly read "science majors who don't have a date on a Friday") have tried at various times, mostly for purposes of amusement, to compute how much it would cost in materials to build a human. Perhaps the most respectable and comprehensive attempt of recent years was done by Britain's Royal Society of Chemistry when, as part of the 2013 Cambridge Science Festival, it calculated how much it would cost to assemble all the elements necessary to build the actor Benedict Cumberbatch. (Cumberbatch was the guest director of the festival that year and was, conveniently, a typically sized human.)

Altogether, according to RSC calculations, fifty-nine elements are needed to construct a human being. Six of these—carbon, oxygen, hydrogen, nitrogen, calcium, and phosphorus—account for 99.1 percent of what makes us, but much of the rest is a bit unexpected. Who would have thought that we would be incomplete without some molybdenum inside us, or vanadium, manganese, tin, and copper? Our requirements for some of these, it must be said, are surpassingly modest and are measured in parts per million or even parts per billion. We need, for instance, just 20 atoms of cobalt and 30 of chromium for every 999,999,999½ atoms of everything else.

The biggest component in any human, filling 61 percent of available space, is oxygen. It may seem a touch counterintuitive that we are almost two-thirds composed of an odorless gas. The reason we are not light and bouncy like a balloon is that the oxygen is mostly bound up with hydrogen (which accounts for another 10 percent of you) to make water—and water, as you will know if you have ever tried to move a wading pool or just walked around in really wet clothes, is surprisingly heavy. It is a little ironic that two of the lightest things in nature, oxygen and hydrogen, when combined form one of the heaviest, but that's nature for you. Oxygen and hydrogen are also two of the cheaper elements within you. All your oxygen will set you back just $14 and your hydrogen a little over $26 (assuming you are about the size of Benedict Cumberbatch). Your nitrogen (2.6 percent of you) is a better value still at just forty cents for a body's worth. But after that it gets pretty expensive.

You need about thirty pounds of carbon, and that will cost you $69,550, according to the Royal Society of Chemistry. (They were using only the most purified forms of everything. The RSC would not make a human with cheap stuff.) Calcium, phosphorus, and potassium, though needed in much smaller amounts, would between them set you back a further $73,800. Most of the rest is even more expensive per unit of volume, but fortunately only needed in microscopic amounts.

Thorium costs over $3,000 per gram but constitutes just 0.0000001 percent of you, so you can buy a body's worth for thirty-three cents. All the tin you require can be yours for six cents, while zirconium and niobium will cost you just three cents apiece. The 0.000000007 percent of you that is samarium isn't apparently worth charging for at all. It's logged in the RSC accounts as costing $0.00.*

Of the fifty-nine elements found within us, twenty-four are traditionally known as essential elements, because we really cannot do without them. The rest are something of a mixed bag. Some are clearly beneficial, some may be beneficial but we are not sure in what ways yet, others are neither harmful nor beneficial but are just along for the ride as it were, and a few are just bad news altogether. Cadmium, for instance, is the twenty-third most common element in the body, constituting 0.1 percent of your bulk, but it is seriously toxic. We have it in us not because our body craves it but because it gets into plants from the soil and then into us when we eat the plants. If you are from North America, you probably ingest about eighty micrograms of cadmium a day, and no part of it does you any good at all.

A surprising amount of what goes on at this elemental level is still being worked out. Pluck almost any cell from your body, and it will have a million or more selenium atoms in it, yet until recently nobody had any idea what they were there for. We now know that selenium makes two vital enzymes, deficiency in which has been linked to hypertension, arthritis, anemia, some cancers, and even, possibly, reduced sperm counts. So, clearly it is a good idea to get some selenium inside you (it is found particularly in nuts, whole wheat bread, and fish), but at the same time if you take in too much you can irremediably poison your liver. As with so much in life, getting the balances right is a delicate business.

Altogether, according to the RSC, the full cost of building a new

* The RSC calculations were done in British pounds and have been converted here into U.S. dollars at the rate that prevailed in the summer of 2013 of £1 = $1.57.

human being, using the obliging Benedict Cumberbatch as a template, would be a very precise $151,578.46. Labor and sales tax would, of course, boost costs further. You would probably be lucky to get a take-home Benedict Cumberbatch for much under $300,000—not a massive fortune, all things considered, but clearly not the meager few dollars that my junior high school teacher suggested. That said, in 2012 *Nova,* the long-running science program on PBS, did an exactly equivalent analysis for an episode called "Hunting the Elements" and came up with a figure of $168 for the value of the fundamental components within the human body, illustrating a point that will become inescapable as this book goes on, namely that where the human body is concerned, the details are often surprisingly uncertain.

But of course it hardly really matters. No matter what you pay, or how carefully you assemble the materials, you are not going to create a human being. You could call together all the brainiest people who are alive now or have ever lived and endow them with the complete sum of human knowledge, and they could not between them make a single living cell, never mind a replicant Benedict Cumberbatch.

That is unquestionably the most astounding thing about us—that we are just a collection of inert components, the same stuff you would find in a pile of dirt. I've said it before in another book, but I believe it's worth repeating: the only thing special about the elements that make you is that they make you. That is the miracle of life.

We pass our existence within this warm wobble of flesh and yet take it almost entirely for granted. How many among us know even roughly where the spleen is or what it does? Or the difference between tendons and ligaments? Or what our lymph nodes are up to? How many times a day do you suppose you blink? Five hundred? A thousand? You've no idea, of course. Well, you blink fourteen thousand times a day—so much that your eyes are shut for twenty-three minutes of every waking day. Yet you never have to think about it, because every second of every day your body undertakes a literally unquantifiable number of tasks—a quadrillion, a nonillion, a quindecillion, a vigintillion (these are actual measures), at all events some number vastly beyond imagining—without requiring an instant of your attention.

In the second or so since you started this sentence, your body has made a million red blood cells. They are already speeding around you, coursing through your veins, keeping you alive. Each of those red blood cells will rattle around you about 150,000 times, repeatedly delivering oxygen to your cells, and then, battered and useless, will present itself to other cells to be quietly killed off for the greater good of you.

Altogether it takes 7 billion billion billion (that's 7,000,000,000, 000,000,000,000,000,000, or 7 octillion) atoms to make you. No one can say why those 7 billion billion billion have such an urgent desire to be you. They are mindless particles, after all, without a single thought or notion between them. Yet somehow for the length of your existence, they will build and maintain all the countless systems and structures necessary to keep you humming, to make you you, to give you form and shape and let you enjoy the rare and supremely agreeable condition known as life.

That's a much bigger job than you realize. Unpacked, you are positively enormous. Your lungs, smoothed out, would cover a tennis court, and the airways within them would stretch nearly from coast to coast. The length of all your blood vessels would take you two and a half times around Earth. The most remarkable part of all is your DNA (or deoxyribonucleic acid). You have a meter of it packed into every cell, and so many cells that if you formed all the DNA in your body into a single strand, it would stretch ten billion miles, to beyond Pluto. Think of it: there is enough of you to leave the solar system. You are in the most literal sense cosmic.

But your atoms are just building blocks and are not themselves alive. Where life begins precisely is not so easy to say. The basic unit of life is the cell—everyone is agreed on that. The cell is full of busy things—ribosomes and proteins, DNA, RNA, mitochondria, and much other cellular arcana—but none of those are themselves alive. The cell itself is just a compartment–a kind of little room: a *cell*–to contain them, and of itself is as nonliving as any other room. Yet somehow when all of these things are brought together, you have life. That is the part that eludes science. I kind of hope it always will.

What is perhaps most remarkable is that nothing is in charge. Each component of the cell responds to signals from other components, all of them bumping and jostling like so many bumper cars, yet somehow all this random motion results in smooth, coordinated action, not just across the cell but across the whole body as cells communicate with other cells in different parts of your personal cosmos.

The heart of the cell is the nucleus. It contains the cell's DNA–three feet of it, as we have already noted, scrunched into a space that we may reasonably call infinitesimal. The reason so much DNA can fit into a cell nucleus is that it is exquisitely thin. You would need twenty billion strands of DNA laid side by side to make the width of the finest human hair. Every cell in your body (strictly speaking, every cell with a nucleus) holds two copies of your DNA. That's why you have enough to stretch to Pluto and beyond.

DNA exists for just one purpose–to create more DNA. A DNA molecule, as you will almost certainly remember from countless television programs if not school biology, is made up of two strands, connected by rungs to form the celebrated twisted ladder known as a double helix. Your DNA is simply an instruction manual for making you. A length of DNA is divided into segments called chromosomes and shorter individual units called genes. The sum of all your genes is the genome.

DNA is extremely stable. It can last for tens of thousands of years. It is nowadays what enables scientists to work out the anthropology of the very distant past. Probably nothing you own right now–no letter or piece of jewelry or treasured heirloom–will still exist a thousand years from now, but your DNA will almost certainly still be around and recoverable, if only someone could be bothered to look for it.

DNA passes on information with extraordinary fidelity. It makes only about one error per every billion letters copied. Still, because your cells divide so much, that is about three errors, or mutations, per cell division. Most of those mutations the body can ignore, but just occasionally they have lasting significance. That is evolution.

All of the components of the genome have one single-minded purpose—to keep the line of your existence going. It's a slightly humbling thought that the genes you carry are immensely ancient and possibly—so far anyway—eternal. You will die and fade away, but your genes will go on and on so long as you and your descendants continue to produce offspring. And it is surely astounding to reflect that not once in the three billion years since life began has your personal line of descent been broken. For you to be here now, every one of your ancestors had to successfully pass on its genetic material to a new generation before being snuffed out or otherwise sidetracked from the procreative process. That's quite a chain of success.

What genes specifically do is provide instructions for building proteins. Most of the useful things in the body are proteins. Some speed up chemical changes and are known as enzymes. Others convey chemical messages and are known as hormones. Still others attack pathogens and are called antibodies. The largest of all our proteins is called titin, which helps to control muscle elasticity. Its chemical name is 189,819 letters long, which would make it the longest word in the English language except that dictionaries don't recognize chemical names. Nobody knows how many types of proteins there are within us, but estimates range from a few hundred thousand to a million or more.

The paradox of genetics is that we are all very different and yet genetically practically identical. All humans share 99.9 percent of their DNA, and yet no two humans are alike. My DNA and your DNA will differ in three to four million places, which is a small proportion of the total but enough to make a lot of difference between us. You also have within you about a hundred personal mutations—stretches of genetic instructions that don't quite match any of the genes given to you by either of your parents but are yours alone.

How all this works in detail is still largely a mystery to us. Only

2 percent of the human genome codes for proteins, which is to say only 2 percent does anything demonstrably and unequivocally practical. Quite what the rest is doing isn't known. A lot of it, it seems, is just there, like freckles on skin. Some of it makes no sense. One particular short sequence, called an Alu element, is repeated more than a million times throughout our genome, including sometimes in the middle of important protein-coding genes. It is complete gibberish, as far as anyone can tell, yet it constitutes 10 percent of all our genetic material. No one has any idea why. The mysterious part was for a while called junk DNA but now is more graciously called dark DNA, meaning that we don't know what it does or why it is there. Some is involved in regulating the genes, but much of the rest remains to be determined.

The body is often likened to a machine, but it is so much more than that. It works twenty-four hours a day for decades without (for the most part) needing regular servicing or the installation of spare parts, runs on water and a few organic compounds, is soft and rather lovely, is accommodatingly mobile and pliant, reproduces itself with enthusiasm, makes jokes, feels affection, appreciates a red sunset and a cooling breeze. How many machines do you know that can do any of that? There is no question about it. You are truly a wonder. But then so, it must be said, is an earthworm.

And how do we celebrate the glory of our existence? Well, for most of us by eating maximally and exercising minimally. Think of all the junk you throw down your throat and how much of your life is spent sprawled in a near-vegetative state in front of a glowing screen. Yet in some kind and miraculous way our bodies look after us, extract nutrients from the miscellaneous foodstuffs we push into our faces, and somehow hold us together, generally at a pretty high level, for decades. Suicide by lifestyle takes ages.

Even when you do nearly everything wrong, your body maintains and preserves you. Most of us are testament to that in one way or another. Five out of every six smokers won't get lung cancer. Most of the people who are prime candidates for heart attacks don't get heart attacks. Every day, it has been estimated, between one and five of your cells turn cancerous, and your immune system captures and kills them. Think of that. A couple of dozen times a week, well over a thousand times a year, you get the most dreaded disease of our age, and each time your body saves you. Of course, very occasionally a cancer develops into something more serious and possibly kills you, but overall cancers are rare: most cells in the body replicate billions and billions of times without going wrong. Cancer may be a common cause of death, but it is not a common event in life.

Our bodies are a universe of 37.2 trillion cells operating in more or less perfect concert more or less all the time.* An ache, a twinge of indigestion, the odd bruise or pimple, are about all that in the normal course of things announces our imperfectability. There are thousands of things that can kill us–slightly more than eight thousand, according to the *International Statistical Classification of Diseases and Related Health Problems* compiled by the World Health Organization–and we escape every one of them but one. For most of us, that's not a bad deal.

We are not perfect by any means, goodness knows. We get impacted molars because we have evolved jaws too small to accommodate all the teeth we are endowed with. We have pelvises too small to pass children without excruciating pain. We are hopelessly susceptible to backache. We have organs that mostly cannot repair themselves. If a zebra fish damages its heart, it grows new tissue. If you damage your heart, well, too bad. Nearly all animals produce their own vitamin C, but we can't. We undertake every part of the process except, inexplicably, the last step, the production of a single enzyme.

The miracle of human life is not that we are endowed with some frailties but that we aren't swamped with them. Don't forget that your genes come from ancestors who most of the time weren't even human. Some of them were fish. Lots more were tiny and furry

* That number is of course an educated guess. Human cells come in a variety of types, sizes, and densities and are literally uncountable. The figure of 37.2 trillion was arrived at in 2013 by a team of European scientists led by Eva Bianconi from the University of Bologna in Italy and was reported in the *Annals of Human Biology*.

and lived in burrows. These are the beings from whom you have inherited your body plan. You are the product of three billion years of evolutionary tweaks. We would all be a lot better off if we could just start fresh and give ourselves bodies built for our particular *Homo sapien* needs—to walk upright without wrecking our knees and backs, to swallow without the heightened risk of choking, to dispense babies as if from a vending machine. But we weren't built for that. We began our journey through history as unicellular blobs floating about in warm, shallow seas. Everything since then has been a long and interesting accident, but a pretty glorious one, too, as I hope the following pages make clear.

2 THE OUTSIDE: SKIN AND HAIR

Beauty is only skin deep, but ugly
goes clean to the bone.

—DOROTHY PARKER

I

IT MAY BE slightly surprising to think it, but our skin is our largest organ, and possibly the most versatile. It keeps our insides in and bad things out. It cushions blows. It gives us our sense of touch, bringing us pleasure and warmth and pain and nearly everything else that makes us vital. It produces melanin to shield us from the sun's rays. It repairs itself when we abuse it. It accounts for such beauty as we can muster. It looks after us.

The formal name for the skin is the cutaneous system. Its size is about two square meters (approximately twenty square feet), and all told your skin will weigh somewhere in the region of ten to fifteen pounds, though much depends, naturally, on how tall you are and how much buttock and belly it needs to stretch across. It is thinnest on the eyelids (just one-thousandth of an inch thick) and thickest on the heels of our hands and feet. Unlike a heart or a kidney, skin never fails. "Our seams don't burst, we don't spontaneously sprout leaks," says Nina Jablonski, professor of anthropology at Penn State University, who is the doyenne of all things cutaneous.

The skin consists of an inner layer called the dermis and an outer epidermis. The outermost surface of the epidermis, called the stratum corneum, is made up entirely of dead cells. It is an arresting thought that all that makes you lovely is deceased. Where body meets air, we are all cadavers. These outer skin cells are replaced every month. We shed skin copiously, almost carelessly: some twenty-five thousand flakes a minute, over a million pieces every hour. Run a finger along a dusty shelf, and you are in large part clearing a path through fragments of your former self. Silently and remorselessly we turn to dust.

Skin flakes are properly called squamae (meaning "scales"). We each trail behind us about a pound of dust every year. If you burn the contents of a vacuum cleaner bag, the predominant odor is that unmistakable scorched smell that we associate with burning hair. That's because skin and hair are made largely of the same stuff: keratin.

Beneath the epidermis is the more fertile dermis, where reside all the skin's active systems—blood and lymph vessels, nerve fibers, the roots of hair follicles, the glandular reservoirs of sweat and sebum. Beneath that, and not technically part of the skin, is a subcutaneous layer where fat is stored. Though it may not be part of the cutaneous system, it's an important part of your body because it stores energy, provides insulation, and attaches the skin to the body beneath.

Nobody knows for sure how many holes you have in your skin, but you are pretty seriously perforated. Most estimates suggest you have somewhere in the region of two to five million hair follicles and perhaps twice that number of sweat glands. The follicles do double duty: they sprout hairs and secrete sebum (from sebaceous glands), which mixes with sweat to form an oily layer on the surface. This helps to keep skin supple and to make it inhospitable for many foreign organisms. Sometimes the pores become blocked with little plugs of dead skin and dried sebum in what is known as a blackhead. If the follicle additionally becomes infected and inflamed, the result is the adolescent dread known as a pimple. Pimples plague young people simply because their sebaceous glands—like all their glands—are highly active. When the condition becomes chronic, the result is acne, a word of very uncertain derivation. It appears to be related to the Greek *acme,* denoting a high and admirable achievement, which a faceful of pimples most assuredly is not. How the two became twinned is not at all clear. The term first appeared in English in 1743 in a British medical dictionary.

Also packed into the dermis are a variety of receptors that keep us literally in touch with the world. If a breeze plays lightly on your cheek, it is your Meissner's corpuscles that let you know.* When you put your hand on a hot plate, your Ruffini corpuscles cry out. Merkel cells respond to constant pressure, Pacinian corpuscles to vibration.

Meissner's corpuscles are everyone's favorites. They detect light touch and are particularly abundant in our erogenous zones and other areas of heightened sensitivity: fingertips, lips, tongue, clitoris, penis, and so on. They are named after a German anatomist, Georg Meissner, who is credited with discovering them in 1852, though his colleague Rudolf Wagner claimed that he in fact was the discoverer. The two men fell out over the matter, proving that there is no detail in science too small for animosity.

All are exquisitely fine-tuned to let you feel the world. A Pacinian corpuscle can detect a movement as slight as 0.00001 millimeter, which is practically no movement at all. More than this, they don't even require contact with the material they are interpreting. As David J. Linden points out in *Touch,* if you sink a spade into gravel or sand, you can feel the difference between them even though all you are touching is the spade. Curiously, we don't have any receptors for wetness. We have only thermal sensors to guide us, which is why when you sit down on a wet spot, you can't generally tell whether it really is wet or just cold.

Women are much better than men at tactile sensitivity with fingers, but possibly just because they have smaller hands and thus

* "Corpuscle," from the Latin, meaning "little body," is a somewhat vague term anatomically speaking. It can signify either unattached, free-floating cells, as in blood corpuscles, or it can signify clumps of cells that function independently, as with Meissner's corpuscles.

a more dense network of sensors. An interesting thing about touch is that the brain doesn't just tell you how something feels, but how it *ought* to feel. That's why the caress of a lover feels wonderful, but the same touch by a stranger would feel creepy or horrible. It's also why it is so hard to tickle yourself.

One of the most memorably unexpected events I experienced in the course of doing this book came in a dissection room at the University of Nottingham in England when a professor and surgeon named Ben Ollivere (about whom much more in due course) gently incised and peeled back a sliver of skin about a millimeter thick from the arm of a cadaver. It was so thin as to be translucent. "That," he said, "is where all your skin color is. That's all that race is—a sliver of epidermis."

I mentioned this to Nina Jablonski when we met in her office in State College, Pennsylvania, soon afterward. She gave a nod of vigorous assent. "It is extraordinary how such a small facet of our composition is given so much importance," she said. "People act as if skin color is a determinant of character when all it is is a reaction to sunlight. Biologically, there is actually no such thing as race—nothing in terms of skin color, facial features, hair type, bone structure, or anything else that is a defining quality among peoples. And yet look how many people have been enslaved or hated or lynched or deprived of fundamental rights through history because of the color of their skin."

A tall, elegant woman with silvery hair cut short, Jablonski works in a very tidy office on the fourth floor of the anthropology building on the Penn State campus, but her interest in skin came about almost thirty years ago when she was a young primatologist and paleobiologist at the University of Western Australia in Perth. While preparing a lecture on the differences between primate skin color and human skin color, she realized there was surprisingly little information on the subject and embarked on what has become a lifelong study. "What began as a small, fairly innocent project ended up taking over a big part of my professional life," she says. In 2006, she produced the highly regarded *Skin: A Natural History* and followed that six years later with *Living Color: The Biological and Social Meaning of Skin Color.*

Skin color turned out to be more scientifically complicated than anyone imagined. "Over 120 genes are involved in pigmentation in mammals," says Jablonski, "so it is really hard to unpack it all." What we can say is this: skin gets its color from a variety of pigments, of which the most important by far is a molecule formally called eumelanin but known universally as melanin. It is one of the oldest molecules in biology and is found throughout the living world. It doesn't just color skin. It gives birds the color of their feathers, fish the texture and luminescence of their scales, squid the purply blackness of their ink. It is even involved in making fruits go brown. In us, it also colors our hair. Its production slows dramatically as we age, which is why older people's hair tends to turn gray.

"Melanin is a superb natural sunscreen," says Jablonski. "It is produced in cells called melanocytes. All of us, whatever our race, have the same number of melanocytes. The difference is in the amount of melanin produced." Melanin often responds to sunlight in a literally patchy way, resulting in freckles, which are technically known as ephelides.

Skin color is a classic example of what is known as convergent evolution—that is, similar outcomes that have evolved in two or more locations. The people of, say, Sri Lanka and Polynesia have light brown skin not because of any direct genetic link but because they independently evolved brown skin to deal with the conditions of where they lived. It used to be thought that depigmentation probably took perhaps ten thousand to twenty thousand years, but now thanks to genomics we know it can happen much more quickly—in probably just two or three thousand years. We also know that it has happened repeatedly. Light-colored skin—"de-pigmented skin," as Jablonski calls it—has evolved at least three times on Earth. The lovely range of hues humans boast is an ever-changing process. "We are," as Jablonski puts it, "in the middle of a new experiment in human evolution."

It has been suggested that light skin may be a consequence of human migration and the rise of agriculture. The argument is that hunter-gatherers got a lot of their vitamin D from fish and game and that these inputs fell sharply when people started growing crops, especially as they moved into northern latitudes. It therefore became a great advantage to have lighter skin, to synthesize extra vitamin D.

Vitamin D is vital to health. It helps to build strong bones and teeth, boosts the immune system, fights cancers, and nourishes the heart. It is thoroughly good stuff. We can get it in two ways–from the foods we eat or through sunlight. The problem is that too much UV exposure damages DNA in our cells and can cause skin cancer. Getting the right amount is a tricky balance. Humans have addressed the challenge by evolving a range of skin tones to suit sunshine intensity at different latitudes. When a human body adapts to altered circumstances, the process is known as phenotypic plasticity. We alter our skin color all the time–when we tan or burn beneath a bright sun or blush from embarrassment. The red of sunburn is because the tiny blood vessels in the affected areas become engorged with blood, making the skin hot to the touch. The formal name for sunburn is erythema. Pregnant women frequently undergo a darkening of the nipples and areolae, and sometimes of other parts of the body such as the abdomen and face, as a result of increased production of melanin. The process is known as melasma, but its purpose is not understood. The flush we get when angry is a little counterintuitive. When the body is poised for a fight, it mostly diverts blood flow to where it is really needed–namely, the muscles–so why it would send blood to the face, where it confers no obvious physiological benefit, remains a mystery. One possibility suggested by Jablonski is that it helps in some way to mediate blood pressure. Or it could just serve as a signal to an opponent to back off because one is really angry.

At all events, the slow evolution of different skin tones worked fine when people stayed in one place or migrated slowly, but nowadays increased mobility means that lots of people end up in places where sun levels and skin tones don't get along at all. In regions like northern Europe and Canada, it isn't possible in the winter months to extract enough vitamin D from weakened sunlight to maintain health no matter how pale one's skin, so vitamin D must be consumed as food, and hardly anyone gets enough—and not surprisingly. To meet dietary requirements from food alone, you would have to eat fifteen eggs or six pounds of swiss cheese every day, or, more plausibly if not more palatably, swallow half a tablespoon of cod liver oil. In America, milk is helpfully supplemented with vitamin D, but that still provides only a third of daily adult requirements. In consequence, some 50 percent of people globally are estimated to be vitamin D deficient for at least part of the year. In northern climes, it may be as much as 90 percent.

As people evolved lighter skin, they also developed lighter-colored eyes and hair—but only pretty recently. Lighter-colored eyes and hair evolved somewhere around the Baltic Sea about six thousand years ago. It's not obvious why. Hair and eye color don't affect vitamin D metabolism, or anything else physiological come to that, so there seems to be no practical benefit. The supposition is that these traits were selected for as tribal markers or because people found them more attractive. If you have blue or green eyes, it's not because you have more of those colors in your irises than other people but because you simply have less of other colors. It is the paucity of other pigments that leaves the eyes looking blue or green.

Skin color has been changing over a much longer period—at least sixty thousand years. But it hasn't been a straightforward process. "Some people have de-pigmented; some have re-pigmented," Jablonski says. "Some people have altered skin tones a lot in moving to new latitudes, others hardly at all."

Indigenous populations in South America, for instance, are lighter-skinned than would be expected at the latitudes they inhabit. That is because in evolutionary terms they are recent arrivals. "They were able to get to the tropics quite quickly and had lots of gear, including some clothing," Jablonski told me. "So in effect they thwarted evolution." Rather harder to explain have been the KhoeSan people of southern

Africa. They have always lived under a desert sun and have never migrated any great distance, yet have 50 percent lighter skin than would be predicted by their environment. It now appears that a genetic mutation for lighter skin was introduced to them sometime in the last two thousand years by outsiders—but who these mysterious light-skinned outsiders were and how they came to be in southern Africa are unknown.

The development in recent years of techniques for analyzing ancient DNA means that we are learning more all the time and much of it is surprising—and some is confusing and some disputed. Using DNA analysis, in early 2018 scientists from University College London and Britain's Natural History Museum announced to widespread astonishment that an ancient Briton known as Cheddar Man had had "dark to black" skin. He seems also to have had blue eyes. Cheddar Man was among the first people to return to Britain after the end of the last ice age some ten thousand years ago. His forebears had been in Europe for thirty thousand years, more than sufficient time to have evolved light skin, so if he was truly dark-skinned, it would be a real surprise. However, other authorities have suggested that the DNA used in the analysis was too degraded and our understanding of the genetics of pigmentation too uncertain to allow any conclusions about the color of Cheddar Man's skin and eyes. If nothing else, it was a reminder of how much we have still to learn.

"Where skin is concerned, we are still in many ways at the very beginning," Jablonski told me.

Skin comes in two varieties: with hair and without. Hairless skin is called glabrous, and there isn't much of it. Our only truly hairless parts are lips, nipples and genitalia, and the bottoms of our hands and feet. The rest of the body is covered with either conspicuous hair, called terminal hair, as on your head, or vellus hair, which is the downy stuff you find on a child's cheek. We are actually as hairy as our cousins the apes. It's just that our hair is much wispier and fainter. Altogether we are estimated to have five million hairs, but the number varies with age and circumstances, and is only a guess anyway.

Hair is unique to mammals. Like the underlying skin, it serves a multitude of purposes: it provides warmth, cushioning, and camouflage, shields the body from ultraviolet light, and allows members of a group to signal to each other that they are angry or aroused. But some of these features clearly don't work so well when you are nearly hairless. In all mammals, when they are cold, the muscles around their hair follicles contract in a process known formally as horripilation but more commonly as getting goose bumps. In furry mammals, it adds a useful layer of insulating air between the hair and the skin, but in humans it has absolutely no physiological benefit and merely reminds us how comparatively bald we are. Horripilation also makes mammalian hair stand up (to make animals look bigger and more ferocious), which is why we get goose bumps when we are frightened or on edge, but of course that doesn't work very well for humans either.

The two most enduring questions with respect to human hair are when did we become essentially hairless and why did we retain conspicuous hair on the few places we did? As to the first, it isn't possible to state categorically when humans lost their hair, because hair and skin aren't preserved in the fossil record, but it is known from genetic studies that dark pigmentation dates from between 1.2 and 1.7 million years ago. Dark skin wasn't necessary when we were still furry, so that would strongly suggest a time frame for hairlessness. Why we retained hair on some parts of our bodies is fairly straightforward with respect to the head but not so clear elsewhere. Hair on the head acts as a good insulator in cold weather and a good reflector of heat in hot weather. According to Nina Jablonski, tightly curled hair is the most efficient kind "because it increases the thickness of the space between the surface of the hair and the scalp, allowing air to blow through." A separate but no less important reason for the retention of head hair is that it has been a tool of seduction since time immemorial.

Pubic and underarm hair are more problematic. It is not easy to think of a way that armpit hair enriches human existence. One line of supposition is that secondary hair is used to trap or disperse (depending on theory) sexual scents, or pheromones. The one problem with this theory is that humans don't seem to have pheromones. A study published in 2017 in *Royal Society Open Science* by researchers from Australia concluded that human pheromones probably don't exist and certainly play no detectable role in attraction. Another hypothesis is that secondary hair somehow protects the skin beneath it from chafing, though clearly a lot of people remove hair from all around their bodies without a notable increase in skin irritation. A more plausible theory, perhaps, is that secondary hair is for display—that it announces sexual maturity.

Every hair on your body has a growth cycle, with a growing phase and a resting phase. For facial hair a cycle is normally completed in four weeks, but a scalp hair may be with you for as much as six or seven years. A hair in your armpit is likely to last about six months, a leg hair for two months. Removing hair, whether through cutting, shaving, or waxing, has no effect on what happens at the root. We each grow about twenty-five feet of hair in a lifetime, but because all hair falls out at some point, no single strand can ever get longer than about three feet. Hair grows by one third of a millimeter a day, but the rate of hair growth depends on your age and health and even the season of the year. Our hair cycles are staggered, so we don't usually much notice as our hair falls out.

II

IN OCTOBER 1902, police in Paris were called to an apartment at 157 rue du Faubourg Saint-Honoré, in a wealthy neighborhood a few hundred yards from the Arc de Triomphe in the 8th arrondissement. A man had been murdered and some works of art stolen. The murderer left behind no obvious clues, but luckily detectives were able to call upon Alphonse Bertillon, a wizard at identifying criminals.

Bertillon had invented a system of identification that he called anthropometry but that became known to an admiring public as Bertillonage. The system introduced the concept of the mug shot and the practice, still universally observed, of recording every arrested person full face and in profile. But it was in the fastidiousness of its measurements that Bertillonage stood out. Subjects were measured for eleven oddly specific attributes–height when seated, length of left little finger, cheek width–which Bertillon had chosen because they would not change with age. Bertillon's system was developed not to convict criminals but to catch recidivists. Because France gave stiffer sentences to repeat offenders (and often exiled them to distant, steamy outposts like Devil's Island), many criminals tried desperately to pass themselves off as first-time offenders. Bertillon's system was designed to identify them, and it did that very well. In the first year of operation, he unmasked 241 fraudsters.

Fingerprinting was actually only an incidental part of Bertillon's system, but when he found a single fingerprint on a window frame at 157 rue du Faubourg Saint-Honoré and used that to identify the murderer as one Henri Léon Scheffer, it caused a sensation not just in France but around the world. Quickly, fingerprinting became a fundamental tool of police work everywhere.

The uniqueness of fingerprints was first established in the West by the nineteenth-century Czech anatomist Jan Purkinje, though in fact the Chinese had made the same discovery more than a thousand years earlier and for centuries Japanese potters had identified their wares by pressing a finger into the clay before baking. Charles Darwin's cousin Francis Galton had suggested using fingerprints to catch criminals years before Bertillon came up with the notion, as did a Scottish missionary in Japan named Henry Faulds. Bertillon wasn't even the first to use a fingerprint to catch a murderer–that happened in Argentina ten years earlier–but it is Bertillon who gets the credit.

What evolutionary imperative led us to get whorls on the ends of our fingers? The answer is that nobody knows. Your body is a universe of mystery. A very large part of what happens on and within it happens for reasons that we don't know–very often, no doubt, because there are no reasons. Evolution is an accidental process, after all. The idea that all fingerprints are unique is actually a supposition. No one can say for absolute certain that no one else has fingerprints to match yours. All that can be said is that no one has yet found two sets of fingerprints that precisely match.

The textbook name for fingerprints is dermatoglyphics. The plow lines that make up our fingerprints are papillary ridges. They are assumed to aid in gripping, in the way tire treads improve traction on roads, but no one has ever actually proved that. Others have suggested that the whorls of fingerprints drain water better, make the skin of the fingers more stretchy and supple, or improve sensitivity, but again no one really knows what they are there for. Similarly, no one has ever come close to explaining why our fingers wrinkle when we have long baths. The explanation most often given is that wrinkling helps them to drain water better and improves grip. But that doesn't really make a great deal of sense. Surely the people who most urgently need a good grip are those who have just fallen in water, not those who have been in it for some time.

Very, very occasionally, people are born with completely smooth fingertips, a condition known as adermatoglyphia. They also have slightly fewer sweat glands than normal. This would seem to suggest a genetic connection between sweat glands and fingerprints, but what that connection is has yet to be determined. As cutaneous features go, fingerprints are frankly pretty trivial. Far more important are your sweat glands. You might not think it, but sweating is a crucial part of being human. As Nina Jablonski has put it, "It is plain old unglamorous sweat that has made humans what they are today."

Chimpanzees have only about half as many sweat glands as we have, and so can't dissipate heat as quickly as humans can. Most quadrupeds cool by panting, which is incompatible with sustained running and simultaneous heavy breathing, especially for furry creatures in hot climates. Much better to do as we do and seep watery fluids onto nearly bare skin, which cools the body as it evaporates, turning us into a kind of living air conditioner. As Jablonski has written, "The loss of most of our body hair and the gain of the ability to dissipate excess body heat through eccrine sweating helped to make possible the dramatic enlargement of our most temperature-sensitive organ, the brain." That, she says, is how sweat helped to make you brainy.

Even at rest we sweat steadily, if inconspicuously, but if you add in vigorous activity and challenging conditions, we drain off our water supplies very quickly. According to Peter Stark in *Last Breath: Cautionary Tales from the Limits of Human Endurance,* a man who weighs 155 pounds will contain a little over forty-two quarts of water. If he does nothing at all but sit and breathe, he will lose about one and a half quarts of water per day through a combination of sweat, respiration, and urination. But if he exerts himself, that rate of loss can shoot up to one and a half quarts per hour. That can quickly become dangerous. In grueling conditions—walking under a hot sun, say—you can easily sweat away ten and a half to twelve and a half quarts of water in a day. No wonder we need to keep hydrated when the weather is hot.

Unless the loss is halted or replenished, the victim will begin to suffer headaches and lethargy after losing just three to five quarts of fluid. After six or seven quarts of unrestored loss, mental impairment starts to become likely. (That is when dehydrated hikers leave a trail and wander into the wilderness.) If the loss gets much above ten and a half quarts for a 155-pound man, the victim will go into shock and die. During World War II, scientists studied how long soldiers could walk in a desert without water (assuming they were adequately hydrated at the outset) and concluded that they could go forty-five miles in 80-degree heat, fifteen miles in 100-degree heat, and just seven miles in 120-degree heat.

Your sweat is 99.5 percent water. The rest is about half salt and half other chemicals. Although salt is only a tiny part of your overall sweat, you can lose as much as three teaspoonfuls of it in a day in hot weather, which can be a dangerously high amount, so it is important to replenish salt as well as water. Sweating is activated by the release of adrenaline, which is why when you are stressed, you break into a sweat. Unlike the rest of the body, the palms don't sweat in response to physical exertion or heat, but only from stress. Emotional sweating is what is measured in lie detector tests.

Sweat glands come in two varieties: eccrine and apocrine. Eccrine glands are much the more numerous and produce the watery sweat that dampens your shirt on a sweltering day. Apocrine glands are confined mostly to the groin and armpits (technically the axilla) and produce a thicker, stickier sweat.

It is eccrine sweat in your feet—or more correctly the chemical breakdown by bacteria of the sweat in your feet—that accounts for their lush odor. Sweat on its own is actually odorless. It needs bacteria to create a smell. The two chemicals that account for the odor—isovaleric acid and methanediol—are also produced by bacterial actions on some cheeses, which is why feet and cheese can often smell so very alike.

Your skin microbes are exceedingly personal. The microbes that live on you depend to a surprising degree on what soaps or laundry detergents you use, whether you favor cotton clothing or wool, whether you shower before work or after. Some of your microbes are permanent residents. Others camp out on you for a week or a month and then, like a wandering tribe, quietly vanish.

You have about 100,000 microbes per square centimeter of your skin, and they are not easily eradicated. According to one study, the number of bacteria on you actually rises after a bath or shower because they are flushed out from nooks and crannies. But even when you try scrupulously to sanitize yourself, it isn't easy. To make one's hands safely clean after a medical examination requires thorough washing with soap and water for at least a full minute—a standard that is, in practical terms, all but unattainable for anyone dealing with lots of patients. It is a big part of the reason why every year some two million Americans pick up a serious infection in the hospital (and ninety thousand of them die of it). "The greatest difficulty," Atul Gawande has written, "is getting clinicians like me to do the one thing that consistently halts the spread of infections: wash our hands."

A study at New York University in 2007 found that most people had about 200 different species of microbes on their skin, but the species load differed dramatically from person to person. Only four types appeared on everyone tested. In another widely reported study, the Belly Button Biodiversity Project, conducted by researchers at North Carolina State University, sixty random Americans had their belly buttons swabbed to see what was lurking there microbially. The study found 2,368 species of bacteria, 1,458 of which were unknown to science. (That is an average of 24.3 new-to-science microbes in every navel.) The number of species per person varied from 29 to 107. One volunteer harbored a microbe that had never been recorded outside Japan—where he had never been.

The problem with antibacterial soaps is that they kill good bacteria on your skin as well as bad. The same is true of hand sanitizers. In 2016, the Food and Drug Administration banned nineteen ingredients commonly used in antibacterial soaps on the grounds that manufacturers had not proved them to be safe over the long term.

Microbes are not the only inhabitants of your skin. Right now, grazing in the divots on your head (and elsewhere on your oily surface, but above all on your head) are tiny mites called *Demodex folliculorum*. They are generally harmless, thank goodness, as well as invisible. They have lived with us for so long that according to one study their DNA can be used to track the migrations of our ancestors from hundreds of thousands of years ago. At their scale, your skin to them is like a giant crusty bowl of cornflakes. If you close your eyes and use your imagination, you can almost hear the crunching.

One other thing the skin does a lot, for reasons not always understood, is itch. Although a great deal of itching is easily explained (mosquito bites, rashes, encounters with poison ivy), an awful lot of it is beyond explanation. As you read this passage, you may feel an urge to scratch yourself in various places that didn't itch at all a moment ago simply because I have raised the matter. No one can say why we are so suggestible with respect to itches or even why in the absence of obvious irritants we have them at all. No single location in the brain is devoted to itching, so it is all but impossible to study neurologically.

Itching (the medical term for the condition is pruritus) is confined to the outer layer of skin and a few moist outposts—eyes, throat, nose, and anus primarily. No matter how else you suffer, you will never have an itchy spleen. Studies of scratching showed that the most prolonged relief comes from scratching the back but the most pleasurable relief comes from scratching the ankle. Chronic itching occurs in all kinds of conditions—brain tumors, strokes, autoimmune disorders, as a side effect of medications, and many more. One of the most maddening forms is phantom itching, which often accompanies an amputation and provides the miserable sufferer with a constant itch that simply cannot be satisfied. But perhaps the most extraordinary case of unappeasable suffering concerned a patient known as M., a Massachusetts woman in her late thirties who developed an irresistible itch on her upper forehead following a bout of shingles. The itch became so maddening that she rubbed the skin completely away over a patch of scalp about an inch and a half in diameter. Medications didn't help. She rubbed the spot especially furiously while asleep—so much so that one morning she awoke to find a trickle of cerebrospinal fluid running down her face. She had scratched through the skull bone and into her own brain. Today, more than a dozen years later, she is reportedly able to manage the scratch without doing severe damage to herself, but the itch has never gone away. What is most puzzling is that she has destroyed virtually all the nerve fibers in that patch of skin, yet the maddening itch remains.

Probably no mystery of the outer surface causes greater consternation, however, than our strange tendency to lose our hair as we age. We have about 100,000 to 150,000 hair follicles on our heads, though clearly not all follicles are equal among all people. You lose, on average, between fifty and a hundred head hairs every day, and sometimes they don't grow back. About 60 percent of men are "substantially bald" by the age of fifty. One man in five achieves that condition by thirty.

Little is understood about the process, but what is known is that a hormone called dihydrotestosterone tends to go slightly haywire as we age, directing hair follicles on the head to shut down and more reserved ones in the nostrils and ears to spring to dismaying life. The one known cure for baldness is castration. Ironically, considering how easily some of us lose it, hair is pretty impervious to decay and has been known to last in graves for thousands of years.

Perhaps the most positive way to look at it is that if some part of us must yield to middle age, the hair follicles are an obvious candidate for sacrifice. No one ever died of baldness, after all.

3 MICROBIAL YOU

And we are not at the end of the penicillin story.
Perhaps we are only just at the beginning.

—ALEXANDER FLEMING, NOBEL PRIZE
ACCEPTANCE SPEECH, DECEMBER 1945

I

TAKE A DEEP breath. You probably suppose that you are filling your lungs with rich, life-giving oxygen. Actually, not really. Eighty percent of the air you breathe is nitrogen. It is the most abundant element in the atmosphere and it is vital to our existence, but it doesn't interact with other elements. When you take a breath, the nitrogen in the air goes into your lungs and straight back out again, like an absentminded shopper who has wandered into the wrong store. For nitrogen to be useful to us, it must be converted into more sociable forms, like ammonia, and it is bacteria that do that job for us. Without their help, we would die. Indeed, we could never have existed. It is time to say thank you to your microbes.

You are home to trillions and trillions of tiny living things, and they do you a surprising amount of good. They provide you with about 10 percent of your calories by breaking down foods that you couldn't otherwise make use of, and in the process extract beneficial nutriments like vitamins B_2 and B_{12} and folic acid. Humans produce twenty digestive enzymes, which is a pretty respectable number in the animal world, but bacteria produce ten thousand, or five hundred times as many, according to Christopher Gardner of Stanford University. "Our lives would be vastly less well nourished without them," he says.

Individually they are infinitesimally small and their lives are fleeting—the average bacterium weighs about one-trillionth of the weight of a dollar bill and lives for no more than twenty minutes—but collectively they are formidable indeed. The genes you are born with are all you are ever going to have. You can't buy or trade for better ones. But bacteria can swap genes among themselves, as if they were Pokémon cards, and they can pick up DNA from dead neighbors. These horizontal gene transfers, as they are known, massively accelerate their capacity to adapt to whatever nature and science throw at them. The DNA of bacteria is less scrupulous in its proofreading, too, so they mutate more often, giving them even greater genetic nimbleness.

We can't begin to compete with them for speed of change. *E. coli* can reproduce seventy-two times in a day, which means that in three days they can rack up as many new generations as we have managed in the whole of human history. A single parent bacterium could in theory produce a mass of offspring greater than the weight of Earth in less than two days. In three days, its progeny would exceed the mass of the observable universe. Clearly that could never happen, but they are with us already in numbers beyond imagining. If you put all Earth's microbes in one heap and all the other animal life in another, the microbe heap would be twenty-five times greater than the animal one.

Make no mistake. This is a planet of microbes. We are here at their pleasure. They don't need us at all. We'd be dead in a day without them.

We know surprisingly little about the microbes in and on and around us because overwhelmingly they will not grow in a lab, which makes them exceedingly difficult to study. What can be said is that as you sit here now, you are likely to have something like 40,000 species of microbes calling you home—900 in your nostrils, 800 more on your inside cheeks, 1,300 next door on your gums, as many as 36,000 in your gastrointestinal tract, though such numbers must constantly be adjusted as new discoveries are made. In early 2019, a study of just twenty people by the Wellcome Sanger Institute in England found 105 new species of gut microbes whose existence had been quite unsuspected. Precise numbers will vary from person to person and within individuals over time depending on whether you are an infant or elderly, where and with whom you've been sleeping, whether you have been taking antibiotics, or whether you are fat or thin. (Thin people have more gut microbes than fat people; having hungry microbes may at least partly account for their thinness.) That is of course just the numbers of species. In terms of individual microbes, the number is beyond imagining, never mind counting: it's in the trillions. Altogether your private load of microbes weighs roughly three pounds, about the same as your brain. People have even begun describing our microbiota as one of our organs.

For years, it was commonly stated that we each contain ten times as many bacterial cells as human ones. It turns out that that confident-sounding figure came from a paper written in 1972 that was little more than a guess. In 2016, researchers from Israel and Canada did a more careful assessment and concluded that each of us contains about thirty trillion human cells and between thirty and fifty trillion bacterial cells (depending on a lot of factors like health and diet), so the numbers are much closer to being equal—though it should also be noted that 85 percent of our own cells are red blood cells, which aren't true cells at all, because they don't have any of the usual machinery of cells (like nuclei and mitochondria), but are really just containers for hemoglobin. A separate consideration is that bacterial cells are tiny, whereas human cells are comparatively gigantic, so in terms of massiveness, not to mention the complexity of what they do, human cells are unquestionably more consequential. Then again, looked at genetically, you have about twenty thousand genes of your own within you, but perhaps as many as twenty million bacterial genes, so from that perspective you are roughly 99 percent bacterial and not quite 1 percent you.

Microbial communities can be surprisingly specific. Although you and I will each have several thousand bacterial species within us, we may have only a fraction in common. Microbes are ferocious housekeepers, it seems. Have sex and you and your partner will perforce exchange a lot of microbes and other organic material. Passionate kissing alone, according to one study, results in the transfer of up to one billion bacteria from one mouth to another, along with about 0.7 milligrams of protein, 0.45 milligrams of salt, 0.7 micrograms of fat, and 0.2 micrograms of "miscellaneous organic compounds" (that is, bits of food). But as soon as the party is over, the host microorganisms in both participants will begin a kind of giant sweeping-out process, and within only a day or so the microbial profile for both parties will be more or less fully restored to what it was before they locked tongues. Occasionally, some pathogens sneak through, and that is when you get herpes or a head cold, but that is the exception.*

Luckily, most microbes have nothing to do with us. Some live benignly inside us and are known as commensals. Only a tiny portion of them make us ill. Of the million or so microbes that have been identified, just 1,415 are known to cause disease in humans—very few, all things considered. On the other hand, that is still a lot of ways to be unwell, and together those 1,415 tiny, mindless entities cause one-third of all the deaths on the planet.

As well as bacteria, your personal repertoire of microbes consists of fungi, viruses, protists (amoebas, algae, protozoa, and so on), and archaea, which for a long time were thought to be just more bacteria

* According to Dr. Anna Machin of Oxford University, something you are doing when you are kissing another person is sampling his or her histocompatibility genes, which are involved in immune response. Though it may not be the matter uppermost on your mind at that moment, you are essentially testing whether the other person would make a good mate from an immunological perspective.

but actually represent a whole other branch of life. Archaea are very like bacteria in that they are quite simple and have no nucleus, but they have the great benefit to us that they cause no known diseases in humans. All they give us is a little gas, in the form of methane.

It's worth bearing in mind that all these microbes have almost nothing in common in terms of their history and genetics. All that unites them is tininess. To all of them, you are not a person but a world—a vast and jouncing wealth of marvelously rich ecosystems with the convenience of mobility thrown in, along with the very helpful habits of sneezing, petting animals, and not always washing quite as fastidiously as you really ought to.

II

A VIRUS, IN the immortal words of the British Nobel laureate Peter Medawar, is "a piece of bad news wrapped up in a protein." Actually, a lot of viruses are not bad news at all, at least not to humans. Viruses are a little weird, not quite living but by no means dead. Outside living cells, they are just inert things. They don't eat or breathe or do much of anything. They have no means of locomotion. We must go out and collect them—off door handles or handshakes or drawn in with the air we breathe. They do not propel themselves; they hitchhike. Most of the time, they are as lifeless as a mote of dust, but put them into a living cell, and they will burst into animate existence and reproduce as furiously as any living thing.

Like bacteria, they are incredibly successful. The herpes virus has endured for hundreds of millions of years and infects all kinds of animals—even oysters. They are also terribly small—much smaller than bacteria and too small to be seen under conventional microscopes. If you blew one up to the size of a tennis ball, a human would be five hundred miles high. A bacterium on the same scale would be about the size of a beach ball.

In the modern sense of a very small microorganism, the term "virus" dates only from 1900, when a Dutch botanist, Martinus Beijerinck, found that the tobacco plants he was studying were susceptible to a mysterious infectious agent even smaller than bacteria. At first he called the mysterious agent *contagium vivum fluidum* but then changed it to "virus," from a Latin word for "toxin." Although he was the father of virology, the importance of his discovery wasn't appreciated in his lifetime, so he was never honored with a Nobel Prize, as he really should have been.

It used to be thought that all viruses cause disease—hence the Peter Medawar quotation—but we now know that most viruses infect only bacterial cells and have no effect on us at all. Of the hundreds of thousands of viruses reasonably supposed to exist, just 586 species are known to infect mammals, and of these only 263 affect humans.

We know very little about most other, nonpathogenic viruses because only the ones that cause disease tend to get studied. In 1986, a student at the State University of New York at Stony Brook named Lita Proctor decided to look for viruses in seawater—which was considered a highly eccentric thing to do because it was universally assumed that the oceans have no viruses except perhaps for a transient few introduced through sewage outfall pipes and the like. So it was a slight astonishment when Proctor found that the average quart of seawater contains up to 100 *billion* viruses. More recently, Dana Willner, a biologist at San Diego State University, looked into the number of viruses found in healthy human lungs—somewhere else that viruses were not thought to lurk much. Willner found that the average person harbored 174 species of virus, 90 percent of which had never been seen before. Earth, we now know, is aswarm with viruses to a degree that until recently we barely suspected. According to the virologist Dorothy H. Crawford, ocean viruses alone if laid end to end would stretch for ten million light-years, a distance essentially beyond imagining.

Something else viruses do is bide their time. A most extraordinary example of that came in 2014 when a French team found a previously unknown virus, *Pithovirus sibericum,* in Siberia. Although it had been locked in permafrost for thirty thousand years, when injected into an amoeba, it sprang into action with the lustiness of youth. Luckily,

P. sibericum proved not to infect humans, but who knows what else may be out there waiting to be uncovered? A rather more common manifestation of viral patience is seen in the varicella-zoster virus. This is the virus that gives you chicken pox when you are small, but then may sit inert in nerve cells for half a century or more before erupting in that horrid and painful indignity of old age known as shingles. It is usually described as a painful rash on the torso, but in fact shingles can pop up almost anywhere on the body surface. A friend of mine had it in his left eye and described it as the worst experience of his life. (The word, incidentally, has nothing to do with the tiles of a roof. Shingles as a medical condition comes from the Latin *cingulus,* meaning a kind of belt; as a roofing material, it is from the Latin *scindula,* meaning a stepped tile. It is just by chance that they ended up in English with the same spellings.)

The most regular of unwelcome viral encounters is the common cold. Everyone knows that if you get chilled, you are more likely to catch a cold (that is why we call it a cold, after all), yet science has never been able to prove why—or even, come to that, *if* that is actually so. Colds unquestionably are more frequent in winter than in summer, but that may only be because we spend more time indoors then and are more exposed to others' leakages and exhalations.

The common cold is not a single illness but rather a family of symptoms generated by a multiplicity of viruses, of which the most pernicious are the rhinoviruses. These alone come in a hundred varieties. There are, in short, lots of ways to catch a cold, which is why you never develop enough immunity to stop catching them all.

For years, Britain operated a research facility called the Common Cold Unit, but it closed in 1989 without ever finding a cure. It did, however, conduct some interesting experiments. In one, a volunteer was fitted with a device that leaked a thin fluid at his nostrils at the same rate that a runny nose would. The volunteer then socialized with other volunteers, as if at a cocktail party. Unknown to any of them, the fluid contained a dye visible only under ultraviolet light. When that was switched on after they had been mingling for a while, the participants were astounded to discover that the dye was everywhere—on the hands, head, and upper body of every participant and on glasses, doorknobs, sofa cushions, bowls of nuts, you name it. The average adult touches his face sixteen times an hour, and each of those touches transferred the pretend pathogen from nose to snack bowl to innocent third party to doorknob to innocent fourth party and so on until pretty much everyone and everything bore a festive glow of imaginary snot. In a similar study at the University of Arizona, researchers infected the metal door handle to an office building and found it took only about four hours for the "virus" to spread through the entire building, infecting over half of employees and turning up on virtually every shared device like photocopiers and coffee machines. In the real world, such infestations can stay active for up to three days. Surprisingly, the least effective way to spread germs (according to yet another study) is kissing. It proved almost wholly ineffective among volunteers at the University of Wisconsin who had been successfully infected with cold virus. Sneezes and coughs weren't much better. The only really reliable way to transfer cold germs is physically by touch.

A survey of subway trains in Boston found that metal poles are a fairly hostile environment for microbes. Where microbes thrive is in the fabrics on seats and on plastic handgrips. The most efficient method of transfer for germs, it seems, is a combination of folding money and nasal mucus. A study in Switzerland in 2008 found that flu virus can survive on paper money for two and a half weeks if it is accompanied by a microdot of snot. Without snot, most cold viruses could survive on folding money for no more than a few hours.

The two other forms of microbe that commonly lurk within us are fungi and protists. Fungi for a long time were a kind of scientific bewilderment, classified as just slightly strange plants. In fact, at a cellular level, they aren't very like plants at all. They don't photosynthesize, so they have no chlorophyll and thus are not green. They are actually more closely related to animals than to plants. It wasn't until 1959 that they were recognized as quite separate and given their own kingdom. They essentially divide into two groups—molds and yeasts. By and large fungi leave us alone. Only about three hundred out of several million species affect us at all, and most of those mycoses, as they are known, don't make you really ill, but rather cause only mild discomfort or irritation, as with athlete's foot, say. A few, however, are much nastier than that, and the number of nasty ones is growing.

Candida albicans, the fungus behind thrush, until the 1950s was found only in the mouth and genitals, but now it sometimes invades the deeper body, where it can grow on the heart and other organs, like mold on fruit. Similarly, *Cryptococcus gattii* was for decades known to exist in British Columbia in Canada, mostly on trees or in the soil around them, but it never harmed a human. Then, in 1999, it developed a sudden virulence, causing serious lung and brain infections among a scattering of victims in western Canada and the United States. Exact figures are impossible to come by because the disease is often misdiagnosed and, remarkably, is not reportable in California, one of the main sites of occurrence, but something over three hundred cases in western North America have been confirmed since 1999, with about a third of victims dying.

Rather better reported are figures for coccidioidomycosis, which is more commonly known as valley fever. It occurs almost entirely in California, Arizona, and Nevada, infecting about ten thousand to fifteen thousand people a year and killing about two hundred, though the actual number is probably higher because it can be confused with pneumonias. The fungus is found in soils, and the number of cases rises whenever soils are disturbed, as with earthquakes and dust storms. Altogether fungi are thought to be responsible for about a million deaths globally every year, so hardly inconsequential.

Finally, protists. A protist is anything that isn't obviously plant, animal, or fungus; it is a category reserved for all those life-forms that don't fit anywhere else. Originally, in the nineteenth century, all single-celled organisms were called protozoa. It was assumed that all were closely related, but over time it became evident that bacteria and archaea were separate kingdoms. Protists is a huge category and includes amoebas, parameciums, diatoms, slime molds, and many others that are mostly obscure to all but people working in biological fields. From a human health perspective, the most notable protists are those from the genus *Plasmodium.* They are the evil little creatures that transfer from mosquitoes into us and give us malaria. Protists are also responsible for toxoplasmosis, giardiasis, and cryptosporidiosis.

There is, in short, an astounding array of microbes all around us, and we have barely begun to understand their effects on us, for good and ill. A most arresting illustration of that arose in 1992 in the north of England in the old mill town of Bradford, West Yorkshire, when Timothy Rowbotham, a government microbiologist, was sent to try to track down the source of an outbreak of pneumonia. In a sample of water he took from a storage tower, he found a microbe unlike anything he or anyone else had ever seen before. He tentatively identified it as a new bacterium, not because it was particularly bacterial in nature, but because it couldn't be anything else. He dubbed it the Bradford coccus for want of a better term. Though he had no idea of it, Rowbotham had just changed the world of microbiology.

Rowbotham saved the samples in a freezer for six years before sending them on to colleagues when he took early retirement. Eventually, they came into the hands of Richard Birtles, an English biochemist working in France. Birtles realized that the Bradford coccus was not a bacterium but a virus—but one that didn't fit any definitions of what viruses should be. For a start, this one was massively bigger—by a factor of more than a hundred—than any virus previously known. Most viruses have only a dozen or so genes. This one had over a thousand. Viruses aren't considered living things, but its genetic code contained a stretch of sixty-two letters that has been found in all living things since the dawn of creation, making it not only arguably alive but as ancient as anything else on Earth.*

Birtles named the new virus mimivirus, for "microbe-mimicking." When Birtles and his colleagues wrote up their findings, they couldn't at first find any journal that would publish them, because they were too bizarre. The cooling tower was knocked down in the late 1990s, and it appears that the only colony of this odd and ancient virus was lost with it.

Since then, however, other colonies of even more enormous viruses have been found. In 2013, a team of French researchers led by Jean-Michel Claverie from Aix-Marseille University in France (the institution to which Birtles was attached when he characterized mimivirus) found a new giant virus that they called pandoravirus, which contains no fewer than twenty-five hundred genes, 90 percent of which are found nowhere else in nature. They then found a third group, pithovirus, which is even bigger and at least as strange. Altogether as of this writing there are now five groups of giant viruses, which are all not only different from everything else on Earth but also very different from one another. Such strange and foreign bioparticles, it has been argued, are evidence for the existence of a fourth domain of life, in addition to bacteria, archaea, and eukaryotes, the latter of which include complex life like us. Where microbes are concerned, we are really just at the beginning.

III

WELL INTO THE modern age, the idea that something as small as a microorganism could cause us serious harm was thought self-evidently preposterous. When the German microbiologist Robert Koch reported in 1884 that cholera was wholly caused by a bacillus (a rod-shaped

* For the record: GTGCCAGCAGCCGCGGTAATTCAGCTCCAATAGCGTATAT-TAAAGTTGCTGCAGTTAAAAAG.

bacterium), an eminent but skeptical colleague named Max von Pettenkofer was so vehemently offended by the thought that he made a great show of swallowing a vial of the bacilli to prove Koch wrong. This would be a much better anecdote if Pettenkofer had thereupon fallen gravely ill and recanted his ill-founded objections, but in fact he didn't become ill at all. Sometimes that happens. It is now believed that Pettenkofer had suffered from cholera earlier in his life and enjoyed some residual immunity. What is less well publicized is that two of his students also drank cholera extract and both grew very ill. At all events, the episode served to delay even further general acceptance of the germ theory, as it was known. In a sense, it didn't matter terribly much what caused cholera or many other common maladies, because there weren't any treatments for them anyway.*

Before penicillin, the closest thing to a wonder drug that existed was Salvarsan, developed by the German immunologist Paul Ehrlich in 1910, but Salvarsan was effective against only a few things, principally syphilis, and had a lot of drawbacks. For a start, it was made from arsenic, so was toxic, and treatment consisted in injecting roughly a pint of solution into the patient's arm once a week for fifty weeks or more. If it wasn't administered exactly right, fluid could seep into

* Koch's discoveries are of course extremely well known, and he is justly celebrated for them. What is often overlooked, however, is what a difference small, incidental contributions can make to scientific progress, and nowhere was that better illustrated than in Koch's own productive lab. Culturing lots and lots of different bacterial samples took up a great deal of lab space and raised the constant risk of cross-contamination. But luckily Koch had a lab assistant named Julius Richard Petri who devised the shallow dish with a protective lid that bears his name. Petri dishes took up very little space, provided a sterile and uniform environment, and effectively eliminated the risk of cross-contamination. But there was still a need for a growing medium. Various gelatins were tried, but all proved unsatisfactory. Then Fanny Hesse, the American-born wife of another junior researcher, suggested that they try agar. Fanny had learned from her grandmother to use agar to make jellies because it didn't melt in the heat of an American summer. Agar worked perfectly for lab purposes, too. Without these two developments, Koch might have taken years longer, or possibly never succeeded, in making his breakthroughs.

muscle, causing painful and sometimes serious side effects, including the need for amputation. Doctors who could administer it safely became celebrated. Ironically, one of the most highly regarded was Alexander Fleming.

The story of Fleming's accidental discovery of penicillin has been told many times, but hardly any two versions are quite the same. The first thorough account of the discovery was not published until 1944, a decade and a half after the events it describes, by which time details were already blurring, but as best as can be said, the story seems to be this: In 1928, while Alexander Fleming was away on a holiday from his job as a medical researcher at St. Mary's Hospital in London, some spores of mold from the genus *Penicillium* drifted into his lab and landed on a petri dish that he had left unattended. Thanks to a sequence of chance events—that Fleming hadn't cleaned up his petri dishes before departing on holiday, that the weather was unusually cool that summer (and thus good for spores), that Fleming remained away long enough for the slow-growing mold to act—he returned to find that the bacterial growth in the petri dish had been conspicuously inhibited.

It is often written that the type of fungus that landed on his dish was a rare one, making the discovery practically miraculous, but this appears to have been a journalistic invention. The mold was in fact *Penicillium notatum* (now called *Penicillium chrysogenum*), which is very common in London, so it was hardly momentous that a few spores should drift into his lab and settle on his agar. It has also become a commonplace that Fleming failed to exploit his discovery and that years passed before others finally converted his findings into a useful medicine. That is, at the very least, an ungenerous interpretation. First, Fleming deserves credit for perceiving the significance of the mold; a less alert scientist might simply have tossed the whole lot out. Moreover, he dutifully reported his discovery, and even noted the antibiotic implications of it, in a respected journal. He also made some effort to turn the discovery into a usable medicine, but it was a technically tricky proposition—as others would later discover—and he had more pressing research interests to pursue, so he didn't stick with it. It is often overlooked that Fleming was a distinguished and busy scientist already. He had in 1923 discovered lysozyme, an antimicrobial enzyme found in saliva, mucus, and tears as part of the body's first line of defense against invading pathogens, and was still preoccupied with exploring its properties. He was hardly foolish or slapdash, as is sometimes implied.

In the early 1930s, researchers in Germany produced a group of antibacterial drugs known as sulfonamides, but they didn't always work well and often had serious side affects. At Oxford, a team of biochemists led by the Australian-born Howard Florey began searching for a more effective alternative and in the process rediscovered Fleming's penicillin paper. The principal investigator at Oxford was an eccentric German émigré named Ernst Chain, who bore an uncanny resemblance to Albert Einstein (right down to the bushy mustache) but had a far more challenging disposition. Chain had grown up in a wealthy Jewish family in Berlin but had decamped to England with the rise of Adolf Hitler. Chain was gifted in many fields and considered a career as a concert pianist before settling on science. But he was also a difficult man. He had a volatile temperament and slightly paranoid instincts, though it seems fair to say that if there was ever a time when a Jew might be excused paranoia it was the 1930s. He was an unlikely candidate to make any discoveries because he had a pathological fear of being poisoned in a lab. Despite his dread, he persevered and found to his astonishment that penicillin not only killed pathogens in mice but had no evident side effects. It appeared to be the perfect drug: one that could devastate its target without wreaking collateral damage. The problem, as Fleming had seen, was that it was very hard to produce penicillin in clinically useful quantities. Under Florey's command, Oxford gave over a significant amount of resources and research space to growing mold and patiently extracting from it tiny amounts of penicillin.

By early 1941, they had just enough to trial the drug on a policeman named Albert Alexander, who was a tragically ideal demonstration of how vulnerable humans were to infections before antibiotics. While pruning roses in his garden, Alexander had scratched his face on a thorn. The scratch had grown infected and spread. Alexander had lost an eye and now was delirious and close to death. The effect of penicillin was miraculous. Within two days, he was sitting up and looking almost back to normal. But supplies quickly ran short. In desperation the scientists filtered and reinjected all they could from Alexander's urine, but after four days the supplies were exhausted. Poor Alexander relapsed and died.

With Britain preoccupied by World War II and the United States not yet in it, the quest to produce bulk penicillin moved to a U.S. government research facility in Peoria, Illinois. Scientists and other interested parties all over the Allied world were secretly asked to send in soil and mold samples. Hundreds responded, but nothing they sent proved promising. Then, two years after testing had begun, a lab assistant in Peoria named Mary Hunt brought in a cantaloupe from a local grocery store. It had a "pretty golden mold" growing on it, she recalled later. That mold proved to be two hundred times more potent than anything previously tested. The name and location of the store where Mary Hunt shopped are now forgotten, and the historic cantaloupe itself was not preserved: after the mold was scraped off, it was cut into pieces and eaten by the staff. But the mold lived on. Every bit of penicillin made since that day is descended from that single random cantaloupe.

Within a year, American pharmaceutical companies were producing 100 billion units of penicillin a month. The British discoverers found to their chagrin that the production methods had been patented by the Americans and that they were now required to pay royalties to make use of their own discovery.

Alexander Fleming didn't become famous as the father of penicillin until the closing days of the war, some twenty years after his serendipitous discovery, but then he became very famous indeed. He received 189 honors of all types from around the world, and even had a crater on the moon named for him. In 1945, he shared the Nobel

Prize in Physiology or Medicine with Ernst Chain and Howard Florey. Florey and Chain never enjoyed the popular acclaim they deserved, partly because they were much less gregarious than Fleming and partly because his story of accidental discovery made better copy than their story of dogged application. Chain, despite sharing the Nobel Prize, became convinced that Florey had not given him sufficient credit, and their friendship, such as it was, dissolved.

As early as 1945, in his Nobel acceptance speech, Fleming warned that microbes could easily evolve resistance to antibiotics if they were carelessly used. Seldom has a Nobel speech been more prescient.

IV

THE GREAT VIRTUE of penicillin—that it scythes its way through all manner of bacteria—is also its elemental weakness. The more we expose microbes to antibiotics, the more opportunity they have to develop resistance. What you are left with after a course of antibiotics, after all, are the most resistant microbes. By attacking a broad spectrum of bacteria, you stimulate lots of defensive action. At the same time, you inflict unnecessary collateral damage. Antibiotics are about as nuanced as a hand grenade. They wipe out good microbes as well as bad. Increasing evidence shows that some of the good ones may never recover, to our permanent cost.

Most people in the Western world, by the time they reach adulthood, have received between five and twenty courses of antibiotics. The effects, it is feared, may be cumulative, with each generation passing on fewer microorganisms than the one before. Few people are more aware of this than an American scientist named Michael Kinch. In 2012, when he was director of the Yale Center for Molecular Discovery in Connecticut, Kinch's twelve-year-old son, Grant, developed severe abdominal pains.

"He'd been at the first day of a summer camp and he'd eaten some cupcakes," Kinch recalls, "so we thought at first it was just a combination of excitement and overindulgence, but the symptoms got worse." Eventually, Grant ended up in Yale New Haven Hospital, where a number of alarming things happened quickly. It was found that he had a ruptured appendix and that his intestinal microbes had escaped into the abdomen, giving him peritonitis. Then the infection developed into septicemia, which meant it had spread to his blood and could go anywhere in his body. Dismayingly, four of the antibiotics Grant was given didn't have any effect on the marauding bacteria.

"That was really astounding," Kinch recalls now. "This was a kid who had been on antibiotics just once in his life, for an ear infection, and yet he had gut bacteria that were resistant to antibiotics. That shouldn't have happened." Fortunately, two other antibiotics did work and Grant's life was saved.

"He was lucky," Kinch says. "The day is fast approaching when the bacteria inside us may not be resistant to two-thirds of the antibiotics we hit them with, but to all of them. Then we really are in trouble."

Today Kinch is the director of the Center for Research Innovation in Business at Washington University in St. Louis. He works in a once derelict, now stylishly renovated telephone factory that is part of a neighborhood salvation project undertaken by the university. "This used to be the best place in St. Louis to score crack," he says with a hint of ironic pride. A cheerful man of early middle years, Kinch was brought to Washington University to foster entrepreneurship, but one of his central passions remains the future of the pharmaceutical industry and where new antibiotics will come from. In 2016, he wrote an alarming book on the matter, *A Prescription for Change: The Looming Crisis in Drug Development*.

"From the 1950s through the 1990s," he says, "roughly three antibiotics were introduced into the U.S. every year. Today it's roughly one new antibiotic every other year. The rate of antibiotic withdrawals—because they don't work anymore or have become obsolete—is twice the rate of new introductions. The obvious consequence of this is that the arsenal of drugs we have to treat bacterial infections has been going down. There is no sign of it stopping."

What makes this much worse is that a great deal of our antibiotic use is simply crazy. Almost three-quarters of the forty million antibiotic prescriptions written each year in the United States are for conditions that cannot be cured with antibiotics. According to Jeffrey Linder, professor of medicine at Northwestern University, antibiotics are prescribed for 70 percent of acute bronchitis cases, even though guidelines explicitly state that they are of no use there.

Even more appallingly, in the United States 80 percent of antibiotics are fed to farm animals, mostly to fatten them. Fruit growers can also use antibiotics to combat bacterial infections in their crops. In consequence, most Americans consume secondhand antibiotics in their food (including even some foods labeled as organic) without knowing it. Sweden banned the agricultural use of antibiotics in 1986. The European Union followed in 1999. In 1977, the Food and Drug Administration ordered a halt to the use of antibiotics for purposes of fattening farm animals, but backed off when there was an outcry from agricultural interests and the congressional leaders who supported them.

In 1945, the year that Alexander Fleming won the Nobel Prize, a typical case of pneumococcal pneumonia could be knocked out with forty thousand units of penicillin. Today, because of increased resistance, it can take more than twenty million units per day for many days to achieve the same result. On some diseases, penicillin now has no effect at all. In consequence, the death rate for infectious diseases has been climbing and is back to the level of about forty years ago.

Bacteria really are not to be trifled with. They not only have grown steadily more resistant but have evolved into a fearsome new class of pathogen commonly known, with scarcely a hint of hyperbole, as superbugs. *Staphylococcus aureus* is a microbe found commonly on human skin and in nostrils. Generally it does no harm, but it is an opportunist, and when the immune system is weakened, it can slip in and wreak havoc. By the 1950s, it had evolved resistance to penicillin, but luckily another antibiotic called methicillin had become available and it stopped *S. aureus* infections in their tracks. But just two years after methicillin's introduction, two people at the Royal Surrey County

Hospital in Guildford, near London, developed *S. aureus* infections that would not respond to methicillin. *S. aureus* had, almost overnight, evolved a new drug-resistant form. The new strain was dubbed Methicillin-resistant Staphylococcus aureus, or MRSA. Within two years, it had spread to mainland Europe. Soon after that, it leaped to the United States.

Today, MRSA and its cousins kill an estimated 700,000 people around the world annually. Until recently a drug called vancomycin was effective against MRSA, but now resistance has begun to emerge to it. At the same time, we are facing the formidable-sounding carbapenem-resistant Enterobacteriaceae (CRE) infections, which are immune to virtually everything we can throw at them. CRE kills about half of all those it sickens. Luckily, so far, it doesn't usually infect healthy people. But watch out if it does.

Yet as the problem has grown, the pharmaceutical industry has retreated from trying to create new antibiotics. "It's just too expensive for them," Kinch says. "In the 1950s, for the equivalent of a billion dollars in today's money, you could develop about ninety drugs. Today, for the same money, you can develop on average just one-third of a drug. Pharmaceutical patents last only for twenty years, but that includes the period of clinical trials. Manufacturers usually have just five years of exclusive patent protection." In consequence, all but two of the eighteen largest pharmaceutical companies in the world have given up the search for new antibiotics. People take antibiotics for only a week or two. Much better to focus on drugs like statins or antidepressants that people can take more or less indefinitely. "No sane company will develop the next antibiotic," Kinch says.

The problem needn't be hopeless, but it does need to be addressed. At the current rate of spread, antimicrobial resistance is forecast to lead to ten million preventable deaths a year—that's more people than die of cancer now—within thirty years, at a cost of perhaps $100 trillion in today's money.

What nearly everyone agrees is that we need a more targeted approach. One interesting possibility would be to disrupt bacteria's lines of communication. Bacteria never mount an attack until they have assembled sufficient numbers—what is known as a quorum—to make it worthwhile to do so. The idea would be to produce quorum-sensing drugs that wouldn't kill all bacteria but would just keep their numbers permanently below the threshold, the quorum, that triggers an attack.

Another possibility is to enlist bacteriophages, a kind of virus, to hunt down and kill harmful bacteria for us. Bacteriophages—often shortened to just phages—are not well known to must of us, but they are the most abundant bioparticles on Earth. Virtually every surface on the planet, including us, is covered in them. They do one thing supremely well: each one targets a particular bacterium. That means clinicians would have to identify the offending pathogen and select the right phage to kill it, a more costly and time-consuming process, but it would make it much harder for bacteria to evolve resistance.

What is certain is that something must be done. "We tend to refer to the antibiotics crisis as a looming one," Kinch says, "but it is not that at all. It's a current crisis. As my son showed, these problems are with us now—and it is going to get much worse."

Or as a doctor put it to me, "We are looking at a possibility where we can't do hip replacements or other routine procedures because the risk of infection is too high."

The day when people die once again from the scratch of a rose thorn may not be far away.

4 THE BRAIN

The brain is wider than the sky,
For, put them side by side,
The one the other will include
With ease, and you beside.

—EMILY DICKINSON

THE MOST EXTRAORDINARY thing in the universe is inside your head. You could travel through every inch of outer space and very possibly nowhere find anything as marvelous and complex and high functioning as the three pounds of spongy mass between your ears.

For an object of pure wonder, the human brain is extraordinarily unprepossessing. It is, for one thing, 75 to 80 percent water, with the rest split mostly between fat and protein. Pretty amazing that three such mundane substances can come together in a way that allows us thought and memory and vision and aesthetic appreciation and all the rest. If you were to lift your brain out of your skull, you would almost certainly be surprised at how soft it is. The consistency of the brain has been variously likened to tofu, soft butter, or a slightly overcooked Jell-O pudding.

The great paradox of the brain is that everything you know about the world is provided to you by an organ that has itself never seen that world. The brain exists in silence and darkness, like a dungeoned prisoner. It has no pain receptors, literally no feelings. It has never felt warm sunshine or a soft breeze. To your brain, the world is just a stream of electrical pulses, like taps of Morse code. And out of this bare and neutral information it creates for you—quite literally creates—a vibrant, three-dimensional, sensually engaging universe. Your brain is you. Everything else is just plumbing and scaffolding.

Just sitting quietly, doing nothing at all, your brain churns through more information in thirty seconds than the Hubble Space Telescope has processed in thirty years. A morsel of cortex one cubic millimeter in size—about the size of a grain of sand—could hold two thousand terabytes of information, enough to store all the movies ever made, trailers included, or about 1.2 billion copies of this book. Altogether, the human brain is estimated to hold something on the order of two hundred exabytes of information, roughly equal to "the entire digital content of today's world," according to *Nature Neuroscience*.* If that is not the most extraordinary thing in the universe, then we certainly have some wonders yet to find.

The brain is often depicted as a hungry organ. It makes up just 2 percent of our body weight but uses 20 percent of our energy. In newborn infants, it's no less than 65 percent. That's partly why babies sleep all the time—their growing brains exhaust them—and have a lot of body fat, to use as an energy reserve when needed. Your muscles actually use even more of your energy, about a quarter, but you have a lot of muscle; per unit of matter, the brain is by far the most expensive of our organs. But it is also marvelously efficient. Your brain requires only about four hundred calories of energy a day—about the same as you get in a blueberry muffin. Try running your laptop for twenty-four hours on a muffin and see how far you get.

Unlike other parts of the body, the brain burns its four hundred calories at a steady rate no matter what you are doing. Hard thinking

* I am much indebted to Dr. Magnus Bordewich, director of research in the Department of Computer Science at Durham University, for some of these calculations.

doesn't help you slim. In fact, it doesn't seem to confer any benefit at all. An academic at the University of California at Irvine named Richard Haier used positron emission tomography scanners to find that the hardest-working brains are usually the least productive. The most efficient brains, he found, were those that could solve a task quickly and then go into a kind of standby mode.

For all its powers, nothing about your brain is distinctively human. We use exactly the same components–neurons, axons, ganglia, and so on–as a dog or hamster. Whales and elephants have much larger brains than we have, though of course they also have much larger bodies. But even a mouse scaled up to the size of a human would have a brain just as big, and many birds would do even better. It also turns out that the human brain is a little less imposing than we had long assumed. For years, it was written that it has 100 billion nerve cells, or neurons, but a careful assessment by the Brazilian neuroscientist Suzana Herculano-Houzel in 2015 found that the number is more like 86 billion–a pretty substantial demotion.

Neurons are not like other cells, which are typically compact and spherical. Neurons are long and stringy, the better to pass on electrical signals from one to another. The main strand of a neuron is called an axon. At its terminal end, it splits into branch-like extensions called dendrites, as many as 400,000 of them. The tiny space between nerve cell endings is called a synapse. Each neuron connects with thousands of other neurons, giving trillions and trillions of connections–as many connections "in a single cubic centimeter of brain tissue as there are stars in the Milky Way," to quote the neuroscientist David Eagleman. It is in all that complex synaptic entanglement that our intelligence lies, not in the number of neurons, as was once thought.

What is surely most curious and extraordinary about our brain is how largely unnecessary it is. To survive on Earth, you don't need to be able to write music or engage in philosophy–you really only need to be able to outthink a quadruped–so why have we invested so much energy and risk in producing mental capacity that we don't really need? That is just one of the many things about your brain that your brain won't tell you.

As the most complex of our organs, the brain not surprisingly has more named features and landmarks than any other part of the body, but essentially it divides into three sections. At the top, literally and figuratively, is the cerebrum, which fills most of the cranial vault and is the part that we normally think of when we think of "the brain." The cerebrum (from the Latin word for "brain") is the seat of all our higher functions. It is divided into two hemispheres, each of which is principally concerned with one side of the body, but for reasons unknown the wiring is crossed, so that the right side of the cerebrum controls the left side of the body and vice versa. The two hemispheres are connected by a band of fibers called the corpus callosum (meaning "tough material" or literally "calloused body" in Latin). The brain is wrinkled by deep fissures known as sulci and ridges called gyri, which give it more surface area. The exact pattern of grooves and ridges in brains is distinctive to each individual—as distinctive as your fingerprints—but whether it has anything to do with your intelligence or temperament or anything else that defines you is unknown.

Each hemisphere of the cerebrum is further divided into four lobes: frontal, parietal, occipital, and temporal—each broadly specializing in certain functions. The parietal lobe manages sensory inputs like touch and temperature. The occipital lobe processes visual information, and the temporal lobe principally manages auditory information, though it also helps with processing visual information. It has been known for some years that six patches on the temporal lobe, known as face patches, become excited when we look at another face, though which parts of my face excite which of your patches is still largely uncertain, it seems. The frontal lobe is the seat of the higher functions of the brain—reasoning, forethought, problem solving, emotional control, and so on. It is the part responsible for personality, for who we are. Ironically, as Oliver Sacks once noted, the frontal lobes were the last parts of the brain to be deciphered. "Even in my own medical student days, they were called 'the silent lobes,'" he wrote in 2001. That's not because they were thought to lack functions but because those functions do not reveal themselves.

Beneath the cerebrum, at the very back of the head about where it meets the nape of the neck, is the cerebellum (Latin for "little brain"). Although the cerebellum occupies just 10 percent of the cranial cavity, it has more than half the brain's neurons. It has a lot of neurons not because it does a great deal of thinking but because it controls balance and complex movements, and that requires an abundance of wiring.

At the base of the brain, descending from it rather like an elevator shaft connecting the brain to the spine and the body beyond, is the oldest part of the brain, the brain stem. It is the home of our more basic operations: sleeping, breathing, keeping the heart going. The brain stem doesn't get a lot of attention in the popular consciousness, but it is so central to our existence that "brain-stem death" is the fundamental measure of deadness in humans in the United Kingdom.

Scattered through the brain rather like nuts in a fruitcake are many smaller structures—hypothalamus, amygdala, hippocampus, telencephalon, septum pellucidum, habenular commissure, entorhinal cortex, and a dozen or so others—which are collectively known as the limbic system (from the Latin *limbus,* meaning "peripheral"). It's easy to go a lifetime without hearing a word about any of these components unless they go wrong. The basal ganglia, for instance, play an important part in movement, language, and thought, but it is only when they degenerate and lead to Parkinson's disease that they normally attract attention to themselves.

Despite their obscurity and modest dimensions, the structures of the limbic system have a fundamental role in our happiness by controlling and regulating basic processes like memory, appetite, emotions, drowsiness and alertness, and the processing of sensory information. The concept of the limbic system was invented in 1952 by an American neuroscientist, Paul D. MacLean. Not all of today's neuroscientists agree that the components form a coherent system. Many think they are just lots of disparate parts connected only by the fact that they are concerned with bodily performance rather than with thinking.

The most important component of the limbic system is a little powerhouse called the hypothalamus, which isn't really a structure at all but just a bundle of neural cells. The name describes not what it does but where it is: under the thalamus. (The thalamus, meaning "inner chamber," is a kind of relay station for sensory information and is an important part of the brain—there isn't any part of the brain that isn't important, obviously—but is not a component of the limbic system.) The hypothalamus is curiously unimposing. Though only about the size of a peanut and weighing barely a tenth of an ounce, it controls much of the most important chemistry of the body. It regulates sexual function, controls hunger and thirst, monitors blood sugar and salts, decides when you need to sleep. It may even play a part in how slowly or rapidly we age. A large measure of your success or failure as a human being is dependent on this tiny thing in the middle of your head.

The hippocampus is central to the laying down of memories. (The name comes from the Greek for "sea horse" because of its supposed resemblance to that creature.) The amygdala (Greek for "almond") specializes in handling intense and stressful emotions—fear, anger, anxiety, phobias of all types. People whose amygdalae are destroyed are left literally fearless, and often cannot even recognize fear in others. The amygdala grows particularly lively when we are asleep, and thus may account for why our dreams are so often disturbing. Your nightmares may simply be the amygdalae unburdening themselves.*

* You have two of each, one in each hemisphere, so really they ought to be referred to in the plural (thalami, hippocampi, amygdalae, and so on), but they seldom are.

* * *

Considering how exhaustively the brain has been studied, and for how long, it is remarkable how much elemental stuff we still don't know or at least can't universally agree upon. Like what exactly is consciousness? Or what precisely is a thought? It is not something you can capture in a jar or smear on a microscopic slide, and yet a thought is clearly a real and definite thing. Thinking is our most vital and miraculous talent, yet in a profound physiological sense we don't really know what thinking is.

Much the same could be said of memory. We know a good deal about how memories are assembled and how and where they are stored, but not why we keep some and not others. It clearly has little to do with actual value or utility. I can remember the entire starting lineup of the 1964 St. Louis Cardinals baseball team—something that has been of no importance to me since 1964 and wasn't actually very useful then—and yet I cannot recollect the number of my own cell phone, or where I parked my car in any large parking lot, or what was the third of three things my wife told me to get at the supermarket, or any of a great many other things that are unquestionably more urgent and necessary than remembering the starting players for the 1964 Cardinals (who were, incidentally, Tim McCarver, Bill White, Julian Javier, Dick Groat, Ken Boyer, Lou Brock, Curt Flood, and Mike Shannon).

So there is a huge amount we have left to learn and many things we may never learn. But equally some of the things we do know are at least as amazing as the things we don't. Consider how we see—or, to put it slightly more accurately, how the brain tells us what we see.

Just look around you now. The eyes send a hundred billion signals to the brain every second. But that's only part of the story. When you "see" something, only about 10 percent of the information comes from the optic nerve. Other parts of your brain have to deconstruct the signals—recognize faces, interpret movements, identify danger. In other words, the biggest part of seeing isn't receiving visual images; it's making sense of them.

For each visual input, it takes a tiny but perceptible amount of time—about two hundred milliseconds, one-fifth of a second—for the information to travel along the optic nerves and into the brain to be processed and interpreted. One-fifth of a second is not a trivial span of time when a rapid response is required—to step back from an oncoming car, say, or to avoid a blow to the head. To help us deal better with this fractional lag, the brain does a truly extraordinary thing: it continuously forecasts what the world will be like a fifth of a second from now, and *that* is what it gives us as the present. That means that we never see the world as it is at this very instant, but rather as it will be a fraction of a moment in the future. We spend our whole lives, in other words, living in a world that doesn't quite exist yet.

The brain tricks you in a lot of ways for your own good. Sound and light reach you at very different speeds—a phenomenon we experience every time we hear a plane passing overhead and look up to find the sound coming from one part of the sky and a plane moving silently through another. In the more immediate world around you, your brain normally irons out these differences, so that you sense all stimuli as reaching you simultaneously.

In a similar way, the brain manufactures all the components that make up our senses. It is a strange, nonintuitive fact of existence that photons of light have no color, sound waves no sound, olfactory molecules no odors. As James Le Fanu has put it, "While we have the overwhelming impression that the greenness of the trees and the blueness of the sky are streaming through our eyes as through an open window, yet the particles of light impacting on the retina are colourless, just as the waves of sound impacting on the eardrum are silent and scent molecules have no smell. They are all invisible, weightless, subatomic particles of matter travelling through space." All the richness of life is created inside your head. What you see is not what is but what your brain tells you it is, and that's not the same thing at all. Consider a bar of soap. Has it ever struck you that soap lather is always white no matter what color the soap is? That isn't because the soap somehow changes color when it is moistened and rubbed. Molecularly, it's exactly as it was before. It's just that the foam reflects light in a different way. You get the same effect with crashing waves on a beach—greeny-blue water, white foam—and lots of other phenomena. That is because color isn't a fixed reality but a perception.

You have probably at some time or other encountered one of those illusion tests that require you to stare for fifteen or twenty seconds at a red square, then shift your vision to a blank sheet of paper, and for a few moments you will see a ghostly square of greenish blue on the white paper. This "afterimage" is a consequence of tiring some of the photoreceptors in your eyes by making them work extra intently, but what is relevant is that the greenish-blue color is not there and has never existed anywhere but in your imagination. In a very real sense, that is true of all colors.

Your brain is also extraordinarily good at finding patterns and determining order in chaos, as these two well-known illusions show:

In the first illustration, most people see only random smudges until it is pointed out to them that the picture contains a dalmatian dog; then suddenly for nearly everyone the brain fills in the missing edges and makes sense of the whole composition. The illusion dates from the 1960s, but no one seems to have kept a record of who first created it.

The second illustration does have a known history. It is called a Kanizsa triangle, after the Italian psychologist Gaetano Kanizsa, who created it in 1955. There is of course no actual triangle in the picture, except for the one your brain puts there.

Your brain does all these things for you because it is designed to help you in every way it can. Yet paradoxically it is also strikingly unreliable. Some years ago, a psychologist at the University of California at Irvine, Elizabeth Loftus, discovered that it is possible through suggestion to implant entirely false memories in people's heads—to convince them that they were traumatically lost in a department store or shopping mall when they were small or that they were hugged by Bugs Bunny at Disneyland—even though these things never happened. (Bugs Bunny is not a Disney character and has never been at Disneyland.) She could show many people pictures of themselves as a child in which the image had been manipulated to make them look as if they were in a hot-air balloon, and often the subjects would suddenly remember the experience and excitedly describe it, even though in each case it was known that it had never happened.

Now, you might think that you could never be that suggestible, and you would probably be right—only about one-third of people are that gullible—but other evidence shows that we all sometimes completely misrecall even the most vivid events. In 2001, immediately after the 9/11 disaster at the World Trade Center in New York, psychologists at the University of Illinois took detailed statements from seven hundred people about where they were and what they were doing when they learned of the event. One year later, the psychologists asked the same question of the same people and found that nearly half now contradicted themselves in some significant way—put themselves in a different place when they learned of the disaster, believed that they had seen it on TV when in fact they had heard it on the radio, and so on—but without being aware that their recollections had changed.

(I, for my part, vividly recall watching the events live on television in New Hampshire, where we were then living, with two of my children, only to learn later that one of those children was in fact in England at the time.)*

Memory storage is idiosyncratic and strangely disjointed. The mind breaks each memory into its component parts—names, faces, locations, contexts, how a thing feels to the touch, even whether it is living or dead—and sends the parts to different places, then calls them back and reassembles them when the whole is needed again. A single fleeting thought or recollection can fire up a million or more neurons scattered across the brain. Moreover, these fragments of memory move around over time, migrating from one part of the cortex to another, for reasons entirely unknown. It's no wonder we get details muddled.

The upshot is that memory is not a fixed and permanent record, like a document in a filing cabinet. It is something much more hazy and mutable. As Elizabeth Loftus told an interviewer in 2013, "It's a little more like a *Wikipedia* page. You can go in there and change it, and so can other people."

Memories are categorized in many different ways, and no two authorities seem to use quite the same terminologies. The most frequently cited divisions are long-term, short-term, and working (for duration) and procedural, conceptual, semantic, declarative, implicit, autobiographical, and sensual (for type). Fundamentally, however, memories come in two principal varieties: declarative and procedural. Declarative memory is the kind you can put into words—the names

* Another extraordinary example of imaginary memories occurred in an experiment at an unidentified university in Canada where sixty volunteer students were confronted with the accusation that during adolescence they had committed a crime involving theft or assault for which they had been arrested. None of this had actually happened, but after three sessions with a kindly but manipulative interviewer, 70 percent of the volunteers confessed to these imaginary incidents, often adding vivid incriminating details—entirely imaginary but sincerely believed.

of state capitals, your date of birth, how to spell "ophthalmologist," and everything else you know as fact. Procedural memory describes the things you know and understand but couldn't so easily put into words—how to swim, drive a car, peel an orange, identify colors.

Working memory is where short-term and long-term memories combine. Say you are presented with a mathematical problem to solve. The problem resides in short-term memory—you won't, after all, need to remember the problem months from now—but the skills necessary to make the computation are kept in long-term memory.

Researchers also sometimes find it useful to distinguish between recall memory, which is what you can remember spontaneously—the kinds of things you know when you do a general knowledge quiz—and recognition memory, which is where you are a bit hazy on the substance but can recall the context. Recognition memory explains why so many of us struggle to remember the contents of a book but can often recall where we read the book, the color or design of the cover, and other seeming irrelevancies. Recognition memory is actually useful because it doesn't clutter the brain with unnecessary details but does help us to remember where we can find those details if we should need them again.

Short-term memory is really short—no more than half a minute or so for things like addresses and phone numbers. (If you can still remember something after half a minute, it is no longer technically a short-term memory. It's long term.) Most people's short-term memory is pretty abysmal. Six random words or digits is about all that most of us can reliably retain for more than a few moments.

On the other hand, with effort we can train our memories to perform the most extraordinary stunts. Every year the United States has a national memory championship, and the feats performed there are truly astounding. One memory champion could recall 4,140 random digits after looking at them for just thirty minutes. Another was able to remember twenty-seven randomly shuffled decks of cards in the same time period. Yet another could recall a single deck of cards after thirty-two seconds of study. That may not be the most worthwhile use of the human mind, but it is certainly a demonstration of its incredible powers and versatility. Most of the memory champions, by the way, are not spectacularly intelligent. They just are motivated enough to train their memories to do some extraordinary tricks.

It used to be thought that every experience is stored permanently as memory somewhere in the brain but that most of it is locked away beyond our power of immediate recall. The idea arose principally from a series of experiments in Canada from the 1930s to the 1950s by the neurosurgeon Wilder Penfield. While carrying out surgical procedures at the Montreal Neurological Institute, Penfield discovered that when he touched a probe to patients' brains, it often evoked powerful sensations—vivid smells from childhood, feelings of euphoria, sometimes a recollection of a forgotten scene from very early life. From this it was concluded that the brain records and stores every conscious event in our lives, however trivial. Now, however, it is thought that the stimulation was mostly providing the sensation of memory and that what the patients were experiencing was more like a hallucination than a recalled event.

What is certainly true is that we retain a great deal more than we can easily summon to mind. You may not recollect much of a neighborhood you lived in when you were small, but if you went back and walked around it, you would almost certainly remember very particular details you hadn't thought about for years. With sufficient time and prompting, we would probably all be astonished at how much we have stored away inside us.

The person from whom we learned a good deal of what we know about memory was, ironically, a man who had very little of it himself. Henry Molaison was an amiable and good-looking young man of twenty-seven in Connecticut who suffered from crippling episodes of epilepsy. In 1953, inspired by the efforts of Wilder Penfield in Canada, a surgeon named William Scoville drilled into Molaison's head and removed half of the hippocampus from each side of his brain and most of the amygdalae. The procedure greatly reduced Molaison's seizures (though it didn't entirely eliminate them) but at the tragic cost of robbing him of the ability to form new memories—a condition known as anterograde amnesia. Molaison could recall events from his distant past but had almost no capacity to form new memories. Someone who left the room would be immediately forgotten. Even a psychiatrist who saw him almost daily for years was a new person to him each time she came through the door. Molaison always recognized himself in the mirror but was often astounded at how old he had become. Occasionally, and mysteriously, he was able to lay down just a few memories. He could recall that John Glenn was an astronaut and Lee Harvey Oswald an assassin (though he couldn't recall whom Oswald had assassinated) and learned the address and layout of his new house when he moved. But beyond that he was locked in an eternal present that he could never understand. Poor Henry Molaison's plight was the first scientific intimation that the hippocampus has a central role in laying down memories. But what scientists learned from Molaison was not so much how memory works as how difficult it is to understand how it works.

What is surely the most striking feature of the brain is that all its higher processes—thinking, seeing, hearing, and so on—happen right at the surface, in the four-millimeter-thick sheath of the cerebral cortex. The person who first mapped this area was the German neurologist Korbinian Brodmann (1868–1918). Brodmann was one of the most brilliant and least appreciated of modern neuroscientists. In 1909, while working at a research institute in Berlin, he painstakingly identified forty-seven distinct regions of the cerebral cortex, which have been known ever since as Brodmann areas. "Rarely in the history of neuroscience has a single illustration been as influential," wrote Karl Zilles and Katrin Amunts in *Nature Neuroscience* a century later.

Painfully shy, Brodmann was repeatedly overlooked for promotions despite the importance of his work and struggled for years to secure an adequate research position. His career was further sidetracked with the outbreak of World War I, when he was sent to work at a mental asylum in Tübingen. Finally, in 1917, at the age of forty-eight his luck turned. He landed an important job as head of the Department of Topographical Anatomy at an institute in Munich. At last he had the economic security to get married and have a child, both of which he did in short order. Brodmann enjoyed not quite a year of unaccustomed serenity. In the summer of 1918, eleven and a half months after his marriage, two and a half months after the birth of his child, and at the very height of his happiness, he contracted a sudden infection and within five days was dead. He was forty-nine years old.

The area that Brodmann mapped, the cerebral cortex, is the brain's celebrated gray matter. Beneath it is the much greater volume of white matter, which is so called because the neurons are sheathed in a pale fatty insulator called myelin, which greatly accelerates the speed at which signals are transmitted. Both white matter and gray matter are misleadingly named. Gray matter isn't terribly gray in life, but has a pinkish blush. It only becomes conspicuously gray in the absence of blood flow and with the addition of preservatives. White matter is also a posthumous attribute because the pickling process turns the myelin coatings on its nerve fibers a luminous white.

Incidentally, the idea that we use only 10 percent of our brains is a myth. No one knows where the idea came from, but it has never been true or close to true. You may not use it all terribly sensibly, but you employ all your brain in one way or another.

The brain takes a long time to form completely. A teenager's brain is only about 80 percent finished (which may not come as a great surprise to the parents of teenagers). Although most of the growth of the brain occurs in the first two years and is 95 percent completed by the age of ten, the synapses aren't fully wired until a young person is in his or her mid- to late twenties. That means that the teenage years effectively extend well into adulthood. In the meantime, the person in question will almost certainly have more impulsive, less reflective behavior than his elders and will also be more susceptible to the effects of alcohol. "The teenage brain is not just an adult brain with fewer miles on it," Frances E. Jensen, a neurology professor, told *Harvard Magazine* in 2008. It is, rather, a different kind of brain altogether.

The nucleus accumbens, a region of the forebrain associated with pleasure, grows to its largest size in one's teenage years. At the same time, the body produces more dopamine, the neurotransmitter that conveys pleasure, than it ever will again. That is why the sensations you feel as a teenager are more intense than at any other time of life. But it also means that seeking pleasure is an occupational hazard for teenagers. The leading cause of deaths among teenagers is accidents—and the leading cause of accidents is simply being with other teenagers. When more than one teenager is in a car, for instance, the risk of an accident multiplies by 400 percent.

Everybody has heard of neurons, but not so many are familiar with the other main brain cells, glia or glial cells, which is a little odd because they outnumber neurons by ten to one. Glia (the word means "glue" or "putty") are the cells that support neurons in the brain and central nervous system. For a long time, they were assumed to be not too important—their role was thought to be principally to provide a kind of physical support, or extracellular matrix as anatomists put it, for neurons—but now it is known that they engage in a lot of important chemistry, from producing myelin to clearing away wastes.

There is quite a lot of disagreement over whether the brain can make new neurons. A team at Columbia University led by Maura Boldrini announced in early 2018 that the brain's hippocampi definitely produce at least some new neurons, but a team at the University of California at San Francisco came to precisely the opposite conclusion. The difficulty is that there is no certain way of telling whether neurons in the brain are new or not. What is beyond doubt is that even if we do make some new neurons, it is nothing like enough to offset the kind of loss you get from general aging, never mind stroke or Alzheimer's. So either literally or to all intents and purposes, once you pass early childhood, you have all the brain cells you are ever going to have.

On the plus side, the brain is able to compensate for quite severe loss of mass. In one case cited by the British doctor James Le Fanu in his book *Why Us?*, doctors scanning the brain of a middle-aged man of normal intelligence were astounded to discover that two-thirds of the space inside his skull was occupied by a giant benign cyst that he had evidently had since infancy. All of his frontal lobes and some of his parietal and temporal lobes were missing. The remaining third of his brain had simply taken on the duties and functions of the missing two-thirds and had done it so well that neither he nor anyone else had ever suspected that he was operating at a much-reduced capacity.

For all its marvels, the brain is a curiously undemonstrative organ. The heart pumps, the lungs inflate and deflate, the intestines quietly ripple and gurgle, but the brain just sits pudding-like, giving away nothing. Nothing in its structure outwardly suggests that this is an instrument of higher thinking. As Professor John R. Searle of Berkeley once put it, "If you were designing an organic machine to pump blood you might come up with something like a heart, but if you were designing a machine to produce consciousness, who would think of a hundred billion neurons?"

So it is hardly surprising that our understanding of how the brain functions was slow in coming and largely inadvertent. One of the great (and, it must be said, most written about) events in early neuroscience occurred in 1848 in rural Vermont when a young railroad builder named Phineas Gage was packing dynamite into a rock and it exploded prematurely, shooting a two-foot tamping rod through his left cheek and out the top of his head before it clattered back to Earth about fifty feet away. The rod removed a perfect core of brain about an inch in diameter. Miraculously, Gage survived and appears not even to have lost consciousness, though he did lose his left eye and his personality was forever transformed. Previously happy-go-lucky and popular, he was now moody, argumentative, and given to profane outbursts. He was just "no longer Gage," as one old friend reported sadly. As often happens to people with frontal lobe damage, he had no insight into his condition and didn't understand that he had changed. Unable to settle, he drifted from New England to South America and on to San Francisco, where he died aged thirty-six after falling prey to seizures.

Gage's misfortune was the first proof that physical damage to the brain could transform personality, but over the following decades others noticed that when tumors destroyed or impinged upon parts of the frontal lobes, the victims sometimes became curiously placid and serene. In the 1880s, in a series of operations, a Swiss physician named Gottlieb Burckhardt surgically removed eighteen grams of brain from a disturbed woman, in the process turning her (in his own words) from "a dangerous and excited demented person to a quiet demented one." He tried the process on five more patients, but three died and two developed epilepsy, so he gave up. Fifty years later, in Portugal, a professor of neurology at the University of Lisbon, Egas Moniz, decided to try again and began experimentally cutting the frontal lobes of schizophrenics to see if that might quiet their troubled minds. It was the invention of the frontal lobotomy (though it was then often called a leukotomy, particularly in Britain).

Moniz provided an almost perfect demonstration of how not to do science. He undertook operations without having any idea what damage they might do or what the outcomes would be. He conducted no preliminary experiments on animals. He didn't select his patients with particular care and didn't monitor outcomes closely afterward. He didn't actually perform any of the surgeries himself, but supervised his juniors—though freely took credit for any successes. The practice did actually work up to a point. People with lobotomies generally became less violent and more tractable, but they also routinely suffered massive, irreversible loss of personality. Despite the many shortcomings of the procedure and Moniz's lamentable clinical standards, he was feted around the world and in 1949 received the ultimate accolade of a Nobel Prize.

In the United States, a doctor named Walter Jackson Freeman heard of Moniz's procedure and became his most enthusiastic acolyte. Over a period of almost forty years, Freeman traveled the country performing lobotomies on almost anyone brought before him. On one tour, he lobotomized 225 people in twelve days. Some of his patients were as young as four years old. He operated on people with phobias, on drunks picked up off the street, on people convicted of homosexual acts—on anyone, in short, with almost any kind of perceived mental or social aberration. Freeman's method was so swift and brutal that it made other doctors recoil. He inserted a standard household ice pick into the brain through the eye socket, tapping it through the skull bone with a hammer, then wriggled it vigorously to sever neural connections. Here is his breezy description of the procedure in a letter to his son:

> I have been . . . knocking them out with a shock and while they are under the "anesthetic" thrusting an ice pick up between the eyeball and the eyelid through the roof of the orbit actually into the frontal lobe of the brain and making the lateral cut by swinging the thing from side to side. I have done two patients on both sides and another on one side without running into any complications, except a very black eye in one case. There may be trouble later on but it seemed fairly easy, although definitely a disagreeable thing to watch.

Indeed. The procedure was so crude that an experienced neurologist from New York University fainted while watching a Freeman operation. But it was quick: patients generally could go home within an hour. It was this quickness and simplicity that dazzled many in the medical community. Freeman was extraordinarily casual in his approach. He operated without gloves or a surgical mask, usually in street clothes. The method caused no scarring but also meant that he was operating blind without any certainty about which mental capacities he was destroying. Because ice picks were not designed for brain surgery, sometimes they would break off inside the patient's head and have to be surgically removed, if they didn't kill the patient first. Eventually, Freeman devised a specialized instrument for the procedure, but it was essentially just a more robust ice pick.

What is perhaps most remarkable is that Freeman was a psychiatrist with no surgical certification, a fact that horrified many other physicians. About two-thirds of Freeman's subjects received no benefit from the procedure or were worse off. Two percent died. His most notorious failure was Rosemary Kennedy, sister of the future president. In 1941, she was twenty-three years old, a vivacious and attractive girl but headstrong and with a tendency to mood swings. She also had some learning difficulties, though these seem not to have been nearly as severe and disabling as has sometimes been reported. Her father, exasperated by her willfulness, had her lobotomized by Freeman without consulting his wife. The lobotomy essentially destroyed Rosemary. She spent the next sixty-four years in a care home in the Midwest, unable to speak, incontinent, and bereft of personality. Her loving mother did not visit her for twenty years.

Gradually, as it became evident that Freeman and others like him were leaving trails of human wreckage behind them, the procedure fell out of fashion, especially with the development of effective psychoactive drugs. Freeman continued to perform lobotomies well into his seventies before finally retiring in 1967. But the effects that he and others left in their wake lasted for years. I can speak with some experience here. In the early 1970s, I worked for two years at a psychiatric hospital outside London where one ward was occupied in large part by people who had been lobotomized in the 1940s and 1950s. They were, almost without exception, obedient, lifeless shells.*

* In surely its most questionable entry, the 2001 *Oxford Companion to the Body* says, "For many people the term 'lobotomy' conjures up images of disturbed

* * *

The brain is one of our most vulnerable organs. Paradoxically, the very fact that the brain is so snugly encased in its protective skull leaves it susceptible to damage when it swells from infection or when fluid is added to it, as with a bleed, because the additional material has nowhere to go. The result is compression of the brain, which can be fatal. The brain is also easily injured by being dashed against the skull by sudden violence as in a car crash or fall. A thin layer of cerebrospinal fluid in the meninges, the brain's outer membrane, provides a bit of cushioning, but only a bit. These injuries, known as contrecoup injuries, appear on the opposite side of the brain from the point of impact because the brain is flung against its own protective (or in this case not so protective) casing.

Above all, the brain is vulnerable to its own internal storms. Strokes and seizures are peculiarly human frailties. Most other mammals never suffer strokes, and for those that do, it is a rare event. But for humans, it is the second most common cause of death globally, according to the World Health Organization. Why this should be is something of a mystery. As Daniel E. Lieberman observes in *The Story of the Human Body,* we have an excellent blood supply to the brain to minimize stroke and yet we get strokes.

Epilepsy likewise is a perennial mystery, but with the additional burden that sufferers have been shunned and demonized throughout history. Well into the twentieth century, it was commonly believed by medical authorities that seizures were infectious—that just watching someone have a seizure could provoke a seizure in others. Epileptics were often treated as mental defectives and confined to institutions. As recently as 1956, it was illegal in seventeen U.S. states for epileptics to marry; in eighteen states, epileptics could be involuntarily sterilized. The last of these laws was repealed only in 1980. In Britain, epilepsy

---

beings whose brains have been damaged or mutilated extensively, leaving them at best in a vegetative state without a personality or feelings. This was never true." Actually, it was.

remained on the statute books as grounds for annulment until 1970. As Rajendra Kale put it in the *British Medical Journal* some years ago, "The history of epilepsy can be summarised as 4,000 years of ignorance, superstition and stigma followed by 100 years of knowledge, superstition and stigma."

Epilepsy isn't really a single disease but a collection of symptoms that can range from a brief lapse of awareness to prolonged convulsions, all caused by misfiring neurons in the brain. Epilepsy can be brought on by illness or head trauma, but very often there is no clear precipitating event, just a sudden, frightening seizure from out of the blue. Modern drugs have greatly reduced or eliminated seizures for millions of sufferers, but about 20 percent of epileptics do not respond successfully to medications. Every year about one epileptic in a thousand dies during or just after a seizure in a condition known as sudden unexpected death in epilepsy. As Colin Grant noted in *A Smell of Burning: The Story of Epilepsy*, "No one knows what causes it. The heart just stops." (An additional one in a thousand epileptics dies tragically each year from losing consciousness in unfortunate circumstances—in the bath, say, or by striking their head badly in a fall.)

The inescapable fact is that the brain is an unnerving place as well as a marvelous one. There seems to be an almost limitless number of curious or bizarre syndromes and conditions associated with neural disorders. Anton-Babinski syndrome, for instance, is a condition in which people are blind but refuse to believe it. In Riddoch syndrome, victims cannot see objects unless they are in motion. Capgras syndrome is a condition in which sufferers become convinced that those close to them are impostors. In Klüver-Bucy syndrome, the victims develop an urge to eat and fornicate indiscriminately (to the understandable dismay of loved ones). Perhaps the most bizarre of all is Cotard delusion, in which the sufferer believes he is dead and cannot be convinced otherwise.

Nothing about the brain is simple. Even being unconscious is a complicated matter. As well as being asleep, anesthetized, or concussed, you can be in a coma (eyes closed and wholly unaware), a vegetative state (eyes open but unaware), or minimally conscious (occasionally lucid but mostly confused or unaware). Locked-in syndrome is different again. It is being fully alert but paralyzed and often able to communicate only with eye blinks.

No one knows how many people are alive but minimally conscious or worse, but *Nature Neuroscience* suggested in 2014 that the number globally is probably in the hundreds of thousands. In 1997, Adrian Owen, then a young neuroscientist working in Cambridge, England, discovered that some people thought to be in a vegetative state are in fact fully aware but powerless to indicate the fact to anyone.

In his book *Into the Gray Zone,* Owen discusses the case of a patient named Amy who suffered a serious head injury in a fall and for years lay in a hospital bed. Using an fMRI scanner, and carefully watching the woman's neural responses when researchers asked her a series of questions, they were able to determine that she was fully conscious. "She had heard every conversation, recognised every visitor, and listened intently to every decision being made on her behalf." But she was unable to move a muscle–to open her eyes, scratch an itch, express any desire. Owen believes that something in the region of 15 to 20 percent of people thought to be in a permanent vegetative state are in fact fully aware. Even now the only certain way to tell if a brain is working is if its owner says it is.

Perhaps nothing is more unexpected about our brains than that they are much smaller today than they were ten thousand or twelve thousand years ago, and by quite a lot. The average brain has shrunk from 1,500 cubic centimeters then to 1,350 cubic centimeters now. That's equivalent to scooping out a portion of brain about the size of a tennis ball. That's not at all easy to explain, because it happened all over the world at the same time, as if we agreed to reduce our brains by treaty. The common presumption is that our brains have simply become more efficient and able to pack more performance into a smaller space, rather like cell phones, which have grown more sophisticated as they have contracted in size. But no one can prove that we haven't simply grown dimmer.

Over roughly the same period, our skulls have also become thinner. No one can really explain that either. It may be simply that a less robust and active lifestyle means that we don't need to invest in skull bone in the way we used to. But then again it may simply be that we aren't what we once were.

And with that sobering thought to reflect upon, let's look at the rest of the head.

5 THE HEAD

> This was not merely an idea, but a flash of inspiration. At the sight of that skull, I seemed to see all of a sudden, lighted up as a vast plain under a flaming sky, the problem of the nature of the criminal.
>
> —CESARE LOMBROSO

WE ALL KNOW that you can't live without your head, but how long exactly is a question that received rather a lot of attention in the late eighteenth century. It was a good time to wonder because the French Revolution gave inquiring minds a steady supply of freshly lopped heads to examine.

A decapitated head will still have some oxygenated blood in it, so loss of consciousness may not be instantaneous. Estimates of how long the brain can keep working range from two seconds to seven, and that is assuming a clean removal, which was by no means always the case. Heads don't come off easily even with stout blows from a specially sharpened ax wielded by an expert. As Frances Larson notes in her fascinating history of decapitation, *Severed,* Mary, Queen of Scots, needed three hearty whacks before her head hit the basket, and hers was a comparatively delicate neck.

Many observers at executions claimed to have witnessed evidence of consciousness from newly separated heads. Charlotte Corday, guillotined in 1793 for the murder of the radical leader Jean-Paul Marat, was said to wear a look of fury and resentment when the executioner held her head up to the cheering crowd. Others, as Larson notes, were reported to have blinked or moved their lips as if trying to speak. A man named Terier was said to have turned his gaze to a speaker some fifteen minutes after being separated from his body. But how much of this was reflex, or exaggerated in the retelling, no one could say. In 1803, two German researchers decided to bring some scientific rigor to the matter. They pounced on the heads as they fell and examined them immediately for any sign of alertness, shouting, "Do you hear me?" None responded, and the investigators concluded that loss of consciousness was immediate or at least too swift to measure.

No other part of the body has received more misguided attention, or proved more resistant to scientific understanding, than the head. The nineteenth century in particular was something of a golden age in this respect. The period saw the rise of two distinct but often confused disciplines, phrenology and craniometry. Phrenology was the practice of correlating bumps on a skull with mental powers and attributes of character, and it was always a marginal pursuit. Craniometrists virtually without exception dismissed phrenology as crackpot science while promulgating an alternative nonsense of their own. Craniometry focused on more precise and comprehensive measurements of volume, shape, and structure of the head and brain but in pursuit, it must be said, of equally preposterous conclusions.*

The greatest cranial enthusiast of all, now forgotten but once very famous indeed, was Barnard Davis (1801–81), a doctor in the English Midlands. Davis became gripped by craniometry in the 1840s and rapidly made himself into the world's supreme authority. He produced a stream of books with weighty titles like *On the Peculiar Crania of*

* Craniometry is also sometimes referred to as craniology, in which case it needs to be distinguished from the modern, perfectly respectable discipline of the same name. Modern craniology is used by anthropologists and paleontologists to study anatomical differences in ancient peoples, and by forensic scientists to make determinations about the age, sex, and race of recovered skulls.

the Inhabitants of Certain Groups of Islands in the Western Pacific and *Contributions Towards Determining the Weight of the Brain in Different Races of Man.* These were surprisingly popular. *On Synostotic Crania Among Aboriginal Races of Man* went through fifteen editions. The epic *Crania Britannica,* published in two volumes, had thirty-one editions. Davis became so celebrated that people from all over the world, among them the president of Venezuela, left their skulls for him to study. Gradually, he built up the world's largest collection of skulls—1,540 in all, or more than all the skulls in all the world's other institutions combined.

Davis would stop at almost nothing to enlarge his collection. When he wished for skulls from the indigenous people of Tasmania, he wrote to George Robinson, official protector of aborigines, for a selection. Because the plundering of aboriginal graves had by this time become a criminal act, Davis supplied Robinson with detailed instructions on how to remove a skull from an indigenous Tasmanian and replace it with the skull of any convenient surrogate in a way that would avoid arousing suspicions. He was evidently successful in his endeavors, for his collection soon included sixteen Tasmanian skulls and one whole skeleton.

Davis's fundamental ambition was to prove that dark-skinned people were created separately from light-skinned people. He was convinced that a person's intellect and moral compass were indelibly written in the curves and apertures of the skull and that these were exclusively products of race and class. People with "cephalic peculiarities" should be treated "not as criminals but as dangerous idiots," he suggested. In 1878, at the age of seventy-seven, he married a woman fifty years his junior. What her cranium was like is unknown.

This instinct on the part of European authorities to prove all other races inferior was widespread, if not universal. In England, in 1866 the eminent physician John Langdon Haydon Down (1828–96) first described the condition that we now know as Down's syndrome in a paper called "Observations on an Ethnic Classification of Idiots," but he referred to it as "Mongolism" and its victims as "Mongoloid

SMITH/DAVID

SEAT 12D

EXIT

ATLANTA

BLOOMINGTON

DL5226 21FEB

Operated By:

ENDEAVOR DBA DELTA CNX

idiots" in the belief that they were suffering an innate regression to an inferior, Asiatic type. Down believed, and no one seems to have doubted him, that idiocy and ethnicity were conjoined qualities. He also listed "Malay" and "Negroid" as regressive types.

In Italy, meanwhile, Cesare Lombroso (1835–1909), the country's most eminent physiologist, developed a parallel theory called criminal anthropology. Lombroso believed that criminals were evolutionary throwbacks who betrayed their criminal instincts through a range of anatomical features—slope of the forehead, whether their earlobes were rounded or spade shaped, even the amount of spacing between their toes. (People with a lot of toe space were closer to apes, he explained.) Though his assertions were without the faintest scientific validity, Lombroso was widely esteemed and is even now sometimes referred to as the father of modern criminology. Lombroso was frequently called as an expert witness. In one case, cited by Stephen Jay Gould in *The Mismeasure of Man,* he was asked to determine which of two men had killed a woman. Lombroso declared one man self-evidently guilty because he had "enormous jaws, frontal sinuses and zygomata, thin upper lip, huge incisors, unusually large head [and] tactile obtuseness with sensorial manicinism." Never mind that no one knew what much of that meant and that there was no actual evidence against the poor fellow. He was found guilty.

But the most influential, and unexpected, practitioner of craniometry was the great French anatomist Pierre Paul Broca (1824–80). Broca was without question a brilliant scientist. In 1861, during an autopsy on a stroke victim who hadn't spoken for years except to repeat endlessly the syllable "tan," Broca discovered the brain's speech center in the frontal lobe—the first time that anyone had connected an area of the brain to a specific action. The speech center is still called Broca's area, and the impediment Broca discovered is Broca's aphasia. (Under it, a person can understand speech but can't reply except to utter meaningless noises or sometimes stock phrases like "I'll say" or "Oh, boy.")

Broca was less astute, however, with respect to character traits.

He was convinced, even when all the evidence was against him, that females, criminals, and dark-skinned foreigners had smaller, less agile brains than their white male counterparts. Whenever Broca was presented with evidence that contradicted this, he disregarded it on the grounds that it must be flawed. He was similarly disinclined to believe a study from Germany showing that German brains were on average a hundred grams weightier than French ones. He explained this awkward discrepancy by suggesting that the French subjects were very old when tested and that their brains had shrunk. "The degree of decadence that old age can impose upon a brain is very variable," he observed. He also had problems accounting for why executed criminals sometimes had big brains, and decided that their brains had become artificially engorged by the stress of hanging. The greatest indignity of all came when Broca's own brain was measured upon his death and was found to be smaller than average.

The person who finally put the study of the human head on something like a sound scientific foundation was none other than the great Charles Darwin. In 1872, thirteen years after he published *On the Origin of Species,* Darwin produced another landmark work, *The Expression of the Emotions in Man and Animals,* which looked at expressions reasonably and without prejudice. The book was revolutionary not just for being sensible but for observing that certain expressions appear to be common to all peoples. This was a much bolder utterance than we may realize today because it underlined Darwin's conviction that all people, whatever their race, have a common heritage, and that was a very revolutionary thought in 1872.

What Darwin realized was something that all babies know instinctively—that the human face is highly expressive and instantly captivating. No two authorities seem to agree on quite how many expressions we can make—estimates range from forty-one hundred to ten thousand—but it is clearly a large number. More than forty muscles, a significant portion of the body's total, are involved in facial expression. Babies fresh from the womb are said to prefer a face, or even the general pattern of a face, to any other shape. Whole regions of the brain are devoted solely to recognizing faces. We are exquisitely sensitive to the subtlest alterations of mood or expression, even if we are not always conscious of them. In an experiment related by Daniel McNeill in his book *The Face,* men were shown two photos of women that were identical in every respect except that the pupils had been subtly enlarged in one. Although the change was too slight to be consciously perceived, the test subjects invariably found the women with larger pupils more attractive, though they were at a loss to explain why.

In the 1960s, nearly a century after Darwin wrote *The Expression of the Emotions,* Paul Ekman, a professor of psychology at the University of California at San Francisco, decided to test the universality, or not, of facial expressions by studying remote tribal people in New Guinea who had no acquaintance with Western habits. Ekman concluded that six expressions are universal: fear, anger, surprise, pleasure, disgust, and sorrow. The most universal expression of all is a smile, which is rather a nice thought. No society has ever been found that doesn't respond to smiles in the same way. True smiles are brief–between two-thirds of a second and four seconds. That's why a held smile begins to look menacing. A true smile is the one expression that we cannot fake. As the French anatomist G.-B. Duchenne de Boulogne noticed as long ago as 1862, a genuine, spontaneous smile involves the contraction of the orbicularis oculi muscle in each eye, and we have no independent control over those muscles. You can make your mouth smile, but you can't make your eyes sparkle with feigned joy.

According to Paul Ekman, we all indulge in "microexpressions"– flashes of emotion, no more than a quarter of a second in duration, that betray our true inner feelings regardless of what our more general, controlled expression is conveying. Nearly all of us miss these telltale expressions, according to Ekman, but we can be taught to spot them, assuming we want to know what workmates and loved ones really think of us.*

By primate standards, we have a very odd head. Our faces are flat, our foreheads high, and our noses protuberant. Almost certainly a number of factors are responsible for our distinctive facial arrangements—our upright posture, our biggish brain, our diet and lifestyle, the fact that we are built for sustained running (which affects how we breathe), and the things that we find adorable in a mate. (Dimples, for instance—not something that gorillas look for when feeling frisky.)

Surprisingly, given how central faces are to our existence, quite a lot about them is still a mystery to us. Take eyebrows. All the many species of hominids that preceded us had prominent browridges, but we *Homo sapiens* gave them up in favor of our small, active eyebrows. It's not easy to say why. One theory is that eyebrows are there to keep sweat out of the eyes, but what the eyebrows do really well is convey feelings. Think how many messages you can send with a single arched eyebrow, from "I find that hard to believe" to "Watch your step" to "Care to have sex?" One of the reasons the *Mona Lisa* looks enigmatic is that she has no eyebrows. In one interesting experiment, subjects were shown two sets of digitally doctored photographs of well-known people: one with the eyebrows eliminated and the other with the eyes themselves taken away. Surprisingly, but overwhelmingly, volunteers found it harder to identify the celebrities without eyebrows than without eyes.

* Surely, however, any figure must be largely notional. How on earth would you distinguish, let us say, expression No. 1,013 from No. 1,012 or 1,014? Any such differences would have to be practically microscopic. Even some basic expressions are almost impossible to distinguish. Fear and surprise cannot usually be told apart without knowing the context that prompted the emotion.

Eyelashes are similarly uncertain. There is some evidence to suggest that eyelashes subtly change airflow around the eye, helping to waft away motes of dust and other tiny particulates from landing there, but the main benefit is probably that they add interest and allure to faces. People with long eyelashes are generally rated more attractive than those without.

Even more anomalous is the nose. It is the convention among mammals to have snouts, not round, projecting noses. According to Daniel Lieberman, professor of human evolutionary biology at Harvard, our external nose and intricate sinuses evolved to help with breathing efficiency and with keeping us from becoming overheated on long runs. It is an arrangement that has clearly suited us, for humans and their ancestors have had projecting noses for some two million years.

Most mysterious of all is the chin. The chin is unique to humans, and no one knows why we have one. It doesn't seem to confer any structural benefit to the head, so it may be simply that we find a good chin dashing. Lieberman, in a rare moment of lightness, observed, "Testing this last hypothesis is especially difficult, but the reader is encouraged to think of appropriate experiments." It is certainly the case that we talk about "chinless wonders" and otherwise equate modest chins with deficiencies of character and intellect.

Much as we all appreciate a pert nose or gorgeous eyes, the real purpose of most of our facial features is to help us interpret the world through our senses. It's curious that we always speak of our five senses because we have way more than that. We have a sense of balance, of acceleration and deceleration, of where we are in space (what is known as proprioception), of time passing, of appetite. Altogether (and depending on how you count them) we have as many as thirty-three systems within us that let us know where we are and how we are doing.

We'll explore the sense of taste in the next chapter when we venture into the mouth, but let's look now at the three other most familiar senses of the head: sight, hearing, and smell.

SIGHT

THE EYE IS a thing of wonder, needless to say. About a third of your entire cerebral cortex is engaged with vision. Victorians so marveled at the intricacy of the eye that they often cited it as proof of intelligent design. It was an odd choice because the eye is really rather the reverse—literally so, for it is built back to front. The rods and cones that detect light are at the rear, but the blood vessels that keep it oxygenated are in front of them. There are vessels and nerve fibers and other incidental detritus all over, and your eye has to see through all this. Normally, your brain edits out any interference, but it doesn't always succeed. You might have had the experience of looking at a clear blue sky on a sunny day and seeing little white sparks popping in and out of existence, like the briefest of shooting stars. What you are seeing, amazingly enough, is your own white blood cells, moving through a capillary in front of the retina. Because white blood cells are big (compared with red blood cells), they sometimes get stuck briefly in the narrow capillaries, and that is what you are seeing. The technical name for these disturbances is Scheerer's blue field entoptic phenomena (named for a German ophthalmologist of the early twentieth century, Richard Scheerer), though they are more commonly and poetically known as blue sky sprites. They are especially visible against a bright blue sky simply because of the way the eye absorbs different wavelengths of light.

Floaters are a similar phenomenon. They are clumps of microscopic fibers in the jellylike vitreous humor of your eye, which cast a shadow on the retina. Floaters are a common occurrence as you get older, and are generally harmless, though they can indicate a retinal tear. The technical name for them, if you wish to impress someone, is muscae volitantes, or "hovering flies."

If you held a human eyeball in your hand, you might be surprised by its size because we only see about one-sixth of it when it is embedded in the eye socket. The eye feels like a gel-filled bag, which is not surprising, because it is filled with a gel-like material, the aforementioned vitreous humor. (Humor in its anatomical sense signifies any fluid or semifluid in the body and not, obviously, its ability to generate laughs.)

As you would expect of a complex instrument, the eye has many parts, some of which are well known to us by name (iris, cornea, retina) and others of which are more obscure (fovea, choroid, sclera), but essentially it is a camera. The front part—the lens and cornea—captures passing images and projects them onto the back wall of the eye—the retina—where photoreceptors convert them into electrical signals that are passed on to the brain via the optic nerve.

If there is one part of your visual anatomy that deserves a moment's thanks, it is the cornea. This modest, dome-shaped goggle not only protects the eye from worldly assaults but actually does two-thirds of the eyeball's focusing. The lens, which gets all the credit in the popular mind, does only about a third of the focusing. The cornea could hardly be less imposing. If you were to pop it out and lay it on the tip of your finger (where it would fit very comfortably), it wouldn't seem much at all. But on closer examination, as with almost every part of the body, it is a wonder of complexity. It has five layers—epithelium, Bowman's membrane, stroma, Descemet's membrane, and endothelium—laminated into a space just slightly over half a millimeter thick. In order to be transparent, it has a very modest blood supply—indeed, practically none. The part of the eye that has the most photoreceptors—that really does the seeing—is called the fovea (from a Latin word for "shallow pit"; the fovea inhabits a slight depression). It is interesting that such a crucial part is one that most of us have never heard of.

To keep all this working smoothly (in the most literal sense), we produce tears constantly. Tears not only keep our eyelids gliding smoothly but also even out tiny imperfections on the eyeball surface, making focused vision possible. They also contain antimicrobial chemicals, which successfully keep most pathogens at bay. Tears come in three varieties: basal, reflex, and emotional. Basal are the functional ones that provide lubrication. Reflex tears are those that emerge when the eye is irritated by smoke or sliced onions or similar. And emotional tears are of course self-evident, but they are also unique. We are the only creatures that cry from feeling, as far as we can tell. Why we do so is another of life's many mysteries. We get no physiological benefit from erupting in tears. It is also a little odd surely that this act signifying powerful sadness is also triggered by extreme joy or quiet rapture or intense pride or almost any other potent emotional state.*

Producing tears involves an extraordinary number of tiny glands around the eyes—namely, the Glands of Krause, Wolfring, Moll, and Zeis, as well as nearly four dozen Meibomian glands in the eyelids. Altogether you produce about five to ten ounces of tears a day. The tears drain away through holes known as puncta on the little fleshy knob (known as the papilla lacrimalis) in the corner of each eye beside the nose. When you cry emotionally, the puncta cannot drain the fluid fast enough, so it overflows your eyes and runs down your cheeks.

The iris is what gives the eye its color. It is composed of a pair of muscles that adjust the opening of the pupil, rather like the aperture on a camera, to let in or keep out light as needed. Superficially, the iris looks like a neat ring, encircling the pupil, but closer inspection shows that it is in fact "a riot of spots, wedges, and spokes," in the words of Daniel McNeill, and these patterns are unique to each of us, which is why iris recognition devices are now increasingly used to identify us at security checkpoints.

The white of the eye is formally known as the sclera (from a Greek word for "hard"). Our scleras are unique among primates. They allow us to monitor the gazes of others with considerable precision, as

* Incidentally, twenty-twenty vision means only that you see as well from twenty feet what any other reasonably well-sighted person would see. It doesn't mean that your vision is perfect.

well as to communicate silently. You have only to move your eyeballs slightly to get a companion to look at, let's say, someone at a neighboring table in a restaurant.

Our eyes contain two types of photoreceptors for vision–rods, which help us see in dim conditions but provide no color, and cones, which work when the light is bright and divide the world up into three colors: blue, green, and red. People who are "color-blind" normally lack one of the three types of cones, so they don't see all the colors, just some of them. People who have no cones at all, and are genuinely color-blind, are called achromatopes. Their main problem isn't that their world is pallid but that they really struggle to cope with bright light and can be literally blinded by daylight. Because we were once nocturnal, our ancestors gave up some color acuity–that is, sacrificed cones for rods–to gain better night vision. Much later, primates re-evolved the ability to see reds and oranges, the better to identify ripe fruit, but we still have just three kinds of color receptors compared with four for birds, fish, and reptiles. It's a humbling fact, but virtually all nonmammalian creatures live in a visually richer world than we do.

On the other hand, we make pretty good use of what we have got. The human eye can distinguish somewhere between 2 million and 7.5 million colors, according to various calculations. Even at the lower end of estimates, that is a lot.

Your visual field is surprisingly compact. Look at your thumbnail at arm's length; that's about the area you have in full focus at any given instant. But because your eye is constantly darting–taking four snapshots every second–you have the impression of seeing a much broader area. The movements of the eye are called saccades (from a French word meaning "to pull violently"), and you have about a quarter of a million of them every day without ever being aware of it. (Nor do we notice it in others.)

In addition, all the nerve fibers leave the eye via a single channel at the back, resulting in a blind spot about fifteen degrees off center in our field of vision. The optic nerve is fairly hefty–it is about the thickness of a pencil–which is quite a lot of visual space to lose. You can experience this blind spot by means of a simple trick. First, close your left eye and stare straight ahead with the other. Now hold up one finger from your right hand as far from your face as you can. Slowly move the finger through your field of vision while steadfastly staring straight ahead. At some point, rather miraculously, the finger will disappear. Congratulations. You have found your blind spot.

You don't normally experience the blind spot, because your brain continually fills in the void for you. The process is called perceptual interpolation. The blind spot, it's worth noting, is much more than just a spot; it's a substantial portion of your central field of vision. That's quite remarkable–that a significant part of everything you "see" is actually imagined. Victorian naturalists sometimes cited this as additional proof of God's beneficence, without evidently pausing to wonder why He had given us a faulty eye to begin with.

HEARING

HEARING IS ANOTHER seriously underrated miracle. Imagine being given three tiny bones, some wisps of muscle and ligament, a delicate membrane, and some nerve cells, and from them trying to fashion a device that can capture with more or less perfect fidelity the complete panoply of auditory experience–intimate whispers, the lushness of symphonies, the soothing patter of rain on leaves, the drip of a tap in another room. When you place a set of $800 headphones over your ears and marvel at the rich, exquisite sound, bear in mind that all that that expensive technology is doing is conveying to you a reasonable approximation of the auditory experience that your ears give you for nothing.

The ear consists of three parts. The outermost of these, the floppy shell on the side of our heads that we call "the ear," is formally the pinna (from the Latin for "fin" or "feather," a bit oddly). On the face of it, the pinna would seem ill-designed to do its job. Any engineer, starting from scratch, would design something larger and more rigid–more like a satellite dish, say–and certainly wouldn't allow hair to cascade over it. In fact, however, the fleshy whorls of our outer ears do a surprisingly good job of capturing passing sounds—and, more than that, of stereoscopically working out where they come from and whether they demand attention. That is why you can not only hear someone across the room speak your name at a cocktail party but turn your head and identify the speaker with uncanny accuracy. Your forebears spent eons as prey to endow you with this benefit.

Although all outer ears function in the same way, each set, it appears, is uniquely built and as distinctive as the owner's fingerprints. According to the British scientist and author Desmond Morris, two-thirds of Europeans have free-hanging earlobes and one-third have attached lobes. Whether tethered or flapping, the earlobes make no difference to your hearing or indeed anything else.

The passage beyond the pinna, the ear canal, ends in a taut and sturdy piece of tissue known to science as the tympanic membrane and to the rest of us as the eardrum, which marks the boundary between the outer ear and the middle ear. The tiny quiverings of the eardrum are passed on to the three smallest bones in the body, collectively known as ossicles and individually known as the malleus, incus, and stapes (or hammer, anvil, and stirrup, because of their very vague resemblances to those objects). The ossicles are perfect demonstrations of how evolution is so often a matter of make-do. They were jawbones in our ancient ancestors and only gradually migrated to new positions in our inner ear. For much of their history, those three bones had nothing to do with hearing.

The ossicles exist to amplify sounds and pass them on to the inner ear via the cochlea, a snail-shaped structure (cochlea means "snail") that is filled with twenty-seven hundred delicate hairlike filaments called stereocilia, which wave like ocean grasses as sound waves pass across them. The brain then puts all the signals together and works out what it has just heard. All this is done on a sublimely modest scale—the cochlea is no bigger than a sunflower seed, the three bones of the ossicles would fit on a shirt button—yet it works incredibly well. A pressure wave that moves the eardrum by less than the width of an atom will activate the ossicles and reach the brain as sound. You genuinely cannot improve upon that. As the acoustics scientist Mike Goldsmith has put it, "If we could hear quieter sounds still, we would live in a world of continuous noise, because the omnipresent random motion of air molecules would be audible. Our hearing really could not get any better." From the quietest detectable sound to the loudest is a range of about a million million times of amplitude.

To help protect us from the damage of really loud noises, we have something called an acoustic reflex, in which a muscle jerks the stapes away from the cochlea, essentially breaking the circuit, whenever a brutally intense sound is perceived, and it maintains that posture for some seconds afterward, which is why we are often deafened after an explosion. Unfortunately, the process is not perfect. Like any reflex, it is quick but not instantaneous, and it takes about a third of a second for the muscle to contract, by which point a lot of damage can be done.

Our ears are built for a quiet world. Evolution did not foresee that one day humans would insert plastic buds in their ears and subject their eardrums to a hundred decibels of melodic roar across a span of millimeters. The stereocilia tend to wear out anyway as we age, and they do not, alas, regenerate. Once you disable a stereocilium, it remains lost to you forever. There isn't any particular reason for this. Stereocilia grow back perfectly well in birds. They just don't do it in us. The high-frequency ones are at the front and the low-frequency ones farther in. This means that all sound waves, high and low, pass over the high-frequency cilia, and this heavier traffic means they wear out more quickly.

In order to gauge the power, intensity, and loudness of different sounds, acoustic scientists in the 1920s came up with the concept of the decibel. The term was coined by Colonel Sir Thomas Fortune Purves, chief engineer of the British Post Office (which in those days was in charge of the British telephone system, hence the interest in sound amplification). The decibel is logarithmic, which means that its units of increment are not mathematical in the everyday sense of the term but increase by orders of magnitude. So the sum of two

10-decibel sounds is not 20 decibels but 13 decibels. Volume doubles about every 6 decibels, which means that a 96-decibel noise is not just a bit louder than a 90-decibel noise but twice as loud. The pain threshold for noise is about 120 decibels, and noises above 150 decibels can burst the eardrum. For purposes of comparison, a quiet place like a library or the countryside is about 30 decibels, snoring is 60 to 80 decibels, a really loud nearby thunderclap is 120 decibels, and standing in the wash of a jet engine at takeoff would be 150 decibels.

The ear is also responsible for keeping you balanced thanks to a tiny but ingenious collection of semicircular ducts and two tiny associated sacs called otolith organs, which together are called the vestibular system. The vestibular system does everything that a gyroscope does on an airplane, but in an extremely miniaturized form. Inside the vestibular channels is a gel that acts a little like the bubbles in a carpenter's level, in that the gel's movements from side to side or up and down tell the brain in which direction we are traveling (which is how you can sense whether you are going up or down in an elevator even in the absence of visual clues). The reason we feel dizzy when we jump from a merry-go-round is that the gel keeps moving even though the head has stopped, so the body is temporarily disoriented. That gel thickens as we age and doesn't slosh around as well, which is one reason why the elderly are often not so steady on their feet (and why they especially shouldn't jump from moving objects). When loss of balance is prolonged or severe, the brain doesn't know quite what to make of it and interprets it as poisoning. That is why loss of balance so generally results in nausea.

Another part of the ear that intrudes upon our consciousness from time to time is the Eustachian tube, which forms a kind of escape tunnel for air between the middle ear and the nasal cavity. Everyone knows that uncomfortable feeling you get in your ears when you change heights rapidly, as when coming in to land in an airplane. It is known as the Valsalva effect, and it arises because the air pressure inside your head fails to keep up with the changing air pressure outside it. Making your ears pop by blowing out while keeping your mouth and nose closed is known as the Valsalva maneuver. Both are named for a seventeenth-century Italian anatomist, Antonio Maria Valsalva—who also, not incidentally, named the Eustachian tube, after his fellow anatomist Bartolomeo Eustachi. As your mother doubtless told you, you shouldn't blow too hard. People have ruptured eardrums from doing so.

SMELL

SMELL IS THE sense that nearly everyone says they would give up if they had to give up one. According to one survey, half of people under the age of thirty said they would sacrifice their sense of smell rather than part with a favored electronic device. I hope it isn't necessary for me to observe that that would be a little foolish. Smell is, in fact, a lot more important to happiness and fulfillment than most people appreciate.

At the Monell Chemical Senses Center in Philadelphia, they are devoted to understanding smell, and thank goodness because not very many others are. Housed in an anonymous brick building alongside the campus of the University of Pennsylvania, the Monell is the largest research institution in the world dedicated to the complex and neglected senses of taste and smell.

"Smell is something of an orphan science," said Gary Beauchamp when I visited in the autumn of 2016. A friendly, soft-spoken man with a trim white beard, Beauchamp is president emeritus of the center. "The number of papers published on vision and hearing is in the tens of thousands every year," he told me. "On smell, it is a few hundred at most. It is the same with research money, where funding is at least ten to one in favor of hearing and vision over smell."

One consequence of this is that there is a great deal that we still don't know about smell, including exactly how it works. When we sniff or inhale, odor molecules in the air drift into our nasal passages and come into contact with the olfactory epithelium—a patch of nerve cells containing some 350 to 400 types of odor receptors. If the right kind of molecule activates the right kind of receptor, it sends a signal to the brain, which interprets it as a smell. How exactly this happens is where the controversy lies. Many authorities believe the odor molecules fit into the receptors like a key into a lock. A problem with this theory is that sometimes molecules have different chemical shapes but the same smell, and some have almost matching shapes but dissimilar smells, which suggests that a simple shape explanation is not enough. So there is a competing, rather more complicated theory which is that the receptors are activated by something called resonance. Essentially, the receptors are stimulated not by the shape of molecules but by how they vibrate.

For those of us who are not scientists, it doesn't really matter, because the outcome is the same in either case. What is important is that odors are complex and hard to deconstruct. Aroma molecules typically activate not one type of odor receptor but several, rather like a pianist playing chords—but on an enormous keyboard. A banana, for example, contains three hundred volatiles, as the active molecules in aromas are called. Tomatoes have four hundred, coffee no fewer than six hundred. Working out how and to what degree these contribute to an aroma is not straightforward. Even at the simplest level, results are often wildly counterintuitive. If you combine the fruity odor of ethyl isobutyrate with the caramel-like allure of ethyl maltol and the violet scent of allyl alpha-ionone, you get pineapple, which smells wholly unlike its three principal inputs. Still other chemicals have very different structures but produce the same smell, and no one knows why that happens either. The smell of burned almonds can be produced by seventy-five different chemical combinations that have nothing in common beyond how the human nose perceives them. Because of the complexities, we are still very much at the beginning of an understanding of it all. The smell of licorice, for instance, was decoded only in 2016. Many, many other common odors are still to be deciphered.

For decades, it was universally agreed that humans can discriminate about ten thousand different smells, but then someone decided to look into the origin of the claim and found that it was first suggested way back in 1927 by two chemical engineers in Boston who simply guessed at it. In 2014, researchers at the Université Pierre et Marie Curie in Paris and Rockefeller University in New York reported in the journal *Science* that in fact we can detect vastly more odors than that—at least a trillion and possibly even more than that. At once other scientists in the field called into question the statistical methodology used in the study. "These claims have no basis," Markus Meister, a professor of biological sciences at Caltech, flatly declared.

An interesting and important curiosity of our sense of smell is that it is the only one of the five basic senses not mediated by the hypothalamus. When we smell something, the information, for reasons unknown, goes straight to the olfactory cortex, which is nestled close to the hippocampus, where memories are shaped, and it is thought by some neuroscientists that that may explain why certain odors are so powerfully evocative of memories for us.

Smell is certainly an intensely personal experience. "I think the single most extraordinary aspect of olfaction is that we all smell the world differently," Beauchamp says. "Although we all have 350 to 400 types of odor receptor, only about half of them are common to all people. That means that we don't smell the same things."

He reached into his desk and pulled out a vial, which he uncapped and passed to me to sniff. I could smell nothing at all.

"It's a hormone called androsterone," Beauchamp explained. "About a third of people, like you, can't smell it. One-third smell something like urine, and one-third smell sandalwood." His smile broadened. "If you have three people who cannot even agree on whether something is pleasant, revolting, or simply odorless, you begin to see how complicated the science of smell is."

We are better at detecting odors than most of us realize. In an arresting experiment, researchers at the University of California at Berkeley dragged a chocolate scent around a huge grassy field and had volunteers try to follow the trail as a bloodhound would, on their hands and knees and with their noses to the ground. Amazingly, about two-thirds of the volunteers were able to follow the scent with considerable accuracy. For five of fifteen smells tested, humans actually outperformed dogs. Other tests have shown that people given a selection of T-shirts to sniff can generally identify the one worn by their spouse. Babies and mothers are similarly skillful at identifying each other by odor. Smell, in short, is much more important to us than we appreciate.

Total smell loss is known as anosmia, and partial loss is hyposmia. Somewhere between 2 and 5 percent of people in the world suffer from one or the other, which is a very high proportion. An especially wretched minority experience cacosmia, which is where everything smells like feces, and it is, by all accounts, as horrible as you would imagine. At Monell, they refer to smell loss as "an invisible disability."

"People hardly ever lose their sense of taste," Beauchamp told me. "Taste is supported by three different nerves, so there is quite a lot of backup. Our sense of smell is much more vulnerable." The main cause of smell loss is infectious diseases like flu and sinusitis, but it can also result from a knock to the head or neural degeneration. One of the early symptoms of Alzheimer's is smell loss. Ninety percent of people who lose smell through head injury never get the sense back; a smaller proportion, about 70 percent, who lose smell through infections suffer permanent loss.

"People who lose their sense of smell are usually astounded at how much pleasure it takes out of their lives," says Beauchamp. "We depend on smell for interpreting the world but also, no less crucially, for getting pleasure from it."

This is especially true of food, and for that very important subject we need another chapter.

6 DOWN THE HATCH: THE MOUTH AND THROAT

To lengthen thy life, lessen thy meals.

—BENJAMIN FRANKLIN

IN THE SPRING of 1843, the great British engineer Isambard Kingdom Brunel took a rare break from his labors—at the time he was building the SS *Great Britain,* the largest and most challenging ship ever to come off a drawing board to that time—to amuse his children with a magic trick. Things didn't go quite to plan, however. Midway through the entertainment, Brunel accidentally swallowed a gold half-sovereign coin that he had secreted under his tongue. We may reasonably imagine Brunel's look of surprise followed by consternation and perhaps slight panic as he felt the coin slide down his throat and lodge at the base of his trachea. It caused him no great pain, but it was uncomfortable and unnerving because he knew that if it shifted even slightly it could choke him.

Over the next few days, Brunel, his friends, colleagues, family, and doctors attempted every obvious remedy, from slapping him hard on the back to holding him aloft by the ankles (he was a small man and easily lofted) and shaking him vigorously, but nothing worked. Seeking an engineered solution, Brunel designed a contraption from which he could hang upside down and be swung in wide arcs in the hope that motion and gravity together would make the coin fall out. That didn't work either.

Brunel's plight became the talk of the nation. Suggestions poured in from every quarter of the country and from abroad, but every attempted remedy failed. At length, the eminent physician Sir Benjamin Brodie decided to attempt a tracheotomy, a risky and disagreeable procedure. Without benefit of anesthetic—the first use of anesthetic in Britain was still three years off—Brodie made an incision in Brunel's throat and tried to extract the coin by reaching into his airway with long forceps, but Brunel couldn't breathe, and coughed so violently that the attempt had to be abandoned.

Finally, on May 16, more than six weeks after his ordeal began, Brunel had himself strapped into his swinging contraption once again and set in motion. Almost immediately, the coin fell out and rolled across the floor.

Very shortly afterward, the eminent historian Thomas Babington Macaulay burst into the Athenaeum Club in Pall Mall and shouted, "It's out!" and everyone knew at once what he meant. Brunel lived the rest of his life without complications from the incident and, as far as is known, never put a coin in his mouth again.

I mention all this here to make the point, if it needed making, that the mouth is a place of peril. We choke to death more easily than any other mammal. Indeed, it can reasonably be said that we are built to choke, which is clearly an odd attribute to go through life with—with or without a coin in your trachea.

Look inside your mouth and a good deal of what you find is familiar—tongue, teeth, gums, dark hole at the back presided over by that curious little flap known as the uvula. But behind the scenes, as it were, are lots and lots of very important apparatus that most of us have never heard of: palatoglossus, geniohyoid, vallecula, levator veli palatini. As with every other part of your head, the mouth is a realm of complexity and mystery.

Take the tonsils. We are all familiar with them, but how many of us know quite what they do? In fact, nobody knows quite what they do. They are the two fleshy hummocks that stand sentinel on either side of the throat at the back. (Confusingly, in the nineteenth century they were often called amygdalae, even though that name was already applied to structures in the brain.) Adenoids are similar but lurk out of sight within the nasal cavity. Both are part of the immune system, but not a particularly impressive part, it must be said. Adenoids often shrink away to virtually nothing in adolescence, and both they and tonsils can be removed without making any discernible difference to your overall well-being. The tonsils are part of a somewhat grander structure known as Waldeyer's tonsillar ring, named for the German anatomist Heinrich Wilhelm Gottfried von Waldeyer-Hartz (1836–1921), who is better remembered for coining the term "chromosome" in 1888 and the term "neuron" in 1891. He was, anatomically speaking, all over the place. Among much else, he was the person who first postulated, way back in 1870, that a woman is born with all her eggs, or ova, fully formed and ready to go.*

The anatomist's word for swallowing is "deglutition," and it is something we do quite a lot—about two thousand times a day, or once every thirty seconds on average. When you swallow, food doesn't just drop into your stomach by means of gravity, but is pushed down by muscular contractions. That's why you can eat and drink while

* It is perhaps worth noting that in 2011 a researcher at the Karolinska Institute in Stockholm noticed that people who had had their tonsils removed while young had a 44 percent greater chance of having a heart attack in later life. Of course, the two events may only be coincidentally related, but in the absence of conclusive evidence it suggests that it might be prudent to leave the tonsils alone. The same study also found that people who kept their appendixes had a 33 percent reduced chance of a heart attack in middle age. Taylor, *Body by Darwin*, 180.

upside down if you choose to. Swallowing is a trickier business than you might think. Altogether, fifty muscles can be called into play just to get a piece of food from your lips to your stomach, and they must snap to attention in exactly the right order to ensure that whatever you dispatch into the alimentary system doesn't go down the wrong way and end up lodged in an airway, like Brunel's coin.

The complexity of human swallowing is largely because our larynx is low in the throat compared with other primates. To accommodate our upright posture when we became bipedal, our necks became longer and straighter and moved to a more central position beneath the skull rather than toward the rear as in other apes. By chance, these changes gave us greater aptitude for speech but also the danger of "tracheal obstruction," in the words of Daniel Lieberman. Uniquely among mammals, we send our air and food down the same tunnel. Only a small structure called the epiglottis, a kind of trapdoor for the throat, stands between us and catastrophe. The epiglottis opens when we breathe and closes when we swallow, sending food in one direction and air in another, but occasionally it errs and the results are sometimes dire.

It is pretty amazing when you reflect upon it that you can sit at a dinner party enjoying yourself extravagantly–eating, talking, laughing, breathing, slurping wine–and that your nasopharyngeal guardians will send everything to the right place, in two directions, without you having to give it a moment's consideration. That's quite an accomplishment. But there is even more to it than that. While you are chattering away about work or school catchment zones or the price of kale, your brain is closely monitoring not just the taste and freshness of what you are eating but also its bulk and texture. So, it will allow you to swallow a large "wet" bolus (like an oyster or a lump of ice cream) but insists on more meticulous chewing for small, dry, sharp items like nuts and seeds that might not pass so smoothly.

Meanwhile, you, far from assisting this critical process, just keep pouring more red wine down your throat, destabilizing all your internal systems and seriously compromising your brain's functional capabilities. To say that your body is your long-suffering servant is to put it mildly.

When you consider the precision required, and the number of times in a lifetime the systems are challenged, it is extraordinary that we don't choke more often. According to official sources, about five thousand people in the United States and some two hundred in Britain choke to death on food each year—which is odd because those figures, adjusted for population, indicate that Americans are five times more likely to asphyxiate while eating than Britons.

Even allowing for the gusto with which my fellow Americans chow down, that seems unlikely. It is more probable that a lot of choking deaths are misattributed as heart attacks. Suspecting as much, a coroner in Florida named Robert Haugen many years ago looked into the deaths of people who had supposedly died of heart attacks in restaurants and, without much difficulty, found nine who had in fact choked. In an article for *The Journal of the American Medical Association,* he suggested that choking deaths were much more common than generally thought. But even using the most cautious estimates, choking is the fourth most common cause of accidental death in America today.

The well-known solution to a choking crisis is the Heimlich maneuver, named for Dr. Henry Judah Heimlich (1920–2016), a surgeon from New York who invented it in the 1970s. The Heimlich maneuver consists in embracing a choking victim from behind and giving him or her a series of sharp hugs at the breastbone, to force out the blockage, like a cork from a bottle. (For the record, the burst of air is known as a bechic blast.)

Henry Heimlich was something of a showman. He promoted the procedure, and himself, relentlessly. He appeared on *The Tonight Show* with Johnny Carson, sold posters and T-shirts, and talked to groups large and small across the country. He boasted that his method had saved the lives of Ronald Reagan, Cher, New York's mayor Ed Koch, and several hundred thousand others. He was not always terribly popular with those close to him. A former colleague called Heimlich "a liar and a thief," and one of his own sons accused him of practicing a "wide-ranging, 50-year history of fraud." Heimlich seriously undermined his reputation by championing a treatment called malaria therapy, in which people were purposely infected with low doses of malaria in the belief that it would cure them of cancer, Lyme disease, and AIDS, among much else. His claims for the treatment were not supported by any actual science. Partly because he had become an embarrassment, in 2006 the American Red Cross stopped using the term "Heimlich maneuver" and started calling it "abdominal thrusts."

Heimlich died in 2016 aged ninety-six. Shortly before his death, he saved the life of a woman at his nursing home with his own maneuver—the only time in his life that he had an opportunity to use it. Or possibly not. It emerged afterward that he had claimed to have saved someone else's life on another occasion. Heimlich, it seems, maneuvered the truth as well as trapped lumps of food.

The greatest choking authority of all time was almost certainly a dour American doctor with the luxuriant name of Chevalier Quixote Jackson, who lived from 1865 to 1958. Jackson has been called (by the Society of Thoracic Surgeons) "the father of American bronchoesophagoscopy," and he was most assuredly that, though it must also be said there were not a lot of other contenders. His specialty—his obsession—was with foreign objects that had been swallowed or inhaled. Over a career that lasted almost seventy-five years, Jackson specialized in designing instruments and refining methods for retrieving such objects, and in the process he built up an extraordinary collection of 2,374 imprudently ingested items. Today the Chevalier Jackson Foreign Body Collection is housed in a cabinet in the basement of the Mütter Museum of the College of Physicians of Philadelphia. Each object is fastidiously cataloged by age and sex of the swallower; type of object; whether it lodged in the trachea, larynx, esophagus, bronchus, stomach, pleural cavity, or elsewhere; whether it proved fatal or not; and by what means it was removed. It is presumed to be the world's largest assemblage of the extraordinary things people have put down their throats, whether by accident or bizarre design. Among the objects Jackson retrieved from the gullets of the living or dead were a wristwatch, a crucifix with rosary beads, miniature binoculars, a small padlock, a toy trumpet, a full-sized meat skewer, a radiator key, several spoons, a poker chip, and a medallion that said (perhaps just a touch ironically) "Carry Me for Good Luck."

Jackson was a cold and friendless man by all accounts, but there seems to have been some kindness buried within him. In his autobiography, he recorded how on one occasion he removed from a child's throat "a grayish mass—perhaps food, perhaps dead tissue"—which had kept her from swallowing for some days, then had his assistant give her a glass of water. The girl took a cautious sip and it went down, then took a larger sip. "Then she gently moved aside the glass of water in the nurse's hand, took hold of my hand and kissed it," Jackson recorded in the sole incident in his life that seems to have touched him.

In the seven and a half decades he was active, Jackson saved hundreds of lives and provided the training that allowed others to save countless more. Had he been a touch more charming with patients and colleagues, he would doubtless be better known today.

It will not have escaped your attention that the mouth is a moist and glistening vault. That's because twelve salivary glands are distributed around it. A typical adult secretes a little less than one and a half quarts a day. According to one calculation, we secrete about 31,700 quarts in a lifetime (about as much as you would get in two hundred or so deep baths). Just recently it was discovered that saliva also contains a powerful painkiller called opiorphin. It is six times more potent than morphine, though we have it only in very small doses, which is why you are not perennially high or indeed notably pain-free when you bite your cheek or burn your tongue. Because it is so dilute, no one is sure why it is there at all. It is so unassertive that its existence wasn't even noticed until 2006.

Saliva is almost entirely water. Only 0.5 percent of it is anything else, but that tiny portion is full of useful enzymes—proteins that speed up chemical reactions. Among these are amylase and ptyalin, which begin to break down sugars in carbohydrates while they are still in our mouths. Chew a starchy food like bread or potato for a bit longer than normal and you will soon notice a sweetness. Unfortunately for us, bacteria in our mouths like that sweetness, too; they devour the liberated sugars and excrete acids, which drill through our teeth and give us cavities. Other enzymes, notably lysozyme—which was discovered by Alexander Fleming before he stumbled onto penicillin—attack many invading pathogens, but not the ones that cause tooth decay, alas. We are in the rather strange position that we not only fail to kill the bacteria that give us a lot of trouble but actively nurture them.

We produce very little saliva while we sleep, which is why microbes can proliferate then and give you a foul mouth to wake to. It is also why brushing your teeth at bedtime is a good idea because it reduces the number of bacteria you go to sleep with. If you've ever wondered why no one wants to kiss you first thing in the morning, it is possibly because your exhalations may contain up to 150 different chemical compounds, not all of them as fresh and minty as we might hope. Among the common chemicals that help to create morning mouth are methyl mercaptan (which smells very like old cabbage), hydrogen sulfide (like rotten eggs), dimethyl sulfide (slimy seaweed), dimethylamine and trimethylamine (rank fish), and the self-explanatory cadaverine.

Professor Joseph Appleton of the University of Pennsylvania School of Dental Medicine, in the 1920s, was the first to study bacterial colonies within the mouth and discovered that, microbially speaking, your tongue, teeth, and gums are like separate continents, each with its own colonies of microorganisms. There are even differences in the bacterial colonies that inhabit the exposed part of a tooth and those beneath the gum line. Altogether, about a thousand species of bacteria have been found in human mouths, though at any one time you are unlikely to have more than about two hundred.

The mouth is not only a welcoming home for germs but an excellent way station for those that want to move elsewhere. Paul Dawson, a professor of food science at Clemson University in South Carolina, has made something of a career of studying the ways people spread bacteria from themselves to other surfaces, as when they share a water bottle or engage in "double dipping" with chips and salsa. In a study called "Bacterial Transfer Associated with Blowing Out Candles on a Birthday Cake," Dawson's team found that candle blowing across a cake increased the coverage of bacteria on it by up to 1,400 percent, which sounds pretty horrifying but is in fact probably not much worse than the kinds of exposures we encounter in normal life anyway. There are a lot of germs adrift in the world or wriggling invisibly on surfaces, and those surfaces include a lot of what you put in your mouth and nearly everything you touch.

The most familiar components of the mouth are of course the teeth and the tongue. Our teeth are formidable creations and nicely versatile, too. They come in three varieties: blades (which are pointy), cusps (which are spade-like), and basins, or fossae (which fall somewhere between the other two). The outside of your tooth is the enamel. It is the hardest substance in the human body, but forms just a thin layer and can't be replaced if it is damaged. That's why you have to go to the dentist for cavities. Under the enamel is a much thicker layer of another mineralized tissue called dentin, which can renew itself. At the center of it all is the fleshy pulp with nerves and blood supply. Because they are so hard, teeth have been called "ready-made fossils." When all the rest of you has turned to dust or dissolved away, the last physical trace of your existence on Earth may be a fossilized molar.

We can bite pretty hard. Bite force is measured in units called newtons (in honor of Isaac Newton's second law of motion, not his oral ferocity), and if you are a typical adult male, you can muster about four hundred newtons of force, which is quite a lot, though nothing like as much as an orangutan, which can bite with five times as much vigor. Still, when you consider how well you can demolish, say, an ice cube (try doing that with your fists and see how far you get) and how little space the five muscles of the jaw occupy, you can appreciate that human chomping is pretty capable.

The tongue is a muscle, but quite unlike any other. For one thing, it is exquisitely sensitive–think how adroitly you pick out something in your food that shouldn't be there, like a tiny piece of eggshell or grain of sand–and intimately involved in vital activities like speech articulation and tasting food. When you eat, the tongue darts about like a nervous host at a cocktail party, checking the taste and shape of every morsel in preparation for dispatching it onward to the gullet. As everyone knows, the tongue is coated with taste buds. These are clumps of taste receptor cells found in the bumps on your tongue, which are formally called papillae. They come in three different shapes: circumvallate (or rounded), fungiform (mushroom shaped), and foliate (leaf shaped). They are among the most regenerative of all cells in the body and are replaced every ten days.

For years, even textbooks spoke of a tongue map, with the elemental tastes each occupying a well-defined zone: sweet on the tip of the tongue, sour at the sides, bitter at the back. In fact, that is a myth, traced to a textbook written in 1942 by one Edwin G. Boring, a Harvard psychologist who misinterpreted a paper written by a German researcher forty years before that. Altogether we have about ten thousand taste buds, mostly distributed around the tongue, except in the very middle, where there are none at all. Additional taste buds are found in the roof of the mouth and lower down the throat, which is said to be why some medicines taste bitter as they go down.

As well as the mouth, the body has taste receptors in the gut and throat (to help identify spoiled or toxic substances), but they don't connect to the brain in the same way as the taste receptors on your tongue, and for good reason. You don't want to taste what your stomach is tasting. Taste receptors have also been found in the heart, the lungs, and even the testicles. No one knows quite what they are doing there. They also send signals to the pancreas to adjust insulin output, and it may be connected to that.

It is generally supposed that taste receptors evolved for two deeply practical purposes: to help us find energy-rich foods (like sweet, ripe fruits) and to avoid dangerous ones. But it must also be said that they don't always fulfill either role terribly well. Captain James Cook, the great British explorer, had a salutary demonstration of that in 1774, on his second epic voyage through the Pacific. One of his crew caught a meaty fish, which no one aboard recognized. It was cooked and proudly presented to the captain and two of his officers, but because they had already dined, they merely sampled it and had the remainder put aside for the following day. This was a very lucky thing, for in the middle of the night all three found themselves "seized with an extraordinary weakness and numbness all over our limbs." Cook was for some hours virtually paralyzed and unable to lift anything—even a pencil. The three men were given emetics, to clear their stomachs. They were lucky to survive, for what they had sampled was puffer fish. These contain a poison called tetrodotoxin, which is a thousand times more powerful than cyanide.

Despite its extreme toxicity, puffer fish is a famous delicacy in Japan, where it is called fugu. Preparing fugu is a job entrusted to only a few specially trained chefs, who must carefully remove the fish's liver, intestines, and skin before cooking because they are particularly saturated in poison. Even then, enough toxin remains to numb the mouth and leave the diner feeling pleasantly woozy. In one famous case in 1975, a well-known actor named Bandō Mitsugorō ate four helpings of fugu—despite pleadings to stop—and died wretchedly four hours later of asphyxiation. Fugu still kills about one person a year.

The difficulty with fugu is that by the time the ill effects become evident, it is much too late to do anything about it. The same is true of all kinds of other substances, from belladonna, or deadly nightshade, to a wide range of fungi. In 2008, in a widely publicized case, the British author Nicholas Evans and three members of his family became deathly ill on holiday in Scotland when they mistook a deadly mushroom, *Cortinarius speciosissimus,* for its benign and delicious cousin cèpe. The effects were horrific—Evans needed a kidney transplant, and all members of the party suffered lasting damage—yet nothing in the taste alerted anyone to the perils ahead. The fact is, our putative defenses are far more putative than defensive.

We have about ten thousand taste receptors, but we actually have more pain and other somatosensory receptors than taste receptors in our mouths. Because they exist side by side on the tongue, we sometimes mix them up. When you describe a chili as hot, you are being more literal than you might suppose. Your brain interprets it as being actually burned. As Joshua Tewksbury of the University of Colorado has put it, "Chilies innervate the same neurons that you activate when you touch a 335-degree burner. Essentially, our brain is telling us that we have got our tongue on the stove." In the same way, menthol is perceived as being cool even in the heated smoke of a cigarette.

The active ingredient in all chili peppers is a chemical called capsaicin. When you ingest capsaicin, the body releases endorphins—it's not at all clear why—and that provides us with a literally warm glow of pleasure. As with any warmth, however, it can quickly grow uncomfortable and then intolerable.

The amount of heat in chilies is measured in units called Scovilles, after Wilbur Scoville (1865–1942), an unassuming American pharmacist who had no known interest in hot dishes and very possibly never tasted a genuinely spicy food in his life. Scoville spent much of his career training students at the Massachusetts College of Pharmacy and churning out academic papers with titles like "Some Observations on Glycerin Suppositories," but in 1907 at the age of forty-two, apparently tempted by a big salary, he moved to Detroit to take up a job with a large pharmaceutical company, Parke, Davis & Co. One of his tasks there was to oversee production of a popular muscle salve called Heet. The warmth of Heet came from chili peppers—the same ones used in food—but the heat of peppers varied enormously from one delivery to another, and there was no reliable way of judging how much to put into any given batch. So Scoville came up with something called the Scoville Organoleptic Test, which was a scientific method for measuring the hotness of any pepper. It is still the standard used today.

A bell pepper will have a Scoville rating of between 50 and 100. Jalapeños usually measure in the range of 2,500 to 5,000 Scovilles. Nowadays many people breed peppers specifically to make them as hot as possible. The record holder at the time of writing is the Carolina Reaper at 2.2 million Scovilles. Capsaicin in pure form has 16 million Scovilles. A purified version of a Moroccan spurge plant—a cousin of the innocuous common garden flowering euphorbia—has been measured at 16 billion Scovilles. Such superhot peppers are of no use in foods—they are beyond any human threshold—but they are of interest to manufacturers of pepper sprays, which also use capsaicin.

Capsaicin has been reported to lower blood pressure, fight inflammation, and reduce susceptibility to cancer, among quite a lot else of benefit to the average human. In a study reported in the *British Medical Journal,* Chinese adults who ate a lot of capsaicin were 14 percent less likely to die, from any cause, during the period of the study compared with less adventurous eaters. But, as always with these findings, the fact that the subjects ate a lot of spicy food and were 14 percent better at surviving may only be coincidental.*

* Capsaicin exists in nature because peppers evolved it as a defense against being eaten by small mammals, which would destroy the seeds with their teeth. Birds, however, swallow seeds whole and can't taste capsaicin, so they can eat ripe pepper seeds with abandon. They then fly off and spread the seeds to new locations, bound up in a little white packet of fertilizer, when they defecate, so it is an arrangement that suits birds and seeds alike.

Incidentally, we have pain detectors not only in the mouth but also in the eyes, anus, and vagina, which is why spicy foods can cause discomfort there.

As far as taste goes, our tongue can only identify the familiar basics of sweet, salty, sour, bitter, and umami (a Japanese word meaning "savory" or "meaty"). Some authorities believe we also have taste receptors specifically allocated for metal, water, fat, and another Japanese concept called *kokumi,* meaning "full-bodied" or "hearty," but the only ones that are universally accepted are the five basics.

In the West, umami is still a rather exotic concept. It is actually a comparatively recent term even in Japan, though the taste has been known for centuries. It comes from a popular fish stock called dashi, which is made from seaweed and dried fish scales, and when added to other foods makes them even more delicious and imparts an ineffable but distinctive flavor. In the early twentieth century, a Tokyo chemist named Kikunae Ikeda determined to identify the source of the flavor and to try to synthesize it. In 1909, he published a brief paper in a Tokyo journal, identifying the source of the flavor as the chemical glutamate, an amino acid. He dubbed the flavor umami, meaning "essence of deliciousness."

Ikeda's discovery attracted virtually no attention outside Japan. The word "umami" isn't recorded anywhere in English until 1963, when it appeared in an academic paper. Its first appearance in a more mainstream publication was in 1979 in *New Scientist.* Ikeda's article wasn't translated into English until 2002, after umami taste receptors had been confirmed by Western researchers. But in Japan, Ikeda became celebrated, not as a scientist so much, but rather as a cofounder of a great company, Ajinomoto, created to exploit his patent for making synthetic umami, in the form universally known today as monosodium glutamate, or MSG. Today Ajinomoto is a behemoth, making about one-third of all the world's MSG.

MSG has had a hard time of it in the West since 1968 when *The New England Journal of Medicine* published a letter—not an article or a study, but simply a letter—from a doctor noting that he sometimes felt vaguely unwell after eating in Chinese restaurants and wondered if it was the MSG added to the food that was responsible. The headline on the letter was "Chinese-Restaurant Syndrome," and from this small beginning it became fixed in many people's minds that MSG was a kind of toxin. In fact, it isn't. It appears naturally in lots of foods, like tomatoes, and has never been found to have deleterious effects on anybody when eaten in normal quantities. According to Ole G. Mouritsen and Klavs Styrbaek in their fascinating study, *Umami: Unlocking the Secrets of the Fifth Taste,* "MSG is the food additive that has been subjected to the most thorough scrutiny of all time," and no scientist has ever found any reason to condemn it, yet its reputation in the West as a source of headaches and low-grade malaise now appears to be undimmed and permanent.

The tongue and its taste buds give us just the basic textures and attributes of food—whether they are soft or smooth, sweet or bitter, and so on—but the full sensuousness of it all is dependent on our other senses. It is nearly always wrong to talk about how food tastes, though of course we all do. What we appreciate when we eat is flavor, which is taste plus smell.*

Smell is said to account for at least 70 percent of flavor, and maybe even as much as 90 percent. We appreciate this intuitively without often thinking about it. If someone hands you a pot of yogurt and says, "Is this strawberry?" your response will normally be to sniff it, not taste it. That is because strawberry is actually a smell, perceived nasally, not a taste in the mouth.

When you eat, most of the aroma reaches you not through your nostrils but by the back staircase of your nasal passage, what is known

* It isn't just English that does this. At least ten other languages use the words "taste" and "flavor" interchangeably.

as the retronasal route, as opposed to the orthonasal route up your nose. An easy way to experience the limitations of your taste buds is to close your eyes, pinch shut your nostrils, and eat a flavored jelly bean collected blindly from a bowl. You will instantly apprehend its sweetness, but you almost certainly won't be able to identify its flavor. But open your eyes and nostrils and its fruity specificity becomes immediately and redolently apparent. Even sound materially influences how delicious we find food. People who are played a range of crunching sounds through headphones while sampling potato chips from various bowls will always rate the crunchier, noisier chips as fresher and tastier, even though all the chips are the same.

Many tests have been done to demonstrate how easily we are fooled with respect to flavor. In a blind taste test at the University of Bordeaux, students in the faculty of enology were given two glasses of wine, one red and one white. The wines were actually identical except that one had been made a rich red with an odorless and flavorless additive. The students without exception listed entirely different qualities for the two wines. That wasn't because they were inexperienced or naive. It was because their sight led them to have entirely different expectations, and this powerfully influenced what they sensed when they took a sip from either glass. In exactly the same way, if an orange-flavored drink is colored red, you cannot help but taste it as cherry.

The fact is that odors and flavors are created entirely inside our heads. Think of something delicious—a moist, gooey, warm chocolate brownie fresh from the oven, say. Take a bite and savor the velvety smoothness, the rich heady waft of chocolate that fills your head. Now consider the fact that none of those flavors or aromas actually exist. All that is really going in your mouth is texture and chemicals. It is your brain that reads these scentless, flavorless molecules and vivifies them for your pleasure. Your brownie is sheet music. It is your brain that makes it a symphony.

As with so much else, you experience the world that your brain allows you to experience.

* * *

There is of course one other remarkable thing we do with our mouths and throats, and that is make meaningful noises. The ability to create and share complex sounds is one of the great wonders of human existence and the characteristic more than any other that sets us apart from all other creatures that have ever lived.

Speech and its evolution "are perhaps more extensively debated than any other topic in human evolution," in the words of Daniel Lieberman. No one knows even approximately when speech began on Earth and whether it is an accomplishment confined to *Homo sapiens* or whether it was a skill mastered by archaic humans like Neanderthals and *Homo erectus*. Lieberman thinks it likely that Neanderthals commanded complex speech based on their large brains and array of tools, but it isn't a provable hypothesis.

What is certain is that the capacity for speech requires a delicate and coordinated balance of tiny muscles, ligaments, bones, and cartilage of exactly the right length, tautness, and positioning in order to expel microbursts of modulated air in just the right measures. The tongue, teeth, and lips must also be nimble enough to take these throaty breezes and turn them into nuanced phonemes. And all of this must be achieved without compromising our ability to swallow or breathe. That's quite a tall order, to put it mildly. It isn't just a big brain that allows us to speak but an exquisite arrangement of anatomy. One reason chimps can't talk is that they appear to lack the ability to make subtle shapes with tongue and lips to form complex sounds.

It may be that all this happened fortuitously in the course of an evolutionary redesign of our upper bodies to accommodate our new posture when we became bipedal, or it may be that some of these features were selected for through the slow, incremental wisdom of evolution, but the bottom line is that we ended up with brains big enough to handle complex thoughts and vocal tracts uniquely able to articulate them.

The larynx is essentially a box about an inch on each side. Within or around it are nine cartilages, six muscles, and a suite of ligaments, including two commonly known as the vocal cords but more properly known as the vocal folds. When air is forced through them, the vocal folds snap and flutter (like flags in a stiff breeze, it has been said), producing a variety of sounds, which are refined by tongue, teeth, and lips working together into the wondrous, resonant, informative exhalations known as speech. The three phases of the process are respiration, phonation, and articulation. Respiration is simply the pushing of air past the vocal ligaments; phonation is the process of turning that air into sound; and articulation is the refinement of sound into speech. If you wish to appreciate what a marvel speech is, try singing a song—"Frère Jacques" serves very well—and notice how effortlessly melodic the human voice is. The fact is, your throat is a musical instrument as well as a sluice and wind tunnel.*

When you consider the complexity, it is hardly surprising that some people struggle to put it all together. Stuttering is one of the cruelest and least understood of everyday maladies. It affects 1 percent of adults and 4 percent of children. For reasons unknown, 80 percent of sufferers are male. The victims have included a great many distinguished figures—Aristotle, Virgil, Charles Darwin, Lewis Carroll, Winston Churchill (when young), Henry James, John Updike, Marilyn Monroe, and King George VI of Great Britain, who was sympathetically portrayed by Colin Firth in the 2010 movie *The King's Speech.*

No one knows what provokes it or why different sufferers stumble over different letters or words in different positions in a sentence. It is more common among left-handers than right-handers, especially those who have been made to write right-handed. For many, the stammering miraculously ceases when they sing the words or speak another language or talk to themselves. The majority of speakers recover from the condition by their teenage years (which is why the proportion of

* Very strictly speaking, the vocal folds consist of the two vocal ligaments plus associated muscles and membranes.

child sufferers is so much higher than adult ones). Females seem to recover more easily than males.

There is no reliable cure for stuttering. Johann Dieffenbach, one of Germany's most eminent surgeons in the nineteenth century, thought stuttering was entirely a muscular complaint and believed he could cure it by cutting out some of his patients' tongue muscles. Although the process was wholly ineffectual, it was widely copied throughout Europe and the United States for a while. Many patients died; all suffered mightily. Today, mercifully, most sufferers are helped significantly with speech therapy and a patient, compassionate approach.

Before we leave the throat and descend further into the body, we should take a moment to consider the strange little fleshy appendage that stands guard at the point where all becomes darkness and with which we began this tour of our largest opening. I refer to the small and permanently mysterious uvula. (The name incidentally comes from the Latin for "little grape," even though it is not especially like a grape at all.)

For a long time, nobody knew what it was for. We are still not completely sure, but it seems to be a sort of mud flap for the mouth. It directs food down the throat and away from the nasal passage (when you cough while eating, for instance). It also helps with the production of saliva, which is always useful, and appears to have a role in triggering the gag reflex. It may also play a part in speech, though this conclusion is based on little more than that we are the only mammals that have a uvula and the only ones that speak. It is a fact that people who have had their uvula removed do lose some control over guttural sounds and sometimes report that they feel they don't sing as melodically as before. The rattling of the uvula in sleep appears to be a significant component of snoring, and is often the reason uvulas are taken out, but the removal of a uvula is a very rare event. For the overwhelming majority of us, the uvula does nothing to draw attention to itself over the course of a lifetime.

The uvula, in short, is a curious thing. Considering its position at the very center of our largest orifice, at the point of no return, it seems to be oddly inconsequential. There is perhaps a kind of strange double comfort in knowing that you will almost certainly never lose your uvula but that it wouldn't matter too much anyway if you did.

7 THE HEART AND BLOOD

Stopped.

—LAST WORD OF THE BRITISH SURGEON AND ANATOMIST JOSEPH HENRY GREEN (1791–1863) WHILE FEELING HIS OWN PULSE

I

THE HEART IS the most misperceived of our organs. For a start, it looks nothing like the traditional symbol associated with Valentine's Day and lovers' initials carved into tree trunks and the like. (That symbol first appeared, as if from out of nowhere, in paintings from northern Italy in the early fourteenth century, but no one knows what inspired it.) Nor is the heart where we place our right hand during patriotic moments; it is more centrally located in the chest than that. Most curious of all, perhaps, is that we make it the emotional seat of our being, as when we declare that we love someone with all our heart or profess a broken heart when they abandon us. Don't misunderstand me. The heart is a wondrous organ and fully deserving of our praise and gratitude, but it is not invested even slightly in our emotional well-being.

That's a good thing. The heart has no time for distractions. It is the most single-minded thing within you. It has just one job to do, and it does it supremely well: it beats. Slightly more than once every second, about 100,000 times a day, as many as 3.5 billion times in a lifetime, it rhythmically pulses to push blood through your body—and these aren't gentle thrusts. They are jolts powerful enough to send blood spurting up to three meters if the aorta is severed.

With such an unrelenting work rate, it is a miracle that most hearts last as long as they do. Every hour your heart dispenses around 70 gallons of blood. That's 1,680 gallons in a day—more gallons pushed through you in a day than you are likely to put in your car in a year. The heart must pump with enough force not merely to send blood to your outermost extremities but to help bring it all the way back again. If you are standing, your heart is roughly four feet above your feet, so there's a lot of gravity to overcome on the return trip. Imagine squeezing a pump the size of a grapefruit with enough force to move a fluid four feet up a tube. Now do that again once every second or so, around the clock, unceasingly, for decades, and see if you don't feel a bit tired. It has been calculated (and goodness knows how, it must be said) that during the course of a lifetime the heart does an amount of work sufficient to lift a one-ton object 150 miles into the air. It is a truly remarkable implement. It just doesn't care about your love life.

For all it does, the heart is a surprisingly modest thing. It weighs less than a pound and is divided into four simple chambers: two atria and two ventricles. Blood enters through the atria (Latin for "entry rooms") and exits via the ventricles (from another Latin word for "chambers"). The heart is not really one pump but two: one that sends blood to the lungs and one that sends it around the body. The output of the two must be in balance, every single time, for it all to work correctly. Of all the blood pumped out of your heart, the brain takes 15 percent, but actually the greatest amount, 20 percent, goes to the kidneys. The journey of blood around your body takes about fifty seconds to complete. Curiously, the blood passing through the chambers of the heart does nothing for the heart itself. The oxygen that nourishes it arrives via the coronary arteries, in exactly the way oxygen reaches other organs.

The two phases of a heartbeat are known as the systole (when the heart contracts and pushes blood out into the body) and diastole (when it relaxes and refills). The difference between these two is your blood pressure. The two numbers in a blood pressure reading–let's say 120/80, or "120 over 80" when spoken–simply measure the highest and lowest pressures your blood vessels experience with each heartbeat. The first, higher number is the systolic pressure; the second, the diastolic. The numbers specifically measure how many millimeters of mercury is pushed up a calibrated tube.

Keeping every part of the body supplied with sufficient quantities of blood at all times is a tricky business. Every time you stand up, roughly a pint and a half of your blood tries to drain downward, and your body has to somehow overcome the dead pull of gravity. To manage this, your veins contain valves that stop blood from flowing backward, and the muscles in your legs act as pumps when they contract, helping blood in the lower body get back to the heart. To contract, however, they need to be in motion. That's why it's important to get up and move around regularly. On the whole, the body manages these challenges pretty well.

"For healthy people there is a less than 20 percent difference between blood pressure at the shoulder and at the ankle," Siobhan Loughna, a lecturer in anatomy at the University of Nottingham Medical School, told me one day. "It's really quite remarkable how the body sorts that out."

As you may gather from this, blood pressure isn't a fixed figure, but changes from one part of the body to another, and across the body as a whole throughout the day. It tends to be highest during the day when we are active (or ought to be active) and to fall at night, reaching its lowest point in the small hours. It has long been known that heart attacks are more common in the dead of night, and some authorities think the nightly change in blood pressure may somehow act as a trigger.

Much of the early research on blood pressure was done in a series of decidedly gruesome experiments on animals conducted by the Reverend Stephen Hales, an Anglican curate of Teddington, Middlesex, near London, in the early eighteenth century. In one experiment, Hales tied down an aged horse and attached a nine-foot-long glass tube to its carotid artery by means of a brass cannula. Then he opened the artery and measured how high blood shot up the tube with each dying pulse. He killed quite a number of helpless creatures in his pursuit of physiological knowledge and was roundly condemned for it—the poet Alexander Pope, who lived locally, was especially vocal on the matter—but among the scientific community his achievements were celebrated. Hales thus had the double distinction of advancing science while at the same time giving it a bad name. Though Hales was denounced by animal lovers, the Royal Society awarded him its very highest honor, the Copley Medal, and for a century or so Hales's book *Haemastaticks* was the last word on blood pressure in animals and man.

Well into the twentieth century, many medical authorities believed that high blood pressure was a good thing because it indicated vigorous flow. We now know, of course, that chronically elevated blood pressure very seriously raises the risk of a heart attack or stroke. A more difficult question is, What exactly constitutes high blood pressure? For a long time, a reading of 140/90 was generally considered the baseline for hypertension, but in 2017 the American Heart Association surprised nearly everyone by abruptly pushing the number downward to 130/80. That small reduction tripled the number of men and doubled the number of women aged forty-five or under who were deemed to have high blood pressure and lifted practically all people over sixty-five into the danger zone. Almost half of all American adults—103 million people—are on the wrong side of the new blood pressure threshold, up from 72 million previously. At least 50 million Americans, it is thought, are not receiving appropriate medical attention for the condition.

Heart health has been one of the success stories of modern medicine. The death rate from heart diseases has fallen from almost 600 per 100,000 in 1950 to just 168 per 100,000 today. As recently as 2000, it was 257.6 per 100,000. But it is still the leading cause of death. In the United States alone, more than eighty million people suffer from cardiovascular disease, and the cost to the nation of treating heart disease has been put as high as $300 billion a year.

There are lots of ways the heart can falter. It can skip a beat, or more usually have an extra beat, because an electrical impulse misfires. Some people can have as many as ten thousand of these palpitations a day without being aware of it. For others, an arrhythmic heart is an endless discomforting ordeal. When the heart's rhythm is too slow, the condition is called bradycardia; when too fast, it is tachycardia.

A heart attack and a cardiac arrest, though usually confused by most of us, are in fact two different things. A heart attack occurs when oxygenated blood can't get to heart muscle because of a blockage in a coronary artery. Heart attacks are often sudden—that's why they are called attacks—whereas other forms of heart failure are often (though not always) more gradual. When heart muscle downstream of a blockage is deprived of oxygen, it begins to die, usually within about sixty minutes. Any heart muscle we lose in this way is gone forever, which is a bit galling when you consider that other creatures much simpler than we are—zebra fish, for instance—can regrow damaged heart tissue. Why evolution deprived us of this useful facility is yet another of the body's many imponderables.

Cardiac arrest is when the heart stops pumping altogether, usually because of a failure in electrical signaling. When the heart stops pumping, the brain is deprived of oxygen and unconsciousness swiftly follows, with death not far behind unless treatment is quickly applied. A heart attack will often lead to cardiac arrest, but you can suffer cardiac arrest without having a heart attack. The distinction between the two is medically important because they require different treatments, though the distinction may be a touch academic to the sufferer.

All forms of heart failure can be cruelly sneaky. For about a quarter of victims, the first (and, more unfortunately, last) time they know they have a heart problem is when they suffer a fatal heart attack. No less appallingly, more than half of all first heart attacks (fatal or otherwise) occur in people who are fit and healthy and have no known obvious risks. They don't smoke or drink to excess, are not seriously overweight, and do not have chronically high blood pressure or even bad cholesterol readings, but they get a heart attack anyway. Living a virtuous life doesn't guarantee that you will escape heart problems; it just improves your chances.

No two heart attacks are quite the same, it seems. Women and men have heart attacks in different ways. A woman is more likely to experience abdominal pain and nausea than a man, which makes it more likely that the problem will be misdiagnosed. Partly for this reason, women who have heart attacks before their mid-fifties are twice as likely to die as a man. Women have more heart attacks than is generally supposed. Twenty-eight thousand women suffer fatal heart attacks in the U.K. each year; about twice as many die of heart disease as die of breast cancer. Some people who are about to experience catastrophic heart failure suffer a sudden, terrifying premonition of impending death. The condition is commonly enough observed that it has a medical name: *angor animi,* or "anguish of the soul." For a lucky few victims (insofar as good fortune can be attached to a fatal event), death comes so swiftly that they appear to feel no pain. My own father went to bed one night in 1986 and never woke up. As far as could be told, he died without pain or distress or indeed awareness. For reasons unknown, the Hmong people of Southeast Asia are particularly susceptible to a condition known as sudden unexplained nocturnal death syndrome. In it, victims' hearts simply stop beating while they are asleep. Autopsies nearly always show the hearts to look normal and healthy.

Hypertrophic cardiomyopathy is the condition that makes athletes die suddenly on playing fields. It arises from an unnatural (and nearly always undiagnosed) thickening of one of the ventricles and causes eleven thousand sudden unexpected deaths a year among people under forty-five in the United States. The heart has more named conditions than just about any other organ, and they are all bad news. If you can go through life without experiencing Prinzmetal angina, Kawasaki disease, Ebstein's anomaly, Eisenmenger syndrome, Takotsubo cardiomyopathy, or many, many others, you may consider yourself fortunate indeed.

Heart disease is now such a common complaint that it is a little surprising to learn that it is largely a modern preoccupation. Until the 1940s, the principal focus of health care was with conquering infectious diseases like diphtheria, typhoid fever, and tuberculosis. Only after many of those were cleared out of the way did it become evident that we had another, growing epidemic on our hands in the form of cardiovascular disease. The triggering event for public awareness seems to have been the death of Franklin Delano Roosevelt. In early 1945, his blood pressure soared to 300/190, and it was clear that this was not a sign of vigor but quite the opposite. When he died soon afterward, aged just sixty-three, the world seemed suddenly to realize that heart disease had become a serious and widespread problem and that it was time to try to do something about it.

The result was the celebrated Framingham Heart Study, conducted in the town of Framingham, Massachusetts, just west of Boston. Starting in the autumn of 1948, the Framingham study recruited five thousand local adults and followed them carefully for the rest of their lives. Though the study has been criticized for being almost entirely composed of white people (a deficiency since corrected), it did at least include women, which was unusually farsighted for the time, particularly because women were not thought to suffer unduly from heart problems then. The study is now in its third generation of volunteers. The idea from the outset was to determine the factors that led some people to have heart problems and others to escape them. It was thanks to the Framingham study that most of the major risks for heart disease were identified or confirmed—diabetes, smoking, obesity, poor diet, chronic indolence, and so on. In fact, the term "risk factor" is said to have been coined in Framingham.

The twentieth century could with some justification be called the Century of the Heart, for no other area of medicine experienced more rapid and revolutionary technical progress. In a single lifetime, we have gone from barely being able to touch a beating heart to operating on them routinely. As with any complicated and risky medical procedure, it took years of patient work by lots of people to perfect the techniques and devise the apparatus to make it all possible. The daring and personal risk that some researchers took on is sometimes quite extraordinary. Consider the case of Werner Forssmann. In 1929, Forssmann was a young, newly qualified doctor working in a hospital near Berlin when he became curious to know if it would be possible to gain direct access to the heart by means of a catheter. Without any idea what the consequences would be, he fed a catheter into an artery in his arm and cautiously pushed it up toward his shoulder and on into his chest until it reached his heart, which, he was gratified to discover, didn't go into arrest when a foreign object invaded it. Then, realizing he needed proof of what he had done, Forssmann walked to the hospital's radiology department, on another floor of the building, and had himself X-rayed to show the shadowy and startling image of the catheter in situ in his heart. Forssmann's procedure would eventually revolutionize heart surgery, but it attracted almost no attention at the time, largely because he reported it in a minor journal. Forssmann would be a rather more sympathetic figure except that he was an early and ardent supporter of the Nazi Party and the National Socialist German Physicians' League, which was behind the purging of Jews in the quest for German racial purity. It's not entirely clear how much personal evil he engaged in during the Holocaust, but at the very least he was philosophically despicable. After the war, partly to escape retribution, Forssmann worked in obscurity as a family physician in a small town in the Black Forest. He would have been forgotten altogether in the wider world except that two academics from Columbia University in New York, Dickinson Richards and André Cournand, whose work was directly reliant on Forssmann's original breakthrough, tracked him down and publicized his contribution to cardiology. In 1956, all three men were awarded the Nobel Prize in Physiology or Medicine.

Far more personally noble than Forssmann, and no less stoic in his capacity for experimental discomfort, was Dr. John H. Gibbon of the University of Pennsylvania. In the early 1930s, Gibbon began a long and patient quest to build a machine that could oxygenate blood artificially, to make open-heart surgery possible. To test the capacity of blood vessels deep within the body to dilate or constrict, Gibbon stuck a thermometer up his rectum, swallowed a stomach tube, and then had icy water poured down it to determine its effect on his internal body temperature. After twenty years of refinements, and much heroic swallowing of iced water, Gibbon unveiled the world's first heart-lung machine at the Jefferson College Hospital in Philadelphia in 1953 and successfully patched a hole in the heart of an eighteen-year-old woman who would otherwise have died. Thanks to his efforts, the woman lived another thirty years.

Unfortunately, the next four patients died, and Gibbon gave up on the machine. It then fell to a surgeon in Minneapolis, Walton Lillehei, to improve both the technology and the surgical techniques. Lillehei introduced a refinement known as controlled cross-circulation in which the patient was hooked up to a temporary donor (usually a close family member) whose blood was circulated through the patient during the period of surgery. The technique worked so well that Lillehei became widely known as the father of open-heart surgery and enjoyed a great deal of acclaim and financial success. Unfortunately, he wasn't quite as impeccable in his private affairs as he might have been. In 1973, he was convicted of five counts of tax evasion and a great deal of very imaginative bookkeeping. Among much else, he had claimed a $100 payment to a prostitute as a charitable tax deduction.

Although open-heart surgery allowed surgeons to correct many faults they previously couldn't get at, it couldn't solve the problem of a heart that wouldn't beat right. That required the device now universally known as a pacemaker. In 1958, a Swedish engineer named Rune Elmqvist, working in collaboration with the surgeon Åke Senning of the Karolinska Institute in Stockholm, built a pair of experimental cardiac pacemakers at his kitchen table. The first was inserted into the chest of Arne Larsson, a forty-three-year-old patient (and himself an engineer) who was very near death from a heart arrhythmia as a result of a viral infection. The device failed after just a few hours. The backup was inserted and it lasted for three years, though it kept breaking down and the batteries had to be recharged every few hours. As technology improved, Larsson was routinely fitted with new pacemakers and lived another forty-three years. When he died in 2002 at the age of eighty-six, he was on his twenty-sixth pacemaker and had outlived both his surgeon Senning and his fellow engineer Elmqvist. The first pacemaker was about the size of a pack of cigarettes. Today's are no bigger than one American quarter and can last up to ten years.

The coronary bypass, which involved taking a length of healthy vein from a person's leg and transplanting it to direct blood flow around a diseased coronary artery, was devised in 1967 by René Favaloro at the Cleveland Clinic in Ohio. Favaloro's was a story at once inspiring and tragic. He grew up poor in Argentina and became the first member of his family to attain a higher education. Upon qualifying as a doctor, he spent twelve years working among the poor but came to the United States in the 1960s to improve his skills. At the Cleveland Clinic, he was little more than a trainee at first but quickly proved himself adept at heart surgery and in 1967 invented the bypass. It was a comparatively simple but ingenious procedure, and it worked brilliantly. Favaloro's first patient, a man too ill to walk up a flight of stairs, recovered completely and lived another thirty years. Favaloro grew wealthy and celebrated and in the twilight of his career decided to return home to Argentina to build a heart clinic and teaching hospital, where doctors could be trained and needy people treated whether they could afford payment or not. All of this he achieved, but because of challenging economic conditions in Argentina, the hospital got into financial difficulties. Unable to see a way out, in 2000 he killed himself.

The great dream was to transplant a heart, but in many places it faced a seemingly insuperable obstacle: a person could not be declared dead until his heart had been stopped for a specified period, but that was all but certain to render the heart unusable for transplant. To remove a beating heart, no matter how far gone the owner was in all other respects, was to risk prosecution for murder. One place where that law did not apply was South Africa. In 1967, at exactly the time that

René Favaloro was perfecting bypass surgery in Cleveland, Christiaan Barnard, a surgeon in Cape Town, attracted far more of the world's attention by transplanting the heart of a young woman fatally injured in a car accident into the chest of a fifty-four-year-old man named Louis Washkansky. It was hailed as a great medical breakthrough, though in fact Washkansky died after just eighteen days. Barnard had much better luck with his second transplant patient, a retired dentist named Philip Blaiberg, who survived for nineteen months.*

Following Barnard, other nations moved to let brain death be used as an alternative measure of irreversible lifelessness, and soon heart transplants were being attempted all over, though nearly always with discouraging results. The main issue was an absence of a wholly reliable immunosuppressive drug to deal with rejection. A drug called azathioprine worked sometimes but couldn't be relied on. Then, in 1969, an employee of the Swiss pharmaceutical company Sandoz named H. P. Frey, while on holiday in Norway, collected soil samples to take back to the Sandoz labs. The company had asked employees to do so when traveling in the hope that they would find potential new antibiotics. Frey's sample contained a fungus, *Tolypocladium inflatum,* which had no useful antibiotic properties but proved excellent at suppressing immune responses—just the thing needed to make organ transplants possible. Sandoz converted Herr Frey's little bag of dirt, and a similar sample subsequently found in Wisconsin, into a best-selling medicine called cyclosporine. Thanks to it and some associated technical improvements, by the early 1980s heart transplant surgeons were managing success rates of 80 percent, an extraordinary achievement in a decade and a half. Today some four to five thousand heart transplants are performed globally each year, with an average survival time of fifteen years. The longest-surviving transplant patient so far

* Barnard's was the first human-to-human heart transplant. The first heart transplant of any type involving a human was in January 1964, when a Dr. James D. Hardy in Jackson, Mississippi, transplanted a chimpanzee's heart into a man named Boyd Rush. The patient died within an hour. Morris, *Heart of the Matter,* 225.

was the Briton John McCafferty, who lived thirty-three years with a transplanted heart before dying in 2016 aged seventy-three.

Incidentally, brain death turned out to be not as straightforward as originally thought. Some peripheral parts of the brain, we now know, may live on after all the rest has grown still. At the time of this writing, that is the issue at the center of a long-running case involving a young woman in the United States who was declared brain-dead in 2013 but who has continued to menstruate, a process that requires a functioning hypothalamus—very much a key part of the brain. The young woman's parents argue that anyone with even part of the brain functioning cannot reasonably be declared brain-dead.

As for Christiaan Barnard, the man who began it all, success rather went to his head. He traveled the world, dated movie stars (Sophia Loren and Gina Lollobrigida notably), and became, in the words of someone who knew him well, "one of the world's great womanizers." Even worse for his reputation, he made a fortune claiming rejuvenative benefits for a range of cosmetics that he most assuredly knew were bogus. He died in 2001, aged seventy-eight, of a heart attack while enjoying himself in Cyprus. His reputation was never again quite what it had been.

Remarkably, even with all the improvements in care, you are 70 percent more likely to die from heart disease today than you were in 1900. That's partly because other things used to kill people first, and partly because a hundred years ago people didn't spend five or six hours an evening in front of a television with a big spoon and a tub of ice cream. Heart disease is far and away the Western world's number one killer. As Michael Kinch has written, "Heart disease kills about the same number of Americans each year as cancer, influenza, pneumonia, and accidents combined. One in three Americans dies of heart disease and more than 1.5 million suffer a heart attack or stroke each year."

Today the problem is as likely to be overtreatment as under, according to some authorities. Balloon angioplasties as a treatment for angina (or chest pains) are a case in point, it seems. With an angioplasty, a balloon is inflated inside a constricted coronary blood vessel to widen it, and a stent, or piece of tubular scaffolding, is left behind to keep the vessel permanently open.* The operation is unquestionably a lifesaver in emergencies, but it has also proven highly popular as an elective procedure. By 2000, a million precautionary angioplasties were being undertaken in the United States every year, but without any proof that they saved lives. When clinical trials were finally undertaken, the results were sobering. According to *The New England Journal of Medicine,* for every one thousand nonemergency angioplasties in America, two patients died on the operating table, twenty-eight suffered heart attacks brought on by the procedure, between sixty and ninety experienced a "transient" improvement in their health, and the rest—about eight hundred people—experienced neither benefit nor harm (unless of course you count the cost, the loss of time, and the anxiety of surgery as harm, in which case there was plenty).

Despite this, angioplasties remain extremely popular. In 2013, the former president George W. Bush had an angioplasty at the age of sixty-seven, even though he was in good shape and had no sign of heart problems. Surgeons don't usually publicly criticize colleagues, but Dr. Steve Nissen, head of cardiology at the Cleveland Clinic, was scathing. "This is really American medicine at its worst," he said. "It's

* The term "stent" has a curious history. It is named after Charles Thomas Stent, a nineteenth-century London dentist who had nothing to do with heart surgery. Stent was the inventor of a compound used to make dental molds, which oral surgeons eventually also found useful when doing repairs to the mouths of soldiers wounded in the Boer War. Over time, the term came to be used for any kind of device used to keep tissue in place during corrective surgery and, in the absence of a better term, gradually took up a position as the word of choice for an arterial support for cardiac surgery. The record for stent insertions, incidentally, appears to be held by a fifty-six-year-old man in New York who, at last report, had had sixty-seven stents inserted for angina in a period of ten years. according to the *Baylor University Medical Center Proceedings*. Charles Stent profile, *Journal of the History of Dentistry,* July 2001; *Baylor University Medical Center Proceedings,* April 2011, 158.

one of the reasons we spend so much on medicine and don't get a lot for it."

II

HOW MUCH BLOOD you have depends, as you might suppose, on how big you are. A newborn baby contains only about eight ounces, whereas a fully grown man will have more like five quarts. What is certain is that you are suffused with the stuff. Prick your skin anywhere and you will draw blood. Within your modest frame are some twenty-five thousand miles of blood vessels (mostly in the form of tiny capillaries), so no part of you is ever far from the refreshment of hemoglobin, the molecule that transports oxygen throughout your body.

We all know that blood carries oxygen to our cells—it is one of the few facts about the human body that everyone does seem to know—but it also does a whole lot more. It transports hormones and other vital chemicals, carries off wastes, tracks down and kills pathogens, makes sure oxygen is directed to the parts of the body where it is most needed, signals our emotions (as when we blush from embarrassment or grow red with fury), helps to regulate body temperature, and even enables the complicated hydraulics of the male erection. It is, in short, a complex material. By one estimate, a single drop of blood may contain four thousand different types of molecules. That's why doctors are so fond of blood tests: your blood is positively packed with information.

Spin a test tube of blood in a centrifuge and it will separate into four layers: red cells, white cells, platelets, and plasma. Plasma is the most abundant, constituting a little over half of blood's volume. It is more than 90 percent water with some salts, fats, and other chemicals suspended in it. That isn't to say plasma is unimportant, however. It is anything but. Antibodies, clotting factors, and other constituent parts can be separated out and used in concentrated form to treat autoimmune diseases or hemophilia—and that is a huge business.

In the United States, plasma sales make up 1.6 percent of all goods exported, more than America earns from the sale of airplanes.

Red blood cells (formally called erythrocytes) are the next most plentiful component, constituting about 44 percent of the total volume of the blood. Red blood cells are exquisitely designed to do one job: deliver oxygen. They are very small but superabundant. A teaspoon of human blood contains about twenty-five billion red blood cells, and each one of those twenty-five billion contains 250,000 molecules of hemoglobin, the protein to which oxygen willingly clings. Red blood cells are biconcave in shape—that is, disk shaped but pinched in the middle on both sides—which gives them the largest possible surface area. To make themselves maximally efficient, they have jettisoned virtually all the components of a conventional cell—DNA, RNA, mitochondria, Golgi apparatus, enzymes of every description. A full red blood cell is almost entirely hemoglobin. It is essentially a shipping container. A notable paradox of red blood cells is that although they carry oxygen to all the other cells of the body, they don't use oxygen themselves. They use glucose for their own energy needs.

Hemoglobin has one strange and dangerous quirk: it vastly prefers carbon monoxide to oxygen. If carbon monoxide is present, hemoglobin will pack it in, like passengers on a rush-hour train, and leave the oxygen on the platform. That's why it kills people. (About 430 of them a year in the United States unintentionally, and a similar number by suicide.)

Each red corpuscle survives for about four months, which is pretty good going considering what a jostling and busy existence it leads. Each will be shot around your body about 150,000 times, logging a hundred miles or so of travel before it is too battered to go on. Then these corpuscles are collected by scavenger cells and sent to the spleen for disposal. You discard about a hundred billion red blood cells every day. They are a big component of what makes your stools brown. (Bilirubin, a by-product of the same process, is responsible for the golden glow of urine as well as the yellow blush of fading bruises.)

* * *

White blood cells (or leukocytes) are vital for fighting off infections. In fact, they are so important that we will treat them separately in chapter 12, on the immune system. For the moment, it is enough to know that they are much less numerous than their red siblings. You have seven hundred times as many red blood cells as white ones, which constitute less than 1 percent of the total.*

Platelets (or thrombocytes), the final part of the blood quartet, also account for less than 1 percent of blood's volume. Platelets were for a long time a mystery to anatomists. They were first seen under a microscope in 1841 by a British anatomist named George Gulliver, but they weren't named or properly understood until 1910 when James Homer Wright, chief pathologist at the Massachusetts General Hospital in Boston, deduced their central role in clotting. Clotting is a tricky business. The blood must be perpetually on alert to clot at a moment's notice, but equally mustn't clot unnecessarily. As soon as a bleed starts, millions of platelets begin to cluster around the wound and are joined by similarly vast numbers of proteins, which deposit a material called fibrin. This agglomerates with the platelets to make a plug. To try to avoid errors, no fewer than twelve fail-safe mechanisms are built into the process. Clotting doesn't work in the principal arteries, because the flow of blood is too fierce; any clot would be swept away, which is why major bleeds must be stopped with the pressure of a tourniquet. In severe bleeding, the body does all it can to keep blood flowing to the vital organs and diverts it away from secondary outposts like muscles and surface tissues. That's why patients who are bleeding heavily turn a cadaverous white and are cold to the touch. Platelets live for only about a week, so must be constantly replenished. In the last decade or

* If our blood is red, incidentally, why do our veins look blue? It is simply a quirk of optics. When light lands on our skin, a higher proportion of the red spectrum is absorbed, but more of the blue light is bounced back, so blue is what we see. Color is not some innate feature that radiates out of an object but rather a marker of the light bouncing off it.

so, scientists have realized that platelets do more than just manage the clotting process. They also play important roles in immune response and in tissue regeneration.

For the longest time, almost nothing was known about the purpose of blood beyond that it was somehow vital to life. The prevailing theory, dating since the time of the venerable but frequently mistaken Greek physician Galen (ca. 129–ca. 210), was that blood was manufactured continuously in the liver and used up by the body as fast as it was made. As you will doubtless recall from your school days, the English physician William Harvey (1578–1657) realized that blood is not endlessly consumed, but rather circulates in a closed system. In a landmark work called *Exercitatio anatomica de motu cordis et sanguinis in animalibus* (*On the Motion of the Heart and Blood in Animals*), Harvey outlined all the details of how the heart and circulatory system work, in more or less the terms we understand today. When I was a schoolboy, this was always presented as one of those eureka moments that changed the world. In fact, in Harvey's day the theory was almost universally ridiculed and rejected. Nearly all Harvey's peers thought him "crack-brained," in the words of the diarist John Aubrey. Harvey was abandoned by most of his clients and died a bitter man.

Harvey didn't understand respiration, so couldn't explain what purpose blood served or why it circulated—two pretty glaring deficiencies, as his critics were quick to point out. Galenists additionally believed that the body contains two separate arterial systems—one in which the blood is bright red and another in which it is much duller. We now know that blood traveling from the lungs is full of oxygen and therefore shiny crimson, while blood returning to the lungs is depleted of oxygen and thus rather duller. Harvey couldn't explain how blood circulating in a closed system could be of two colors, which became yet another reason to scorn his theories.

The secret of respiration was deduced not long after Harvey's death by another Englishman, Richard Lower, who realized that blood dulls in color on its way back to the heart because it has given up its oxygen, or nitrous spirit, as he called it. (Oxygen wouldn't be discovered until the following century.) That, Lower reasoned, was why blood circulated, to continuously pick up and discharge nitrous oxide, which was quite a big insight and one that should have made him famous. In fact, Lower is remembered more now for another aspect of blood. In the 1660s, Lower was one of several eminent scientists who became interested in the possibility of saving lives through blood transfusions, and he became involved in a series of often gruesome experiments. In November 1667 before an audience of "considerable and intelligent persons" at the Royal Society in London, and without having any idea at all what the consequences might be, Lower transfused about half a pint of blood from a live sheep into the arm of an amiable volunteer named Arthur Coga. Then Lower and Coga and all the distinguished onlookers sat keenly for many minutes waiting to see what would happen. Happily, nothing did. One of those present reported that Coga afterward was "well and merry, and drank a glass or two of canary, and took a pipe of tobacco."

Two weeks later, the experiment was repeated, again without ill effect, which is really surprising. Normally, when foreign substances are introduced in volume into the bloodstream, the recipient goes into shock, so why Coga escaped a miserable experience is puzzling. Unfortunately, the results emboldened other scientists across Europe to conduct transfusion tests of their own, and these took on an increasingly inventive, not to say surreal, cast. Volunteers were transfused with milk, wine, beer, and even mercury, as well as the blood of every species of domesticated creature. The results all too often were distressingly agonized, embarrassingly public deaths. Very quickly transfusion experiments were banned or fell into abeyance, and for about a century and a half they remained out of favor.

And then followed a strange thing. Just as the rest of the scientific world was embarking on the outpouring of discovery and insight known to us as the Age of Enlightenment, medicine sank into a kind of dark age. You could hardly imagine more misguided and counterproductive practices than those to which physicians became attached in the eighteenth and even much of the nineteenth centuries. As David Wootton put it in *Bad Medicine: Doctors Doing Harm Since Hippocrates*, "Up until 1865 medicine was almost completely ineffectual where it wasn't positively harmful."

Consider the unfortunate death of George Washington. In December 1799, not long after he had retired as America's first president, Washington spent a long day on horseback in foul weather inspecting Mount Vernon, his plantation in Virginia. Returning home later than expected, he sat through dinner in damp clothes. That night he developed a sore throat. Soon he had difficulty swallowing, and his breathing became labored.

Three physicians were called in. After a hurried consultation, they opened a vein in his arm and drained eighteen ounces of blood, almost enough to fill a British pint glass (or overfill an American one). Washington's condition only worsened, however, so his throat was blistered with a poultice of cantharides—what is more commonly known as Spanish fly—to draw out bad humors. For good measure, he was given an emetic to induce vomiting. When all of this failed to produce any visible benefit, he was bled three times more. Altogether about 40 percent of his blood was removed over two days.

"I die hard," Washington croaked as his well-meaning doctors relentlessly sapped him. No one knows precisely what Washington's complaint was, but it might have been no more than a minor throat infection that required a little rest. As it was, the illness and treatment together left him dead. He was sixty-seven years old.

Upon his death, yet another doctor visited and proposed that they revive—indeed, resurrect—the deceased president by rubbing his skin gently to stimulate blood flow and transfusing him with lamb's blood, to replace the blood he had lost and refresh what remained. His family mercifully decided to leave him to his eternal rest.

It may seem to us self-evidently foolhardy to bleed and pummel a person who is already severely ill, but such practices lasted an extraordinarily long time. Bleeding was thought to be beneficial not just for illness but to instill calm. Frederick the Great of Germany was bled before battle just to soothe his jangled nerves. Bleeding bowls were treasured within families and passed on as heirlooms. The importance of bleeding is recalled by the fact that Britain's venerable medical journal *The Lancet,* founded in 1823, is named for the instrument used for opening veins.

Why did bleeding persist for so long? The answer is that until well into the nineteenth century most doctors approached diseases not as distinct afflictions, each requiring its own treatment, but as generalized imbalances affecting the whole body. They didn't give one drug for headaches and another for, say, ringing in the ears, but rather endeavored to bring the whole body back into a state of equilibrium by purging it of toxins through the administration of cathartics, emetics, and diuretics, or by relieving the victim of a bowl or two of blood. Opening a vein, as one authority put it, "cools and ventilates the blood" and allows it to circulate more freely, "without danger of burning."

The most celebrated bleeder of all, known as the "Prince of Bleeders," was the American Benjamin Rush. Rush trained in Edinburgh and London, where he learned dissecting from the great surgeon and anatomist William Hunter, but his belief that all illnesses arose from a single cause—overheated blood—was largely self-developed during a long career back in Pennsylvania. Rush, it must be said, was a conscientious and learned man. He was a signer of the Declaration of Independence and the most eminent medical practitioner of his day in the New World. But he was a super enthusiast for bleeding. Rush drained up to eighty ounces at a time from his victims and sometimes bled them two or three times in a single day. Part of the problem was that he believed that the human body contains about twice as much blood as it actually does and that one can remove up to 80 percent of that notional amount without ill effect. He was tragically wrong on both counts yet never doubted the rightness of what he did. During a yellow fever epidemic in Philadelphia, he bled hundreds of victims and was convinced that he had saved a great many when in fact all he did was fail to kill them all. "I have observed the most speedy convalescence where the bleeding has been most profuse," he wrote proudly to his wife.

That was the problem with bleeding. If you could tell yourself that those who survived did so because of your efforts while those who died were beyond salvation by the time you reached them, bleeding would always seem a prudent option. Bleeding retained a place in medical treatments right up to the modern age. William Osler, author of *The Principles and Practice of Medicine* (1892), the most influential medical textbook of the nineteenth century, spoke in favor of bleeding well into what we would consider the modern era.

As for Rush, in 1813 at the age of sixty-seven he developed a fever. When it didn't improve, he urged his attending physicians to bleed him, and they did. And then he died.

The beginning of a modern understanding of blood can perhaps be said to date from 1900 and an astute discovery by a young medical researcher in Vienna. Karl Landsteiner noticed that when blood from different people was mixed together, sometimes it clumped and sometimes it did not. By noting which samples joined with which others, he was able to divide the samples into three groups, which he labeled A, B, and 0. Although everybody reads and pronounces the last group as the letter O, Landsteiner in fact meant it to be taken as a zero, because it didn't clump at all. Two other researchers at Landsteiner's lab subsequently discovered a fourth group, which they called AB, and Landsteiner himself, forty years later, co-discovered Rh factor—short for "rhesus," from the type of monkey in which it was found.* The discovery of blood types explained why transfusions often failed: because the donor and the recipient had incompatible types. It was a hugely significant discovery, but unfortunately almost no one paid any

* Rh factor is the name for a kind of surface protein called an antigen. People who have the Rh antigen (about 84 percent of us) are said to be Rh-positive. Those who lack it, the remaining 16 percent, are Rh-negative.

attention to it at the time. Thirty years would pass before Landsteiner's contribution to medical science was recognized with a Nobel Prize in 1930.

The way blood typing works is this: All blood cells are the same inside, but the outsides are covered with different kinds of antigens—that is, proteins that project outward from the cell surface—and that is what accounts for blood types. There are some four hundred kinds of antigens altogether, but only a few have an important effect on transfusion, which is why we have all heard of types A, B, AB, and O, but not, say, Kell, Giblett, and type E, to name just a very few among many. People with blood type A can donate to those with A or AB but not B; people with B can donate to B or AB but not A; people with AB can donate only to other people with AB blood. People with type O blood can donate to all others, and so are known as universal donors. Type A cells have A antigen on their surface, type B have B, and type AB have both A and B. Put A type blood in a B type person and the recipient body sees it as an invasion and attacks the new blood.

We don't actually know why blood types exist at all. Partly it may be because there simply wasn't any reason for them not to. That is to say, there was no reason to suppose that any person's blood would ever end up in someone else's body, so no reason to evolve mechanisms to deal with such issues. At the same time, by favoring certain antigens in our blood, we can gain improved resistance against particular diseases—though often at a price. People with O blood, for instance, are more resistant to malaria but less resistant to cholera. By developing a variety of blood types and spreading them around among populations, we benefit the species, if not always the individuals within it.

Blood typing had a second, unanticipated benefit: establishing parenthood. In a famous case in Chicago in 1930, two sets of parents, the Bambergers and the Watkinses, had babies in the same hospital at the same time. After returning home, they discovered to their dismay that their babies were wearing labels with the other family's name on them. The question became whether the mothers had been sent home with the wrong babies or with the right babies mislabeled. Weeks of uncertainty followed, and in the meantime both sets of parents did what parents naturally do: they fell in love with the babies in their care. Finally, an authority from Northwestern University with a name that might have come out of a Marx Brothers movie, Professor Hamilton Fishback, was called in, and he administered blood tests to all four parents, which at the time seemed the very height of technical sophistication. Fishback's tests showed that both Mr. and Mrs. Watkins had type O blood and therefore could produce only a type O baby, whereas the child in their nursery was type AB. So, thanks to medical science, the babies were swapped back to the right parents, though not without a lot of heartache.

Blood transfusions save a lot of lives every year, but taking and storing blood is an expensive and even risky business. "Blood is a living tissue," says Dr. Allan Doctor of Washington University in St. Louis. "It's as alive as your heart or lungs or any other organ. The moment you take it out of the body, it begins to degrade, and that is where problems begin."

We met in Oxford, where Doctor, a solemn but amiable man with a trim white beard, was attending a conference of the Nitric Oxide Society, a group that was formed as recently as 1996 because before that nobody realized that nitric oxide was worth getting together for. Its importance to human biology was almost entirely unknown. In fact, nitric oxide (not to be confused with nitrous oxide, or laughing gas) is one of our primary signaling molecules and has a central role in all kinds of processes—maintaining blood pressure, fighting infections, powering penile erections, and regulating blood flow, which is where Doctor comes in. His ambition in life is to make artificial blood, but in the meantime he would like to help make real blood safer to use in transfusions. It comes as a shock to most of us to hear it, but transfused blood can kill you.

The problem is that no one knows how long it remains effective in storage. "Legally, in the United States," Doctor says, "blood can be kept for transfusion for forty-two days, but actually it is probably only good for about two and a half weeks. After that, nobody can say to what extent it is working or not." The forty-two-day rule, which comes from the Food and Drug Administration, is based on how long a typical red cell remains in circulation. "It was assumed for a long time that if a red cell is still circulating, it is still functioning, but we now know that that's not necessarily the case," he says.

Traditionally, it was standard practice for doctors to top up any blood that was lost in trauma. Doctor continued, "If you'd lost three pints of blood, they would put three pints back in. But then AIDS and hepatitis C came along and donated blood was sometimes contaminated, so they began to use transfused blood more sparingly, and to their astonishment they found that patients often had better outcomes from *not* receiving transfusions."

It turned out that in some cases it can be better to let patients be anemic than to give them someone else's blood, especially if that blood had been in storage for a while—and that is nearly always the case. When a blood bank receives a call for blood, it normally dispatches the oldest blood first, to use up aging stock before it expires, which means that almost everybody receives old blood. Worse still, it was discovered that even fresh transfused blood actually impedes the performance of existing blood in the recipient's body. This is where nitric oxide comes in.

Most of us think of blood as being more or less equally distributed around the body at all times. Whatever amount is in your arm now is what is always there. In fact, Doctor explained to me, it is not like that at all.

"If you are sitting down, you don't need so much blood in your legs because there is not a great requirement for oxygen in the tissues.

But if you leap up and start running, you are going to need a lot more blood there very quickly. Your red blood cells, using nitric oxide as their signaling molecule, in large part determine where to dispatch blood as the body's requirements change from moment to moment. Transfused blood confuses the signaling system. It impedes function."

On top of all that, real blood has some practical problems. For one thing, it must be kept refrigerated. That makes it difficult to use on battlefields or accident sites, which is a pity because that's where a lot of bleeding takes place. Twenty thousand people die every year in America from bleeding to death before they can get to a hospital. Globally, the number of bleeding deaths a year has been put as high as 2.5 million. Many of those lives would be saved if people could be transfused promptly and safely—hence the desire for an artificial product.

In theory, it ought to be fairly straightforward, particularly because an artificial blood wouldn't need to do most of the many things real blood does except carry hemoglobin.

"In practice, it's proved to be not so simple," says Doctor with a fleeting smile. He explains the problem by likening red blood cells to those magnets that you see picking up cars in junkyards. The magnet has to latch on to an oxygen molecule in the lungs and convey it to a destination cell. In order to do that, it has to know where to take the oxygen and when to release it, and above all it mustn't drop it en route. That has always been the problem with artificial bloods. Even the best-made artificial bloods occasionally drop an oxygen molecule, and in so doing release iron into the bloodstream. Iron is a toxin. Because of the extreme busyness of the circulatory system, even an infinitesimal accident rate will quickly mount up to toxic levels, so the delivery system has to be pretty much perfect. In nature, it is.

For more than fifty years, researchers have been trying to make artificial blood but, despite spending millions of dollars, are still not there yet. Indeed, there have been more setbacks than breakthroughs.

In the 1990s, some blood products made it into trials, but then it became evident that patients enrolled in the trials were having alarming numbers of heart attacks and strokes. In 2006, the FDA temporarily shut down all trials because the results were so bad. Since then, several pharmaceutical companies have abandoned the quest to make a synthetic blood. For now, the best approach is simply to reduce the volume of transfusions. In an experiment at Stanford Hospital in California, clinicians were encouraged to reduce orders for red blood cell transfusions except when absolutely required. In five years, transfusions at the hospital fell by a quarter. The result was not only a $1.6 million saving in costs but fewer deaths, quicker average discharges, and a reduction in posttreatment complications.

Now, however, Doctor and his colleagues in St. Louis think they have nearly cracked the problem. "We have nanotechnology at our disposal now, which wasn't available before," he says. Doctor's team has developed a system that keeps the hemoglobin inside a polymer shell. The shells are shaped like conventional red blood cells but are about fifty times smaller. One of the great virtues of the product is that it can be freeze-dried, enabling it to be stored for up to two years at room temperature.

At the time I met him, he believed they were three years away from trials in humans, and perhaps ten years from using it clinically.

In the meantime, it remains a slightly humbling reflection that about a million times per second our bodies do something that all the science of the world put together so far cannot do at all.

8 THE CHEMISTRY DEPARTMENT

> I hope my disease of the stone may not return to me, but void itself in pissing, which God grant, but I will consult my physitian.
>
> —SAMUEL PEPYS

I

DIABETES IS A horrible disease, but once it was even worse because people could do almost nothing about it. Youngsters with diabetes generally died within a year of diagnosis, and it was a miserable death. The only way to reduce sugar levels in the body, and extend lives even slightly, was to keep victims right on the edge of starvation. One twelve-year-old boy was left so hungry that he was caught eating birdseed from the tray of a canary cage. Eventually he died, as all victims died, famished and wretched. He weighed thirty-nine pounds.

Then, in late 1920, in one of the happiest but most improbable episodes in the history of scientific progress, a struggling young general practitioner in London, Ontario, read an article about the pancreas in a medical journal and got an idea for how he might effect a cure. His name was Frederick Banting, and he knew so little about diabetes that he misspelled it as "diabetus" in his notes. He had no experience of medical research, but he was convinced that he had a notion worth pursuing.

The challenge for anyone tackling diabetes was that the human pancreas has two quite separate functions. Most of it is devoted to making and secreting enzymes that assist in digestion, but the pancreas also contains clusters of cells known as islets of Langerhans. These were discovered in 1868 by a medical student in Berlin, Paul Langerhans, who freely admitted that he had no idea what they were there for. Their function, to produce a chemical that was at first called isletin, was deduced twenty years later by a Frenchman, Édouard Laguesse. We now call that chemical insulin.

Insulin is a small protein that is vital in maintaining a very delicate balance of blood sugar in the body. Too much or too little produces terrible consequences. We get through a lot of insulin. Each molecule only lasts from five to fifteen minutes, so the demand for replenishment is relentless.

The role of insulin in controlling diabetes was well known by Banting's time, but the problem was separating it from the digestive juices. Banting's belief–based on no evidence whatever–was that if you tied off the pancreatic duct and stopped digestive juices from getting to the intestines, the pancreas would stop producing them. There was no reason at all to suppose that this would happen, but he persuaded a professor at the University of Toronto, J. J. R. Macleod, to let him have some lab space, an assistant, and some dogs on which to experiment.

The assistant was a Canadian American named Charles Herbert Best who had grown up in Maine, where his father was a small-town general practitioner. Best was conscientious and willing but, like Banting, knew almost nothing about diabetes and even less about experimental methods. Nonetheless, they set to work, tying off pancreatic ducts in dogs, and, amazingly, got good results. They did almost everything wrong. As one observer put it, their experiments were "wrongly conceived, wrongly conducted, and wrongly interpreted." Yet within weeks they were producing pure insulin.

When given to diabetics, the effect was nothing short of miraculous. Listless, skeletal patients who could barely be called alive were swiftly restored to full vibrancy. It was, to borrow from Michael Bliss, author of the definitive *The Discovery of Insulin*, the closest thing to resurrection modern medicine had ever produced. Another researcher in the lab, J. B. Collip, came up with a more effective method for extracting insulin, and soon it was being produced in vast enough quantities to save lives all over the world. "The discovery of insulin," declared the Nobel laureate Peter Medawar, "may be rated the first great triumph of medical science."

It should have been a happy story for all concerned. In 1923, Banting was awarded the Nobel Prize in Physiology or Medicine along with Macleod, the head of the lab. Banting was appalled. Not only had Macleod not been involved in the experimental work, he hadn't even been in the country when the breakthrough was made, but rather was on an extended annual visit to his native Scotland. Banting clearly thought Macleod did not deserve the honor and announced that he would share the prize money with his trusty assistant Best. Collip, meanwhile, refused to share his improved extraction method with the rest of the team and announced that he intended to patent the procedure in his own name, infuriating the others. Banting, who seems to have had a short fuse in life anyway, on at least one occasion had to be pulled off Collip after physically attacking him.

Best for his part couldn't stand Collip or Macleod and eventually ended up disliking Banting, too. In short, they more or less all ended up loathing one another. But at least the world got insulin.

Diabetes comes in two varieties. Indeed, it is really two diseases, with similar complications and management issues but generally different pathologies. In type 1 diabetes, the body stops producing insulin altogether. In type 2 diabetes, insulin is less effective, usually because of a combination of decreased production and because the cells on which it acts don't respond as they normally would. This is referred to as insulin resistance. Type 1 tends to be inherited; type 2 is usually a consequence of lifestyle. But it's not quite as simple as that. Although type 2 is unequivocally associated with unhealthy living, it also tends to run in families, suggesting a genetic component. Similarly, although type 1 diabetes is associated with a fault in a person's HLA (human leukocyte antigen) genes, only some people with the fault get diabetes, indicating that there is some additional, unrecognized trigger. Many researchers suspect a link to levels of exposure to a range of pathogens in early life. Others have suggested an imbalance in the victim's gut microbes or possibly even a connection to how comfortable and well nourished one was in the womb.

What can be said is that rates everywhere are soaring. Between 1980 and 2014, the number of adults in the world with diabetes of one type or another went from just over 100 million to well over 400 million. Ninety percent of them had type 2 diabetes. Type 2 is growing especially fast in developing countries that have been adopting our bad Western habits of poor diet and inactive lifestyle. Yet type 1 is also growing swiftly. In Finland, it has gone up by 550 percent since 1950. It continues to rise almost everywhere at a rate of about 3 to 5 percent a year, for reasons no one understands.

Although insulin has transformed the lives of millions of diabetics, it is not a perfect solution. For one thing, it cannot be given orally, because it is broken down in the gut before it can be absorbed and put to use, so it must be injected, which is both a tedious process and a crude one. In a healthy body, insulin levels are monitored and adjusted second by second. In a diabetic, they are adjusted only periodically, when the patient self-medicates. That means that insulin levels are still not quite right much of the time, and that has a cumulative negative effect.

Insulin is a hormone. Hormones are the bicycle couriers of the body, delivering chemical messages all around the teeming metropolis that is you. They are defined as any substance that is produced in one part of the body and causes an action somewhere else, but beyond that they are not easy to characterize. They come in different sizes, have different chemistries, go to different places, have different effects when they get there. Some are proteins, some are steroids, some are from a group called amines. They are linked by their purpose, not their chemistry. Our understanding of them is far from complete, and much of what we do know is surprisingly recent.

John Wass, professor of endocrinology at Oxford University, is smitten with hormones. "I love hormones," he likes to say. When we met, in a café in Oxford at the end of a long working day, he was clutching an armful of disorderly papers but looking surprisingly fresh for someone who had flown in that morning from ENDO 2018, the annual conference of the Endocrine Society in the United States.

"It's madness," he tells me in a delighted tone. "You have eight or ten thousand endocrinologists from all over the planet. The meetings start at five thirty in the morning and can go on until nine o'clock at night, so there's a lot to take in and you end up with"—he shakes the papers for me—"a lot of reading. It's very useful but a bit mad."

Wass is a tireless campaigner for a better appreciation of hormones and what they do for us. "They were the last major system in the body to be discovered," he says. "And we are still discovering more all the time. I know I am biased, but it is really a terribly exciting field."

As late as 1958, only about twenty hormones were known. No one seems to know quite how many there are now. "Oh, I think it must be at least eighty," says Wass, "but perhaps as many as a hundred now. We really do keep discovering more all the time."

Until very recently, it was thought that hormones are produced exclusively in the body's endocrine glands (hence the name endocrinology for this branch of medicine). An endocrine gland is one that secretes its products directly into the bloodstream, as opposed to exocrine glands, which secrete onto a surface (like sweat glands onto skin or salivary glands into the mouth). The principal endocrine glands—the thyroid, parathyroid, pituitary, pineal, hypothalamus, thymus, testes (in men), ovaries (in women), pancreas—are scattered all around the body but work together closely. They are mostly tiny and altogether weigh no more than a few ounces but have an importance to your happiness and well-being that is entirely disproportionate to their modest dimensions.

The pituitary gland, for instance, which is buried deep within your brain directly behind your eyes, is only about the size of a baked bean, yet its effects can be—literally—enormous. Robert Wadlow of Alton,

Illinois, the tallest human who ever lived to that point, had a pituitary condition that caused him to grow ceaselessly because of continuous overproduction of growth hormone. A shy and cheerful soul, he was taller than his (normal-sized) father by the age of eight, was 6 feet 11 inches tall at the age of twelve, and over 8 feet tall when he graduated from high school in 1936—all because of a little chemical overexertion by this baked bean in the middle of his skull. He never stopped growing and was just a fraction under 9 feet tall at his greatest eminence. Though not fat, he weighed about five hundred pounds. His shoes were a size 40. By his early twenties, he could walk only with great difficulty. To support himself, he wore leg braces, which caused chafing, and that led to a serious infection that grew septic and killed him as he slept on July 15, 1940. He was just twenty-two. His height at death was 8 feet 11.1 inches. He was much loved and is still celebrated in his hometown.

It is clearly ironic that such a large body resulted from a malfunction in a minuscule gland. The pituitary is often called the master gland because it controls so much. It produces (or regulates the production of) growth hormone, cortisol, estrogen and testosterone, oxytocin, adrenaline, and much else. When you exercise vigorously, the pituitary squirts endorphins into your bloodstream. Endorphins are the same chemicals released when you eat or have sex. They are closely related to opiates. That's why it is called the runner's high. There is barely a corner of your life that the pituitary doesn't touch, yet its functions weren't even broadly understood until well into the twentieth century.

The field of modern endocrinology got off to a somewhat bumpy start, in good measure because of the enthusiastic but misguided endeavors of an otherwise brilliant man named Charles-Édouard Brown-Séquard (1817–94). Brown-Séquard was a man literally of many nations. He was born on the Indian Ocean island of Mauritius, which made him Mauritian and British because Mauritius was then a British colony, but his mother was French and his father was American, so he had claims to four nationalities from the moment of his first breath. He never met his father, a ship's captain who was lost at sea before his son's birth. Brown-Séquard grew up in France and trained as a physician there but then rotated between Europe and America, seldom staying in either long. In one twenty-five-year period, he made sixty Atlantic crossings—this when one trip in a lifetime was exceptional—taking up a variety of posts, many of considerable eminence, in Britain, France, Switzerland, and the United States. During the same period, he wrote nine books and more than five hundred papers; edited three journals; taught at Harvard, the University of Geneva, and the Faculté de Médecine in Paris; lectured widely; and became a leading authority on epilepsy, neurology, rigor mortis, and the secretions of glands. But it was an experiment he conducted in Paris in 1889, at the stately age of seventy-two, that secured his permanent, and somewhat risible, fame.

Brown-Séquard ground up the testes of domesticated animals (dogs and pigs are most often cited, but no two sources seem to quite agree on which animals he favored), injected the extract into himself, and reported feeling as frisky as a forty-year-old. In fact, any improvement he sensed was entirely psychological. Mammalian testes contain almost no testosterone because it is sent out into the body as quickly as it is made, and in any case we manufacture very little of it anyway. If Brown-Séquard ingested any testosterone at all, it was no more than a trace. Even though Brown-Séquard was completely wrong about the rejuvenative effects of testosterone, he was actually right that it is potent stuff—so much so that, when synthesized, it is treated today as a controlled substance.

Brown-Séquard's enthusiasm for testosterone seriously damaged his scientific credibility, and he died soon afterward anyway, but ironically his efforts prompted others to look more closely and systematically at the chemical processes that control our lives. In 1905, a decade after Brown-Séquard's death, the British physiologist E. H. Starling coined the term "hormone" (on advice from a classics scholar at Cambridge University; it comes from a Greek word meaning "to set in motion"), though the science didn't really get going until the following decade. The first journal devoted to endocrinology wasn't founded until 1917, and the umbrella term for the ductless glands of the body, the endocrine system, came even later. It was coined in 1927 by the British scientist J. B. S. Haldane.

Arguably the real father of endocrinology lived a generation before Brown-Séquard. Thomas Addison (1793–1860) was one of a trio of outstanding doctors, known as the Three Greats, at Guy's Hospital in London in the 1830s. The others were Richard Bright, discoverer of Bright's disease (now called nephritis), and Thomas Hodgkin, who specialized in disorders of the lymphatic system and whose name is commemorated in Hodgkin's and non-Hodgkin's lymphomas. Addison was probably the most brilliant, certainly the most productive, of the three. He provided the first accurate account of appendicitis and was a leading authority on all types of anemia. At least five serious medical conditions were named for him, of which the most famous was (and remains) Addison's disease, a degenerative disorder of the adrenal glands that Addison described in 1855, making it the first hormonal disorder to be identified. Despite his fame, Addison was subject to spells of depression, and in 1860, five years after identifying Addison's, he retired to Brighton and killed himself.

Addison's disease is a rare but still-serious illness. It affects about one person in ten thousand. History's most famous sufferer was John F. Kennedy, who was diagnosed with it in 1947, though he and his family always emphatically and untruthfully denied it. In fact, Kennedy not only had Addison's but was lucky to survive it. In those days, before the introduction of glucocorticoids, a type of steroid, 80 percent of sufferers died within a year of diagnosis.

John Wass, at the time we met, was particularly preoccupied with Addison's disease. "It can be a very sad disease because the symptoms—principally loss of appetite and weight loss—are easily misdiagnosed," he told me. "I recently dealt with the case of a really lovely young woman, just twenty-three years old and with a very promising future in front of her, who died of Addison's because her doctor thought she was suffering from anorexia and sent her to a psychiatrist. Addison's in fact arises from an imbalance of cortisol levels—cortisol being a stress hormone that regulates blood pressure. The tragedy of it is that if you correct the cortisol problem, the patient can return to normal health in as little as thirty minutes. She needn't have died at all. A big part of what I do is lecture to general practitioners to try to help them to look out for common hormonal disorders. They are all too often missed."

In 1995, the field of endocrinology experienced a seismic moment when Jeffrey Friedman, a geneticist at Rockefeller University in New York, found a hormone that no one thought could possibly exist. He named it leptin (from a Greek word for "thin"). Leptin was produced not in an endocrine gland but in fat cells. This was a most arresting discovery. No one had ever suspected that hormones could be produced anywhere but in their own dedicated glands. In fact, we now know, hormones are produced all over the place—in the stomach, lungs, kidneys, pancreas, brain, bones, everywhere.

Leptin drew massive and immediate interest not just because of the surprise of where it was produced but even more because of what it does: it helps to regulate appetite. If we could control leptin, then presumably we could help people to control their weight. In studies with rats, scientists discovered that by manipulating leptin levels, they could make rats obese or lean, as they wished. This had the makings of a wonder drug.

Clinical trials with humans were quickly undertaken, amid considerable anticipation. Volunteers with a weight problem received daily leptin injections for a year. At the end of the year, however, they weighed just as much as they had at the beginning. Leptin's effects turned out to be nothing like as straightforward as hoped. Today, nearly a quarter of a century after its discovery, we still haven't figured out exactly how leptin works and are nowhere near being able to use it as an aid for weight control.

A central part of the problem is that our bodies evolved to deal with the challenge of dietary paucity, not overabundance. So leptin isn't programmed to tell you to stop eating. Nothing chemical in your body is. That's a big part of why you tend to just keep on consuming. We are habituated into devouring foods greedily whenever we are able on the assumption that abundance is an occasional condition. When leptin is completely absent, you just keep on eating and eating because your body thinks you are starving. But when it is added to the diet, in normal circumstances it makes no discernible difference to appetite. What leptin is there for essentially is to tell the brain whether you have enough energy reserves to undertake comparatively demanding challenges like getting pregnant or starting puberty. If your hormones think you are starving, those processes will not be allowed to begin. That's why young people who are anorexic often have a very delayed start to puberty. "It's also almost certainly why puberty starts years earlier now than it did in historic times," says Wass. "In Henry VIII's reign, puberty started at sixteen or seventeen. Now it is more commonly eleven. That's almost certainly because of improved nutrition."

Complicating matters further is that bodily processes are nearly always influenced by much more than a single hormone. Four years after leptin's discovery, scientists discovered another hormone involved in appetite regulation. Dubbed ghrelin (the first three letters stand for "growth-hormone related"), it is produced mostly in the stomach but also in several other organs. When we get hungry, our ghrelin levels rise, but it isn't clear whether ghrelin causes hunger or merely accompanies it. Appetite is also influenced by the thyroid gland and by genetic and cultural considerations and by mood and accessibility (a bowl of peanuts on a table is hard to resist), willpower, time of day, season, and much else. No one has figured out how to pack all that into a pill.

On top of all this, most hormones have a multiplicity of functions, which makes it harder to deconstruct their chemistry and riskier to tinker with it. Ghrelin, for instance, doesn't just have a role in hunger, but also helps to control insulin levels and the release of growth hormone. Tampering with one function could destabilize the others.

The range of regulatory jobs any hormone does can be bewilderingly diverse. Oxytocin, to take one example, is well known for its role in generating feelings of attachment and affection–it is sometimes called "the hug hormone"–but it also plays an important part in facial recognition, in directing contractions of the uterus in childbirth, in interpreting the moods of people around us, and in initiating the production of milk by nursing mothers. Why oxytocin accumulated this particular mix of specializations is anyone's guess. Its role in bonding and affection is clearly its most intriguing quality but also its least understood. Oxytocin given to female rats led them to build nests for and fuss over infants that were not their own. Yet in tests where oxytocin has been administered clinically to humans, it has had little or no effect. In some cases, perversely, it has made test subjects more aggressive and less cooperative. Hormones, in short, are complicated molecules. Some of them, like oxytocin, are both hormones and neurotransmitters–signaling molecules for the nervous system. In short, they do a lot, but little of it simply.

Perhaps no one has better understood the boundless complexity of hormones than the German biochemist Adolf Butenandt (1903–95). A native of Bremerhaven, Butenandt studied physics, biology, and chemistry at the Universities of Marburg and Göttingen but also found time for rather more energetic pursuits. He engaged enthusiastically in fencing, without protective equipment, as appears to have been the dashing but not notably prudent convention among young men in Germany of the time, in consequence of which he acquired a jagged scar across his left cheek that he seems to have been rather proud of. His passion in life was biology–animal as well as human–and in particular hormones, which he distilled and synthesized with exceeding patience. In 1931, he took a very large amount of urine donated by the policemen of Göttingen–some sources say 15,850 quarts, some say 26,400, but certainly more than most of us would want to handle–and from it distilled fifteen milligrams of the hormone androsterone. By similar dogged efforts, he distilled several other hormones. To isolate progesterone, for instance, he needed the ovaries of fifty thousand pigs. To isolate the first pheromones—or sexual attractants—required the sex glands of 500,000 Japanese silkworms.

Thanks to his extraordinary focus, his discoveries made possible all kinds of useful products, from synthetic steroids for medical use to birth control pills. He was awarded the Nobel Prize in Chemistry in 1939 when he was just thirty-six years old but wasn't allowed to accept it because Adolf Hitler forbade Germans to receive the award after the Peace Prize was given to a Jew. (Butenandt did finally receive the award in 1949, but not the prize money. According to the terms of Alfred Nobel's will, the cash part of the award expires after a year if not collected.)

For a long time, endocrinologists thought that testosterone was exclusively a male hormone and estrogen exclusively female, but in fact men and women produce and use both. In males, testosterone is produced mostly by the testes, with a little from the adrenal glands, and does three things: it makes a man fertile, it endows him with virile attributes like a deep voice and the need to shave, and it profoundly influences his behavior, giving him not only his sex drive but also a taste for risk and aggression. In women, testosterone is produced about half and half between the ovaries and the adrenal glands, but in much smaller amounts, and boosts libido, but mercifully leaves their common sense undisturbed.

One area where testosterone appears not to be doing us men any good at all is longevity. Many factors determine life span, of course, but it is a fact that men who have been castrated live about as long as women do. In what way exactly testosterone might shorten male lives is not known. Testosterone levels in men fall by about 1 percent a year beginning in their forties, prompting many to take supplements in the hope of boosting their sex drive and energy levels. The evidence that it improves sexual performance or general virility is thin at best; there is much greater evidence that it can lead to an increased risk of heart attack or stroke.

II

NOT ALL GLANDS are tiny, of course. (For the record, a gland is any organ in the body that secretes chemicals.) The liver is a gland and it is, compared with the rest of our glands, gigantic. When fully grown, it weighs about 3.3 pounds, roughly the same as the brain, and fills much of the central abdomen just below the diaphragm. It is disproportionately large in infants, which is why their bellies are so delightfully rounded.

It is also the most multifariously busy organ in the body, with functions so vital that if it shuts down, you will be dead within hours. Among its many jobs, it manufactures hormones, proteins, and the digestive juice known as bile. It filters toxins, disposes of obsolescent red blood cells, stores and absorbs vitamins, converts fats and proteins to carbohydrates, and manages glucose—a process which is so vital for the body that its dilution for even a few minutes can cause organ failure and even brain damage. (Specifically the liver converts glucose into glycogen—a more compact chemical. It is a little bit like shrink-wrapping food so you can pack more of it into your freezer. When energy is needed, the liver converts the glycogen back into glucose and releases it into the bloodstream.) Altogether the liver takes part in some five hundred metabolic processes. It is essentially the body's laboratory. Right now, about a quarter of all your blood is in your liver.

Perhaps the most wondrous feature of the liver is its capacity to regenerate. You can remove two-thirds of a liver and it will grow back to its original size in just a few weeks. "It's not pretty," the Dutch geneticist Professor Hans Clevers told me. "It looks a bit battered and rough compared with the original liver, but it functions well enough. The process is something of a mystery. We don't know how a liver knows to grow back to just the right size and then stop growing, but it is lucky for some of us that it does."

The liver's resilience is not infinite, however. It is subject to more than a hundred disorders, and many of these are grave. Most of us think of liver disease as being caused by excessive alcohol consumption, but in fact alcohol is implicated in only about a third of chronic liver disorders. Nonalcoholic fatty liver disease (NAFLD) is an illness most of us have never heard of, but it is more common than cirrhosis, and far more baffling. It is, for instance, strongly associated with being overweight or obese, and yet a significant proportion of sufferers are fit and lean. No one can explain why. Altogether about a third of us are thought to have early stages of NAFLD, but luckily for most of us it never progresses beyond that. For the unfortunate minority, however, NAFLD means eventual liver failure or other serious diseases. Again, why some people are hit hard and others escape is a mystery. Perhaps the most unnerving aspect is that victims usually suffer no symptoms at all until most of the damage has been done. Even more alarming is that NAFLD is starting to be found in young children–somewhere it had never been seen before until recently. An estimated 10.7 percent of children and adolescents in the United States and 7.6 percent globally are estimated to have fatty livers.

Another risk that many people aren't fully aware of is hepatitis C. According to the Centers for Disease Control and Prevention, about one in thirty people in America born between 1945 and 1965–that's two million people altogether–will have hepatitis C without knowing it. (People born in that period were at greater risk in large part because of contaminated blood transfusions and the sharing of needles by people doing drugs.) Hepatitis C can live within victims for forty years or more, stealthily demolishing their livers, without their being aware of it. The CDC estimates that if all those people could be identified and treated, 120,000 lives would be saved in America alone.

The liver was long thought to be the seat of courage, which is why a cowardly person was deemed "lily-livered." It was also considered the source of two of the four humors–black bile and yellow bile–respectively responsible for melancholy and choler, and thus was considered responsible for both sadness and anger. (The other two humors were blood and phlegm.) The humors were believed to be fluids that circulated within the body and kept everything in balance. For two thousand years, a belief in humors was used to explain people's health, looks, tastes, disposition—everything. In this context, humor has nothing to do with amusement. It comes from a Latin word for "moisture." When we talk today of humoring someone or of people being ill-humored, we are not talking about their capacity for laughter, at least not etymologically.

Packed in beside the liver are two other organs, the pancreas and spleen, which are often paired up because they live side by side and are similarly sized, but actually are quite unalike. The pancreas is a gland and the spleen is not. The pancreas is essential to life; the spleen is expendable. The pancreas is a jellylike organ, about six inches long, shaped roughly like a banana (and about the same size), tucked behind the stomach in the upper abdomen. As well as insulin, it produces the hormone glucagon, which is also involved in regulating blood sugar, and the digestive enzymes trypsin, lipase, and amylase, which help digest cholesterol and fats. Altogether every day it produces over a quart of pancreatic juice, a pretty prodigious amount for an organ of its size.

The pancreas of an animal when cooked for consumption is known as a sweetbread (the word is first recorded in English in 1565), but no one has ever worked out why, because there is nothing sweet or bread-like about it. "Pancreas" isn't recorded in English until late the following decade, so "sweetbread" is actually the older term.

The spleen is roughly the size of your fist, weighs half a pound, and sits fairly high up on the left side of your chest. It does important work monitoring the condition of circulating blood cells and dispatching white blood cells to fight infections. It also acts as a reservoir for blood so that more can be supplied to muscles when suddenly needed, and it aids the immune system. A person who is splenetic is angry or wrathful; we still vent our spleen when angry.

Medical students learn to remember the principal attributes of the spleen by counting upward in odd numbers; 1, 3, 5, 7, 9, 11. That is because the spleen is 1 x 3 x 5 inches in size, weighs 7 ounces, and lies between the 9th and the 11th ribs–though in fact all those numbers but the last two are merely averages.

Just beneath the liver and also closely associated with it is the gallbladder. It is a curious organ in that many animals have gallbladders and many do not. Giraffes, oddly, sometimes have gallbladders and sometimes don't. In humans, the gallbladder stores bile from the liver and passes it on to the intestines. ("Gall" is an old word for "bile.") The chemistry can go wrong for a variety of reasons, resulting in gallstones. Gallstones are a common complaint and were traditionally said to be most often found among women who were "fat, fair, fertile, and forty," according to a well-known but, I'm told, highly inaccurate mnemonic among doctors. As many as a quarter of adults have gallstones, but usually don't know it. Just occasionally a gallstone will block the bladder outlet, leading to abdominal pain.

Surgery for gallstones (which are formally called calculi) is now routine, but once it was often a life-threatening condition. Until late in the nineteenth century, surgeons dared not cut into the upper abdomen because of the dangers of delving amid all the vital organs and arteries up there. One of the very first to attempt an operation on a gallbladder was the great but odd American surgeon William Stewart Halsted (whose extraordinary story we will cover more fully in chapter 21). In 1882, while still a young doctor, Halsted conducted one of the first surgical removals of a gallbladder, on his own mother, on a kitchen table in their family home in upstate New York. What made this all the more remarkable was that there was no certainty at this time that someone could survive without a gallbladder. Whether Mrs. Halsted was quite aware of this as her son pressed a handkerchief of chloroform to her face is not recorded. At all events, she made a full recovery. (In an unfortunate irony, the pioneer Halsted would die following gallbladder surgery on himself forty years later, by which time such surgery had become commonplace.)

Halsted's operation on his mother recalled a procedure undertaken a few years earlier by a German surgeon, Gustav Simon, who removed a diseased kidney from a female patient without having any certain idea what would happen and was delighted to discover–as presumably was the patient–that it didn't kill her. It was the first anyone realized that humans can survive with just one kidney. It remains something of a mystery even now as to why we have two kidneys. It is splendid to have a backup, of course, but we don't get two hearts or livers or brains, so why we have a surplus kidney is a happy imponderable.

The kidneys are invariably called the workhorses of the body. Each day they process about 190 quarts of water–that is the amount a bath holds up to the overflow level–and 3.3 pounds of salt. They are startlingly small for the amount of work they do, weighing just five ounces each. They are not in the small of the back, as everyone thinks, but higher up, about at the bottom of the rib cage. The right kidney is always lower because it is pressed down upon by the asymmetrical liver. Filtering wastes is their principal function, but they also regulate blood chemistry, help maintain blood pressure, metabolize vitamin D, and maintain the vital balance between salt and water levels within the body. Eat too much salt and your kidneys filter out the excess from your blood and send it to the bladder so that you can pee it all away. Eat too little and the kidneys take it back from the urine before it leaves your body. The problem is that if you ask the kidneys to do too much filtering over too long a period, they get tired and stop functioning terribly well. As the kidneys become less efficient, the sodium levels in your blood creep up, pushing your blood pressure dangerously high.

More than most other organs, the kidneys lose function as you age. Between the ages of forty and seventy, their filtration capacity drops by about 50 percent. Kidney stones become more common, as do more life-threatening illnesses. The death rate from chronic kidney disease has jumped by more than 70 percent since 1990 in the United States and by even more in some third world countries. Diabetes is the commonest cause of kidney failure, with obesity and high blood pressure as important contributory factors.

What the kidneys don't return to the body via the bloodstream, they pass on to the second and more familiar of our bladders, the urinary one, for disposal. Each kidney is connected to the bladder by a tube called a ureter. Unlike the other organs discussed here, the urinary bladder doesn't produce hormones (at least none yet found) or have a role in body chemistry, but it does at least possess a kind of venerability. "Bladder" is one of the oldest words in the body, dating from Anglo-Saxon times and predating both "kidney" and "urine" by more than six hundred years. Most other words in Old English with a median *d* sound morphed into a softer *th* sound, so that "feder" became "feather" and "fader" became "father," but "bladder" for some reason resisted the gravitational pull of common usage and has stayed true to its original pronunciation for well over a thousand years, something few other parts of the body could claim.

The urinary bladder is rather like a balloon in that it is designed to swell as we fill it. (In an average-sized man it holds about a British pint, or about six-tenths of a quart; in a woman, rather less.) As we age, the bladder loses elasticity and can't expand as it once did, which is part of the reason old people spend much of their lives scouting for restrooms, according to Sherwin Nuland in *How We Die*. Until very recently, it was thought that the urine and bladder are normally sterile. Occasionally bacteria might sneak in and give us a urinary tract infection, but there were no permanent colonies of bacteria in there. For that reason when the Human Microbiome Project was launched in 2008, with the intention of tracking down and cataloging all the microbes within us, the bladder was excluded from investigation. We now know that the urinary world is at least somewhat microbial, too, if not apparently vastly so.

One unfortunate feature the bladder has in common with both the gallbladder and the kidneys is a tendency to form stones—hardened balls of calcium and salts. For centuries, stones plagued people to a degree almost unimaginable now. Because they were so difficult to deal with, they often grew to a most prodigious size before the victim finally accepted the necessity—and very high risk—of surgery. It was a horrible procedure, combining unsurpassed levels of pain, danger, and indignity in a single mortifying operation. Patients were calmed, to the extent possible, with infusions of opiates and mandragora (a form of mandrake), then placed on their backs on a table with their legs pushed back over their heads, their knees bound to their chests, and their arms bound to the table. Usually four strong men were called upon to hold the patient still while the surgeon scavenged about for stones. Not surprisingly, surgeons who performed the procedure were celebrated for their speed more than any other quality.

Probably history's most famous lithotomy, or stone removal, was that experienced by the diarist Samuel Pepys in 1658, when he was twenty-five years old. This was two years before Pepys started his diary, so we don't have a firsthand account of the experience, but he mentioned it frequently and vividly thereafter (including in the diary's very first entry when he finally started it) and lived in loquacious dread of ever having to undergo anything like it again.

It's not hard to see why. Pepys's stone was the size of a tennis ball (albeit a seventeenth-century tennis ball, which was slightly smaller than a modern tennis ball, though the distinction could fairly be called academic to anyone carrying one). While four men held Pepys down, the surgeon, Thomas Hollyer, inserted an instrument called an itinerarium up his penis and into the bladder to fix the stone in place. Then he took a scalpel and quickly and deftly—but excruciatingly—cut a three-inch-long incision through the perineum (the area between the scrotum and the anus). Peeling back the opening, he gently cut into the exposed and quivering bladder, thrust a pair of duck-billed forceps through the opening, captured the stone, and extracted it. The entire procedure from beginning to end took just fifty seconds but left Pepys bedridden for weeks and traumatized for life.*

* Pepys's complaint is often wrongly described as kidney stones. I regret to say I repeated that error in my book *At Home: A Short History of Private Life*. Pepys had kidney stones aplenty, too—he passed them regularly throughout life—but Dr. Hollyer (sometimes spelled Hollier in other accounts) would not have been able to extract such a large stone from the kidneys without killing him. The

Hollyer charged Pepys twenty-four shillings for the operation, but it was money well spent. Hollyer was famous not just for his speed but also for the fact that his patients very generally survived. In one year, he performed forty lithotomies and lost not a single subject—an extraordinary achievement. Doctors in the past were not always anything like as dangerous and incompetent as we are sometimes led to think them. They might have known nothing of antisepsis, but the best of them did not lack for skill and intelligence.

Pepys for his part marked the anniversary of his survival for some years thereafter with prayers and a special dinner. He kept the stone in a lacquered box, and for the rest of his life showed it off at every opportunity to anyone willing to marvel at it. And who could possibly blame him?

---

experience is recorded fully and memorably in Claire Tomalin's esteemed biography, *Samuel Pepys: The Unequalled Self.*

9 IN THE DISSECTING ROOM: THE SKELETON

Heaven take my soul, and England keep my bones!

—WILLIAM SHAKESPEARE, *THE LIFE AND DEATH OF KING JOHN*

I

THE MOST POWERFUL impression you get in a dissecting room is that the human body is not a wondrous piece of precision engineering. It's meat. It is nothing like the plastic teaching models of torsos lined up on shelves around the perimeter of the room. Those are colorful and shiny, like children's toys. An actual human body in a dissecting room isn't toylike at all. It is just dull flesh and sinew and lifeless organs drained of color. It is slightly mortifying to realize that the only raw flesh we normally see is the meat of animals that we are about to cook and eat. The flesh of a human arm, once the outer skin is removed, looks surprisingly like chicken or turkey. It's only when you see that it ends in a hand with fingers and fingernails that you realize it's human. This is when you think you might be sick.

"Feel this," Dr. Ben Ollivere is saying to me. We are in the dissecting room at the University of Nottingham Medical School in England, and he is directing my attention to a piece of detached tubing in the upper chest of a male body. The tube has been sliced through, evidently for demonstration purposes. Ben instructs me to stick my gloved finger into its interior and feel it. It is stiff, like uncooked pasta—like a cannelloni shell. I have no idea what it is.

"The aorta," Ben says with what seems like pride.

I am frankly amazed. "So that's the heart?" I say, indicating the shapeless lump beside it.

Ben nods. "And the liver, pancreas, kidneys, spleen," he says, pointing out the other organs of the abdomen in turn, sometimes nudging one aside to expose another behind or beneath it. They are not fixed and hard like the plastic teaching models, but move about easily. I am vaguely reminded of water balloons. There is a lot of other stuff in there, too—threaded blood vessels and nerves and tendons, and lots and lots of intestines, all of it just kind of tipped in, as if this poor, anonymous, former person had had to pack himself in a hurry. It was impossible to visualize how any of this disordered interior could ever have conducted the tasks that would allow the very inert body before us to sit up and think and laugh and live.

"You can't mistake death," Ben says to me. "Live people look alive—and even more so on the inside than on the surface. When you open them up in surgery, the organs throb and glisten. They are clearly living things. But in death they lose that."

Ben is an old friend and a distinguished academic and surgeon. He is clinical associate professor of trauma surgery at the University of Nottingham and a consultant trauma surgeon at Queen's Medical Centre in the city. There isn't anything in the human body that doesn't fascinate him. We rather race around this one as he tries to tell me everything about it that interests him, which is everything.

"Just consider all that the hand and wrist do," he says. He tugs gently on an exposed tendon in the cadaver's forearm up near the elbow, and to my surprise the little finger moves. Ben smiles at my startlement and explains that we have so much packed into a small space in the hand that a lot of the work has to be done remotely, like strings on a marionette. "If you make a tight fist, you feel the strain in your forearm. That's because it's the arm muscles that are doing most of the work."

With a blue-gloved hand, he gently swivels the cadaver's wrist, as if conducting an examination. "The wrist is just a thing of beauty," he goes on. "Everything has to go through there—muscles, nerves, blood vessels, everything—and yet it has to be completely mobile at the same time. Think of all the things your wrist has to do—take a lid off a jam jar, wave good-bye, turn a key in a lock, change a lightbulb. It's a magnificent piece of engineering."

Ben's field is orthopedics, so he loves bones and tendons and cartilage—the living infrastructure of the body—the way other people love expensive cars or excellent wines. "See that?" he says, tapping a small, smooth very white obtrusion at the base of the thumb, which I take to be a bit of exposed bone. "No, it's cartilage," he says. "Cartilage is remarkable, too. It is many times smoother than glass: it has a friction coefficient five times less than ice. Imagine playing ice hockey on a surface so smooth that the skaters went sixteen times as fast. That's cartilage. But unlike ice, it isn't brittle. It doesn't crack under pressure as ice would. And you grow it yourself. It's a living thing. None of this has been equaled in engineering or science. Most of the best technology that exists on Earth is right here inside us. And everybody takes it almost completely for granted."

Before we move on, Ben examines the wrist more closely for a moment. "You shouldn't ever try to kill yourself by cutting your wrists, by the way," he says. "All of those things going in are wrapped in a protective band called a fascial sheath, which makes it really hard to get to the arteries. Most people who cut their wrists fail to kill themselves, which is no doubt a good thing." He is briefly thoughtful. "It's also really hard to kill yourself by jumping from a height," he adds. "The legs become a kind of crumple zone. You can make a real mess of yourself, but you are very likely to survive. Killing yourself is actually difficult. We are designed not to die."

This seems a slightly ironic thing to say in a big room full of dead bodies, but I take his point.

* * *

Most of the time the dissecting room at Nottingham is filled with medical students, but it is the summer holidays when Ben Ollivere shows me around. Two other people join us from time to time, Siobhan Loughna, a lecturer in anatomy at the university, and Margaret "Margy" Pratten, head of the anatomy teaching section and associate professor of anatomy.

The dissecting room is a large, well-lit room, clinically clean and slightly chilly, with a dozen anatomical workstations ranged around it. A liniment-like smell of embalming fluid hangs in the air. "We have just changed formulations," Siobhan explains. "It preserves better, but smells a bit more. Embalming fluid is mostly formaldehyde and alcohol."

Most bodies are cut into pieces—or transected, to use the formal term—so that students can focus on a particular area: a leg or shoulder or neck, say. The unit gets through about fifty bodies a year, Margy tells me. I ask her if it is hard to find volunteers.

"No, quite the opposite," she replies. "More bodies are donated than we can accept. Some we have to reject—if the person had Creutzfeldt-Jakob disease, for instance, because there would still be a danger of infection, or if they were morbidly obese." (Very large bodies can be physically challenging to deal with.)

At Nottingham, they have an informal policy that they keep only one-third of a transected body, Margy adds. The retained parts may be kept for years. "The rest is returned to the family so that they can have a funeral service." Whole bodies are generally kept for no more than three years before being sent off for cremation. Members of the staff and medical students often attend the ceremonies. Margy makes a point of always trying to go.

It seems a little strange to say when talking about bodies that have been carefully quartered, then turned over to students to further incise and probe, but at Nottingham they are fastidious about treating the bodies with respect. Not all institutions are quite so rigorous. Not long after my visit to Nottingham, there was a brief scandal in

America after an assistant professor and some graduate students from the University of Connecticut were photographed posing with two severed heads in a selfie taken in a dissecting room in New Haven. By law no photography is allowed in dissecting rooms in Britain. At Nottingham, you cannot bring a phone in.

"These were real people with hopes and dreams and families and all the rest that makes us human, who have given their bodies to help others, and that's incredibly noble, and we try very hard never to lose sight of that," Margy told me.

It took a surprisingly long time for medical science to take a very active interest in what fills the space inside us and how it all works. Up to the Renaissance, human dissection was widely forbidden, and even when it became tolerated, not many people had the stomach for it. A few intrepid souls—Leonardo da Vinci most famously—cut people up for the sake of knowledge, but even Leonardo observed in his notes that a decomposing body was pretty disgusting.

Specimens were nearly always hard to come by. When the great anatomist Andreas Vesalius, as a young man, wanted human remains to study, he stole the body of an executed murderer off a gibbet outside his hometown of Leuven (French Louvain), just east of Brussels in Flanders. William Harvey, in England, was so desperate for subjects that he dissected his own father and sister. No less bizarrely, the Italian anatomist Gabriele Falloppio (after whom the Fallopian tubes are named) was given a criminal who was still alive with instructions to put him down in the manner that best suited his purposes. Falloppio and the criminal together appear to have opted for a comparatively humane overdose of opiates.

In Britain, criminals hanged for murder were distributed to local medical schools for dissection, but there were never enough bodies to meet demand. Because of the shortages, a brisk trade arose in illicit bodies stolen from churchyards. Many people lived in severe dread of having their bodies dug up and violated. A well-known case was that of the celebrated Irish giant Charles Byrne (1761–83). At seven feet seven inches, Byrne was the tallest man in Europe. His skeleton was coveted by the anatomist and collector John Hunter. Terrified of being dissected, Byrne arranged that when he died his coffin would be taken out to sea and dropped in deep water, but Hunter managed to bribe the ship's captain with whom Byrne had made the arrangement, and instead Byrne's body was brought to Hunter's residence in Earl's Court, London, where it was dissected while practically still warm. For decades, Byrne's lanky bones have hung in a display case in the Hunterian Museum of the Royal College of Surgeons in London. However, in 2018 the museum was closed for a three-year-long refurbishment, and there has been talk of allowing Byrne to be buried at sea in fulfillment of his final request.

As medical schools proliferated, the problem of supply steadily worsened. In 1831, London had nine hundred medical students but just eleven executed bodies to share between them. The following year, Parliament passed the Anatomy Act, which made the punishments for grave robbing more severe but also allowed dissecting institutions to take anyone who died penniless in a workhouse, which made a lot of paupers very unhappy but increased the supply considerably.

The rise of scholarly dissection coincided with an improvement in the standard of medical and anatomical textbooks. The most influential anatomical work of the period—and indeed ever since—was *Anatomy: Descriptive and Surgical,* first published in 1858 in London and known ever since as *Gray's Anatomy,* after its author, Henry Gray.

Henry Gray was a rising young demonstrator of anatomy at St. George's Hospital at Hyde Park Corner in London (the building still stands but is now a luxury hotel) when he decided to produce a definitive and modern anatomical guide. Gray was still only in his twenties when he began work on the book in 1855. For the illustrations, he commissioned a medical student at St. George's named Henry Vandyke Carter for a payment of £150 spread over fifteen months. Carter was painfully shy but highly gifted. All of his illustrations had to be drawn in reverse so that they would print the right way around on paper, which must have been an almost unimaginable challenge. Carter did not only all 363 drawings but also nearly all of the preparatory dissection. Although many other anatomy books were available, *Gray's,* in the words of one biographer, "eclipsed all others, partly for its meticulous detail, partly for its emphasis on surgical anatomy, but most of all perhaps for the excellence of the illustrations."

As a collaborator, Gray was spectacularly petty. It is not clear whether he ever paid Carter in full or indeed at all. He certainly never shared royalties. He instructed the printers to reduce the size of Carter's name on the title page and to remove a reference to his medical qualifications, to make him look like a journeyman illustrator. Only Gray's name appeared on the spine, which is why it became known as *Gray's Anatomy* rather than *Gray and Carter's,* as it really should have.

The book was an immediate success, but Gray did not much get to enjoy it. He died in 1861 of smallpox just three years after publication. He was only thirty-four. Carter did somewhat better. In the year of the book's publication, he moved to India, where he became professor of anatomy and physiology (and later principal) at Grant Medical College. He spent thirty years in India before retiring to Scarborough on the Yorkshire coast. He died in 1897 of tuberculosis two weeks before his sixty-sixth birthday.

II

WE ASK A lot of our bodily architecture. The skeleton has to be rigid and yet pliant. We must stand firm but also bend and twist. "We are both floppy and rigid," as Ben Ollivere says. Your knees have to lock into position when you stand but then immediately unlock and bend up to 140 degrees to let us sit and kneel and move about, and we must do all this with a certain grace and fluidity day after day for decades. Think of how jerky and un-lifelike most robots you have ever seen have been—how ploddingly they walk, how tippy they are on stairs or uneven ground, how hopelessly flummoxed they would be in trying to keep up with any three-year-old human at a playground–and you can appreciate what an accomplished creation we are.

It is usually said that we have 206 bones, but the actual number can vary a bit between people. About one person in every eight has an extra, thirteenth pair of ribs, while people with Down's syndrome frequently have a pair missing. So 206 is, for many, an approximate number, and it doesn't include the (mostly) tiny sesamoid bones that are scattered through all of us in our tendons, primarily in the hands and feet. ("Sesamoid" means "like a sesame seed," which is largely an apt description but not always. The kneecap, or patella, is also a sesamoid bone, though hardly sesame-like.)

Your bones are by no means evenly distributed. You have fifty-two in your feet alone, double the number in your spine. The hands and feet together have more than half the bones in the body. Where you have lots of bones isn't necessarily because there is an urgent need for bones to be in one place rather than another, but because that's just where evolution left them.

Our bones do a lot more than keep us from collapsing. As well as providing support, they protect our interiors, manufacture blood cells, store chemicals, transmit sound (in the middle ear), and even possibly bolster our memory and buoy our spirits thanks to the recently discovered hormone osteocalcin. Until the early years of the twenty-first century, no one knew that bones produced hormones at all, but then a geneticist at Columbia University Medical Center, Gerard Karsenty, realized that osteocalcin, which is produced in bones, not only is a hormone but seems to be involved in a large number of important regulatory activities across the body, from helping to manage glucose levels to boosting male fertility to influencing our moods and keeping our memory in working order. Apart from anything else, it could help to explain the long-standing mystery of how regular exercise helps to stave off Alzheimer's disease: exercise builds stronger bones and stronger bones produce more osteocalcin.

Typically about 70 percent of a bone is inorganic material and 30 percent organic. The most fundamental element of bone is collagen. It is the most abundant protein in the body—40 percent of all your proteins are collagens—and it is very adaptable. Collagen makes the white of the eye but also the transparent cornea. In muscle it forms fibers that behave just like rope in that they are strong when stretched but collapse when pushed together. That's great for muscle but wouldn't be so useful in your teeth. So when permanent stiffness is needed, collagen often twins with a mineral called hydroxyapatite, which is strong when compressed and thus allows the body to create good solid structures like bones and teeth.

We tend to think of our bones as inert bits of scaffolding, but they are living tissue, too. They grow bigger with exercise and use just as muscles do. "The bone in a professional tennis player's serving arm may be 30 percent thicker than in his other arm," Margy Pratten told me, and cited Rafael Nadal as an example. Look at bone through a microscope and you will see an intricate array of productive cells just as in any other living thing. Because of the way they are constructed, bones are, to an extraordinary degree, both strong and light.

"Bone is stronger than reinforced concrete," says Ben, "yet light enough to allow us to sprint." All your bones together will weigh no more than about twenty pounds, yet most can withstand up to a ton of compression. "Bone is also the only tissue in the body that doesn't scar," Ben adds. "If you break your leg, after it heals you cannot tell where the break was. There's no practical benefit to that. Bone just seems to want to be perfect."

Even more remarkably, bone will grow back and fill a void. "You can take up to thirty centimeters of bone out of a leg, and with an external frame and a kind of stretcher you can have it grow back," Ben says. "Nothing else in the body will do that." Bone, in short, is amazingly dynamic.

The skeleton is, of course, only a part of the vital infrastructure that keeps you upright and mobile. You also need lots of muscle and a judicious assortment of tendons, ligaments, and cartilage. I think it is safe to say that most of us are not completely clear on what exactly some of these do for us or quite what marks the difference between them. So here is a brief rundown.

Tendons and ligaments are connective tissues. Tendons connect muscles to bone; ligaments connect bone to bone. Tendons are stretchy; ligaments, less so. Tendons are essentially extensions of muscles. When people speak of sinew, they are referring to tendons. If you want to see a tendon, it is easy to do so. Turn your hand palm up. Make a fist and a ridge will form on the underside of your wrist. That's a tendon.

Tendons are strong, and generally it takes a lot of force to tear them, but they also have very little blood supply and therefore take a long time to heal. That at least is better than cartilage, which has no blood supply at all and therefore almost no capacity to heal.

But the bulk of you, no matter how modestly built you are, is muscle. You have more than six hundred muscles altogether. We tend to notice our muscles only when they ache, but of course they are constantly at our service in a thousand unappreciated ways—puckering our lips, blinking our eyelids, moving food through the digestive tract. It takes one hundred muscles just to get us to stand up. You need a dozen to move your eyes over the words you are reading now. The simplest movement of the hand—a twitch of the thumb, say—can involve ten muscles. Many of our muscles we don't even think of as muscles—our tongue and heart, for instance. Anatomists categorize them by what they do. Flexor muscles close joints, and extensor muscles open them; levators lift, and depressors lower; abductors move body parts away, and adductors draw them back; sphincters contract.

Altogether you are about 40 percent muscle if you are a reasonably slender man, slightly less if you are a proportionately similar woman, and just keeping that mass of muscle uses up 40 percent of your energy allowance when you are at rest, and much more when you are active. Because muscle is so expensive to maintain, we sacrifice muscle tone really quickly when we are not using it. Studies by NASA have shown that astronauts even on short missions–from five to eleven days–lose up to 20 percent of muscle mass. (They lose bone density, too.)

All of these things–muscles, bones, tendons, and so on–work together in a deft and splendid choreography. Nowhere is this better demonstrated than in your hands. In each hand you have 29 bones, 17 muscles (plus 18 more that are in the forearm but control the hand), 2 main arteries, 3 major nerves (one of which, the ulnar nerve, is the one you feel in your elbow when you hit your "funny bone") plus 45 other named nerves, and 123 named ligaments, all of which must coordinate their every action with precision and delicacy. Sir Charles Bell, the great nineteenth-century Scottish surgeon and anatomist, thought the hand the most perfect creation in the body–better even than the eye. He called his classic text *The Hand: Its Mechanism and Vital Endowments as Evincing Design,* by which he meant that the hand was proof of divine creation.

The hand is a marvelous creation without question, but not all its parts are equal. If you curl your fingers into a fist, then try to straighten them out one at a time, you will find that the first two pop up obediently enough but the ring finger doesn't seem to want to straighten out at all. Its position on the hand means that it can't really contribute much to fine movement and so it has less in the way of discriminating musculature. Nor, surprisingly, do we all possess the same component parts in our hands. About 14 percent of us lack a muscle called the palmaris longus, which helps to keep the palm tensed. It is rarely missing from top-ranked athletes who need a strong grip to perform, but is otherwise quite dispensable. In fact, the tendon ends of the muscle are sufficiently unneeded that they are frequently used by surgeons when making tendon grafts.

It is often noted that we have opposable thumbs (by which is meant that they can touch the other fingers, giving the capacity for a good grip) as if this were a uniquely human attribute. In fact, most primates have opposable thumbs. Ours are just more pliant and mobile. What we do have in our thumbs are three small but resplendently named muscles not found in any other animals, including chimps: the extensor pollicis brevis, the flexor pollicis longus, and the first volar interosseous of Henle.* Working together, they allow us to grasp and manipulate tools with sureness and delicacy. You might never have heard of them, but these three small muscles are at the heart of human civilization. Take them away and our greatest collective achievement might be maneuvering ants out of their nests with sticks.

"The thumb isn't just a stubbier shape from the other digits," Ben Ollivere told me. "It's actually attached differently. Almost no one ever notices it, but our thumbs are on sideways. The thumbnail faces away from the rest of the fingers. On a computer keyboard you strike the keys with the tips of your fingers but with the side of your thumb. That's what is meant by an opposable thumb. It means we are really good at grasping. The thumb also rotates well—it swings through quite a wide arc—compared with the fingers."

Considering their importance, we have been surprisingly relaxed about naming the digits. Ask most people how many fingers we have and they will say ten. Then ask them which is their first finger and nearly all will unfurl an index finger, thus overlooking the neighboring thumb and relegating it to a separate status. Ask them then to name the next finger along and they will call it the middle finger—but it can only be in the middle if there are five fingers, not four. In the end, even most dictionaries can't decide whether we have eight fingers or ten. Most define fingers as "one of the five terminal members of the hand, or one of the four other than the thumb." Because of the uncertainty, even doctors do not number fingers, because there is no agreement on which is finger number one. Doctors use the usual Latinate technical

* The human body is awash with Henles. We have crypts of Henle in the eye, Henle's ampulla in the uterus, Henle's ligament in the abdomen, Henle's tubules in the kidneys, and several more. All were discovered by a very busy, curiously uncelebrated German anatomist named Jakob Henle (1809–85).

terms for most parts of the hand except, oddly, the fingers, which they call thumb, index, long, ring, and little.

A good deal of what we know about the comparative strengths of the hand and wrist comes from a series of improbable experiments undertaken by a French physician, Pierre Barbet, in the 1930s. Barbet was a surgeon at the Paris Saint-Joseph Hospital who became obsessed with the physical challenges and limitations of human crucifixion. To test how well humans would remain in place on a cross, he nailed real human corpses to wooden crosses using different types of nails driven through different parts of the hands and wrists. He discovered that nails driven through the palm of the hand–the method traditionally depicted in paintings–would not support the weight of a body. The hands would literally tear apart. But if the nails were driven through the wrists, the body would stay in position indefinitely, thus proving that the wrists are much more robust than the hands. And by such means does human knowledge creep forward.

Our other disproportionately bony outposts, the feet, receive a lot less praise and attention when it comes to discussing the things that make us special, but in fact the feet are pretty marvelous, too. The foot has to be three different things: shock absorber, platform, and pushing organ. With every step you take–and in the course of a lifetime you will take probably something in the region of 200 million of them–you execute those three functions in that order. The foot's curved shape, like that of the Roman arch, is immensely strong, but it's also pliant, lending a springy rebound to every step. The combination of arch and springiness gives the foot a recoil mechanism that helps to make our walking rhythmic and bouncy and efficient in comparison with the more lumbering movements of other apes. The average human walks at a pace of about 4.25 feet per second, or 120 steps per minute, though obviously this varies a great deal depending on age, height, urgency, and much else.

Our feet were designed to grasp, which is why you have a lot of bones in them. They were not designed to support a lot of weight, which is one reason they ache at the end of a long day of standing or walking. As Jeremy Taylor points out in *Body by Darwin,* ostriches have eliminated this problem by fusing the bones of their feet and ankle, but then ostriches have had 250 million years to adjust to upright walking, roughly forty times as long as we have had.

All bodies are compromises between strength and mobility. The bulkier an animal is, the more massive its bones must be. So an elephant is 13 percent bone, whereas a tiny shrew needs to devote just 4 percent of itself to skeleton. Humans fall in between at 8.5 percent. If we had stronger scaffolding, we couldn't be as nimble. The price we pay for being able to scamper and sprint is, for many, backache and knee pain in later life—or indeed not so late in life. Such is the pressure on the spine from our upright posture that pathological changes can be detected "as early as the eighteenth year," as Peter Medawar noted.

The problem, of course, is that we come from a long line of beings whose skeletons were designed to take our weight on four legs. We will look at the benefits and consequences of this massive change to our anatomy more closely in the next chapter, but for the moment it's enough to bear in mind that becoming upright meant a wholesale redistribution of our weight load, and with that has come a lot of pain that we would not otherwise have suffered. Nowhere is this more uncomfortably evident in modern humans than in the back. Becoming upright put extra pressure on the cartilage disks that support and cushion the spine, in consequence of which they sometimes become displaced or herniated in what is popularly known as a slipped disk. Between 1 and 3 percent of adults have slipped disks. Back pain is the most common of chronic complaints as we age. An estimated 60 percent of adults have taken at least a week off work at some time with back pain.

Our lower limb joints are also highly vulnerable. Every year in the United States, surgeons perform over 800,000 joint replacements, principally of hips and knees, mostly from wear and tear on the cartilage lining the joints. It is pretty impressive that cartilage lasts as well as it does, especially when you consider that it cannot repair or replenish itself. Think of how many pairs of shoes you have worn out in your life, and you begin to appreciate just how durable your cartilage is.

Because cartilage isn't nourished by blood, the best thing you can do to maintain it is to move around a lot, to keep the cartilage bathed in its own synovial fluid. The worst thing you can do is to pack on a lot of extra body weight. Try walking around all day with a couple of bowling balls tied to your belt and see if you don't feel it in your hips and knees at the end of the day. Well, that's essentially what you are doing already all day every day if you are twenty-five or thirty pounds overweight. It's little wonder that so many of us end up undergoing corrective surgery as the years catch up with us.

For many people, the most problematic part of their infrastructure is their hips. Hips wear out because they have to do two incompatible things: they must provide mobility for the lower limbs, and they must support the weight of the body. This exerts a lot of frictional pressure on the cartilage on both the head of the femur and the hip socket into which it fits. So instead of swiveling smoothly, the two can start to grind painfully, like a pestle in a mortar. Well into the 1950s, there wasn't much medical science could do to relieve the problem. Complications from hip surgery were so great that the usual procedure was to "fuse" the hip, an operation that relieved the pain but left the person with a permanently stiffened leg.

Surgical relief was always short-lived because every synthetic material tried would soon wear down until the bones were grinding painfully again. In some cases, the plastics used in hip replacements squeaked so loudly when people walked that they were embarrassed to go out. Then a dogged orthopedic surgeon in Manchester, England, named John Charnley devoted himself heroically to finding materials and devising methods that would solve all the problems. Essentially, he realized that wear was greatly reduced if the femur was replaced with a stainless steel head and the socket—acetabulum, to use the formal name—was lined with plastic. Almost no one has heard of

Charnley outside orthopedic circles (where he is venerated), but few people have brought relief to greater numbers of sufferers than he did.

Our bones lose mass at a rate of about 1 percent a year from late middle age onward, which is of course why elderly people and broken bones are so unhappily synonymous. Broken hips are especially challenging for the elderly. About 40 percent of people over seventy-five who break their hips are no longer able to care for themselves. For many, it is a kind of last straw. Ten percent die within thirty days, and nearly 30 percent die within twelve months. As the British surgeon and anatomist Sir Astley Cooper liked to quip, "We enter the world through the pelvis and leave it through the hip."

Happily, Cooper was exaggerating. Three-quarters of men and half of women don't break any bones at all in old age, and three-quarters of all people go through the whole of life without any serious problems with their knees, so it is not all bad news. Anyway, as we are about to see, when you consider how many millions of years of risk and hardship our forebears went through to get us comfortably upright, we really don't have much to complain about at all.

10 ON THE MOVE: BIPEDALISM AND EXERCISE

> Not less than two hours a day should be devoted to
> exercise and the weather should be little regarded.
> If the body be feeble, the mind will not be strong.
>
> —THOMAS JEFFERSON

NO ONE KNOWS why we walk. Out of some 250 species of primates, we are the only ones that have elected to get up and move around exclusively on two legs. Some authorities think bipedalism is at least as important a defining characteristic of what it is to be human as our high-functioning brain.

Many theories have been proposed as to why our distant ancestors dropped out of trees and adopted an upright posture—to free their hands to carry babies and other objects; to gain a better line of sight across open ground; to be better able to throw projectiles—but the one certainty is that walking on two legs came at a price. Moving about in the open made our ancient forebears exceedingly vulnerable, for they were not formidable creatures, to say the least. The young and gracile protohuman famously known as Lucy, who lived in what is now Ethiopia some 3.2 million years ago and is often used as a model for early bipedalism, was only about three and a half feet tall and weighed just sixty pounds—hardly the sort of presence to intimidate a lion or cheetah.

It's likely Lucy and her tribal kin had little choice but to take the risk of stepping out into the open. As climate change made their forest habitats shrink, they very probably needed to forage over larger and larger areas to survive, but they almost certainly scampered back to trees when they could. Even Lucy appears to have been only a partial convert to life at ground level. In 2016, anthropologists at the University of Texas concluded that Lucy died after falling out of a tree (or suffered a "vertical deceleration event," as they put it, just a touch drily), the implication being that she spent a great deal of time in the canopy of trees and was probably as much at home up there as on the ground. Or at least she was until the last three or four seconds of her life.

Walking is a more skillful undertaking than we generally appreciate. By balancing on just two supports, we exist in permanent defiance of gravity. As toddlers amusingly demonstrate, walking is essentially a matter of hurling the body forward and letting the legs run to catch up. A pedestrian in motion has one foot or the other off the ground for as much as 90 percent of the time, and thus engages in constant unconscious adjustments of balance. In addition, our center of gravity is high—just above our waists—which adds to our innate tippiness.

In order to proceed from arboreal ape to upright modern human, we had to undertake some pretty profound changes to our anatomy. As noted earlier, our necks became longer and straighter and joined the skull more or less centrally rather than toward the rear as in other apes. We have a supple back that bends, outsized knees, and ingeniously angled thigh bones. You may think your legs drop straight down from your waist—they do in apes—but in fact the femur angles inward as it descends from pelvis to knee. This has the effect of moving our lower legs closer together, giving us a much smoother, more graceful gait. No ape can be trained to walk like a human. They are compelled by their bone structure to waddle, and to do so in a most inefficient way. A chimpanzee uses four times as much energy to move around at ground level as does a human.

To power our forward motion, we have a distinctively gigantic muscle in our buttocks, the gluteus maximus, and an Achilles tendon, something no ape has. We have arches in our feet (for springiness), a sinuous spine (to redistribute weight), and reconfigured pathways for our nerves and blood vessels—all made necessary, or at least advisable, by the evolutionary imperative of putting our head way above our feet. To keep from overheating when we exert ourselves, we became relatively hairless and developed abundant sweat glands.

Above all, we evolved a very different head from other primates. Our faces are flat and conspicuously snoutless. We have a high forehead to accommodate our more impressive brain. Cooking has left us with smaller teeth and a more delicate jaw. Inside, we have a short oral cavity and thus a shorter, more rounded tongue, and a larynx that sits lower in the throat. The changes to our upper anatomy left us by happy accident with vocal tracts uniquely able to make articulate speech. Walking and talking probably went hand in hand. If you are a little creature that hunts big creatures, being able to communicate is obviously an advantage.

At the back of your head is a modest ligament, not found on other apes, that instantly betrays what it is about us that allowed us to thrive as a species. It is the nuchal ligament, and it has just one job: to hold the head steady when running. And running—serious, dogged, long-distance running—is the one thing we do superlatively well.

We are not the speediest of creatures, as anyone who has ever chased a dog or cat or even an escaped hamster will know. The very fastest humans can run about twenty miles an hour, though only for short bursts. But put us up against an antelope or wildebeest on a hot day and allow us to trot after it, and we can run it into the ground. We perspire to keep cool, but quadrupedal mammals lose heat by respiration—by panting. If they can't stop to collect themselves, they overheat and become helpless. Most large animals can't run for more than about nine miles before they drop. That our ancestors could also organize themselves into hunting parties, to harry quarry from different sides or drive prey into confined spaces, made us all the more effective.

These anatomical changes were so monumental that they spawned an entirely new genus (the biological rank above species but below family) called *Homo*. Daniel Lieberman, of Harvard, stresses that the transformation was a two-stage process. First, we became walkers and climbers, but not runners. Then, gradually, we became walkers and runners, but no longer climbers. Running is not just a faster form of locomotion than walking but mechanically quite different. "Walking is a stilt-like gait and involves very different adaptations from running," he says. Lucy was a walker and climber but lacked the physique for running. That came much later, after climate change turned much of Africa into open woodlands and grassy savanna, impelling our vegetarian ancestors to adjust their diets and become carnivores (or really omnivores).

All these changes, in lifestyle and anatomy, happened with exceeding slowness. Fossil evidence suggests that early hominins were walking by about 6 million years ago, but needed an additional 4 million years to acquire the capabilities for endurance running and, with it, persistence hunting. Then a further million and a half years had to pass before they gathered enough cerebral momentum to manufacture tipped spears. That's a long time to wait for a full set of survival capabilities in a hostile, hungry world. Despite these deficiencies, our ancient forebears were successfully hunting large animals 1.9 million years ago.

They were able to do this because of an additional trick in the *Homo* armamentarium: throwing. Throwing required us to change our bodies in three crucial ways. We needed a high and mobile waist (to create a lot of torsion), loose and maneuverable shoulders, and an upper arm capable of flinging in a whiplike fashion. The shoulder joint in humans is not a snug ball and socket, as in our hips, but a more loose and open arrangement. This allows the shoulder to be limber and to rotate freely—exactly what's needed for forceful throwing—but it also means that we dislocate our shoulders easily.

We throw with our whole bodies. Try throwing an object forcefully while standing still and you can hardly do it. A good throw involves a forward step, a brisk rotation of waist and torso, a long backward stretch of the arm at the shoulder, and a powerful hurl. When executed well, a human can throw an object with considerable accuracy at speeds easily in excess of ninety miles an hour, as professional baseball pitchers repeatedly demonstrate. The ability to wound and torment exhausted prey with rocks from a relatively safe distance must have been a highly useful skill among early hunters.

Bipedalism had consequences, too—consequences that we all live with today, as anyone with chronic back pain or knee problems can attest. Above all, the adoption of a narrower pelvis to accommodate our new gait brought a huge amount of pain and danger to women in childbirth. Until recent times, no other animal on Earth was more likely to die in childbirth than a human, and perhaps none even now suffers as much.

For the longest time, the crucial importance to health of just moving around was hardly appreciated. But in the late 1940s a doctor at Britain's Medical Research Council, Jeremy Morris, became convinced that the increasing occurrence of heart attacks and coronary disease was related to levels of activity, and not just to age or chronic stress, as was almost universally thought at the time. Because Britain was still recovering from the war, research funding was tight, so Morris had to think of a low-cost way to conduct an effective large-scale study. While traveling to work one day, it occurred to him that every double-decker bus in London was a perfect laboratory for his purposes because each had a driver who spent his entire working life sitting and a conductor who was on his feet constantly. In addition to moving about laterally, conductors climbed an average of six hundred steps per shift. Morris could hardly have invented two more ideal groups to compare. He followed thirty-five thousand drivers and conductors for two years and found that after he adjusted for all other variables, the drivers—no matter how healthy—were twice as likely to have a heart attack as the conductors. It was the first time that anyone had demonstrated a direct and measurable link between exercise and health.

Study after study since then has shown that exercise produces extraordinary benefits. Going for regular walks reduces the risk of heart attack or stroke by 31 percent. An analysis of 655,000 people in 2012 found that being active for just eleven minutes a day after the age of forty yielded 1.8 years of added life expectancy. Being active for an hour or more a day improved life expectancy by 4.2 years.

As well as strengthening bones, exercise boosts your immune system, nurtures hormones, lessens the risk of getting diabetes and a number of cancers (including breast and colorectal), improves mood, and even staves off senility. As has been noted many times, there is probably not a single organ or system in the body that does not benefit from exercise. If someone invented a pill that could do for us all that a moderate amount of exercise achieves, it would instantly become the most successful drug in history.

And how much exercise should we get? That's not easy to say. The more or less universal belief that we should all walk ten thousand steps a day—that's about five miles—is not a bad idea, but it has no special basis in science. Clearly, any ambulation is likely to be beneficial, but the notion that there is a universal magic number of steps that will give us health and longevity is a myth. The ten-thousand-step idea is often attributed to a single study done in Japan in the 1960s, though it appears that also may be a myth. In the same way, the Centers for Disease Control's recommendations on exercise—namely, 150 minutes per week of moderate activity—are based not on the optimal amount needed for health, because no one can say what that is, but on what the CDC's advisers think people will perceive as realistic goals.

What can be said about exercise is that most of us are not getting nearly enough. Only about 20 percent of people manage even a moderate level of regular activity. Many get almost none at all. Today the average American walks only about a third of a mile a day—and that's walking of all types, including around the house and workplace. Even in an indolent society, it would seem almost impossible to do less. According to *The Economist,* some American companies have begun offering rewards to employees who log a million steps a year on an activity tracker such as a Fitbit. That seems a pretty ambitious number but actually works out to just 2,740 steps a day, or a little over a mile. Even that, however, seems to be beyond many. "Some workers have reportedly strapped their Fitbits to their dogs to boost their activity scores," *The Economist* noted. Modern hunter-gatherers, by contrast, average about nineteen miles of walking and trotting to secure a day's food, and it is reasonable to assume that our ancient forebears would have done about the same.

In short, they worked hard for what they ate and consequently ended up with bodies designed to do two somewhat contradictory things: to be active much of the time, but never to be more active than absolutely necessary. As Daniel Lieberman explains, "If you want to understand the human body, you have to understand that we evolved to be hunter-gatherers. That means being prepared to expend a lot of energy to acquire food, but not wasting energy when you don't need to." So exercise is important, but rest is vital, too. "For one thing," Lieberman says, "you can't digest food while you are exercising because the body shunts blood away from the digestive system in order to meet the increased demand to supply oxygen to the muscles. So you have to rest sometimes just for metabolic purposes and to recover from the exertions of exercise."

Because our ancient ancestors had to survive lean times as well as good, they evolved a tendency to store fat as a fuel reserve—a survival reflex that is now, all too often, killing us. The upshot is that millions of us spend our lives struggling to maintain a balance between paleolithically designed bodies and modern gustatory excess. It's a battle too many of us are losing.

Nowhere in the developed world is that more true than in the United States. According to the World Health Organization, more than 80 percent of American men and 77 percent of American women are overweight, and 35 percent of them are obese—up from just 23 percent as recently as 1988. In roughly the same period, obesity more than doubled in U.S. children and quadrupled in adolescents. If everybody

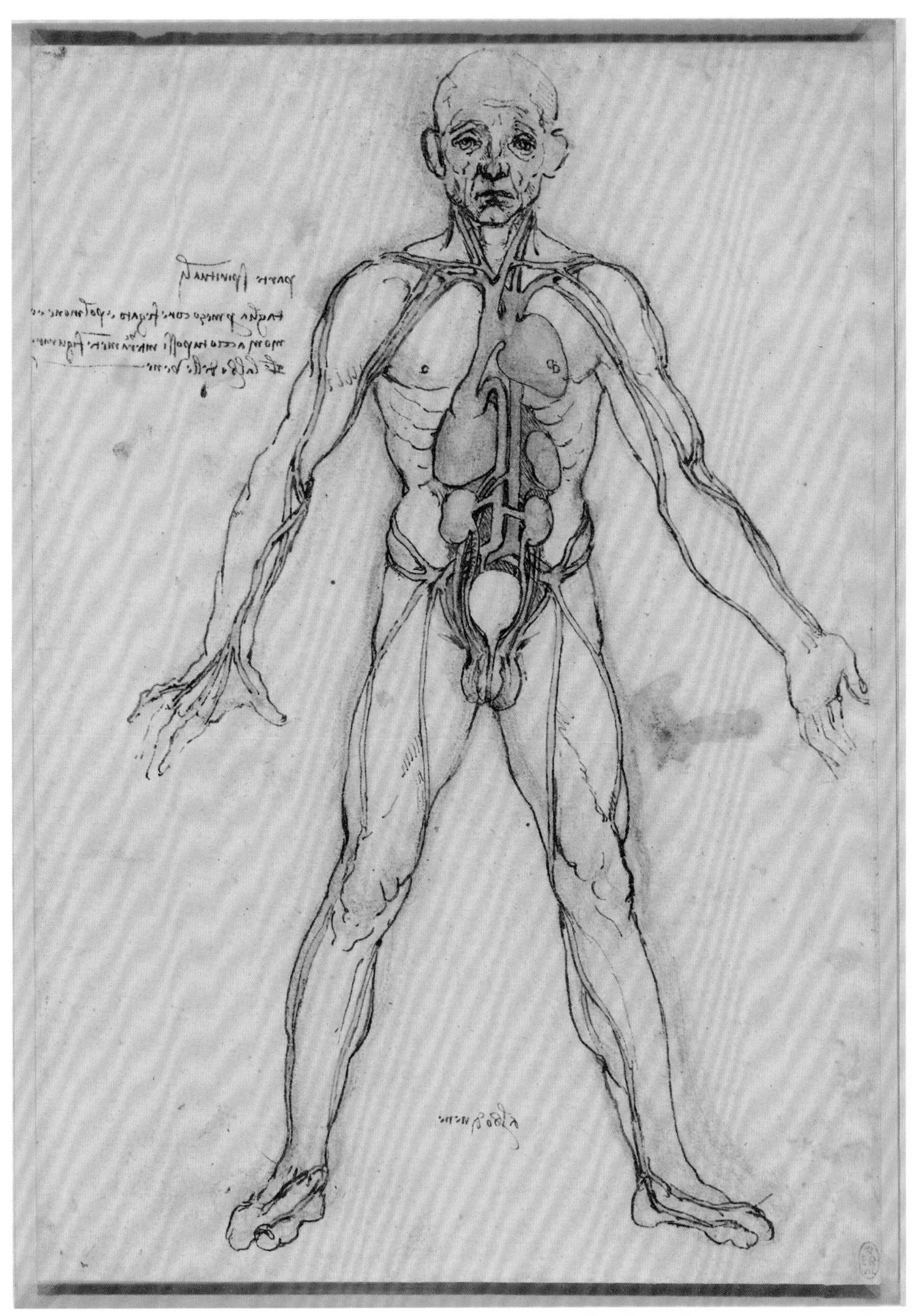

A Leonardo da Vinci drawing of the human body showing blood circulation, c. 1490. It took surprisingly long for medical science to take an active interest in what was inside us and how it worked. Leonardo was one of the first to dissect the human body, but even he noted that he found it disgusting.

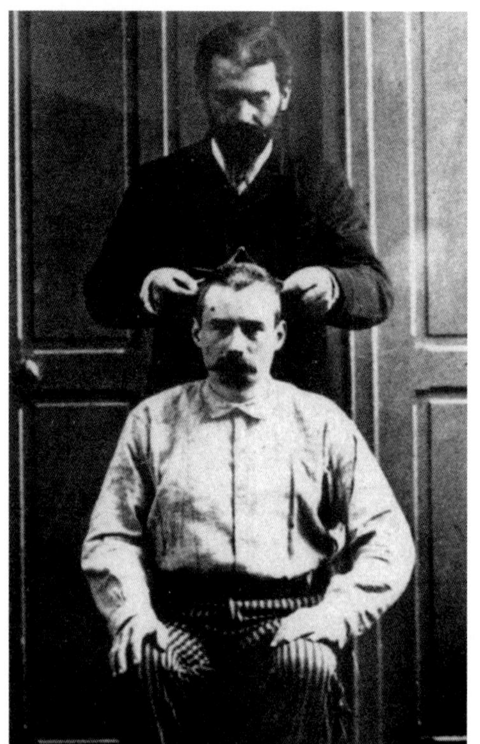

Photograph of Alphonse Bertillon, 1893. The French policeman Bertillon invented the system of identification that became known as Bertillonage, which involved measuring the body parts and individual markings of every arrested person.

Alexander Fleming, photographed in 1945, the year he, Ernst Chain, and Howard Florey shared a Nobel Prize for Physiology or Medicine. By then the Scottish biologist and physician had become famous as the father of penicillin.

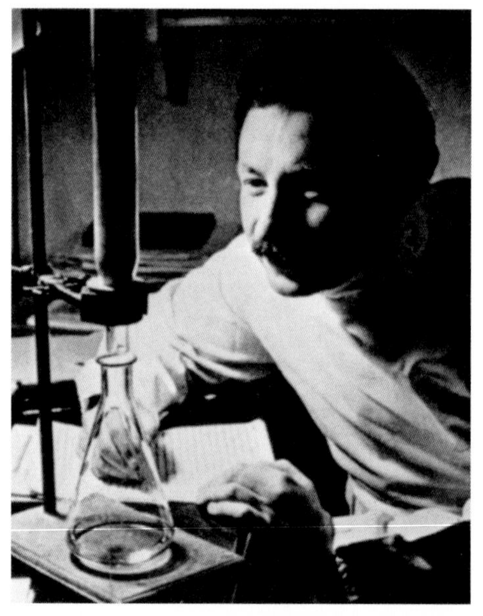

Ernst Chain, German-born biochemist based at Oxford, pictured here in 1944. Pathologically terrified of being poisoned in his lab, he nevertheless went on to discover that penicillin not only killed pathogens in mice but had no evident side effects.

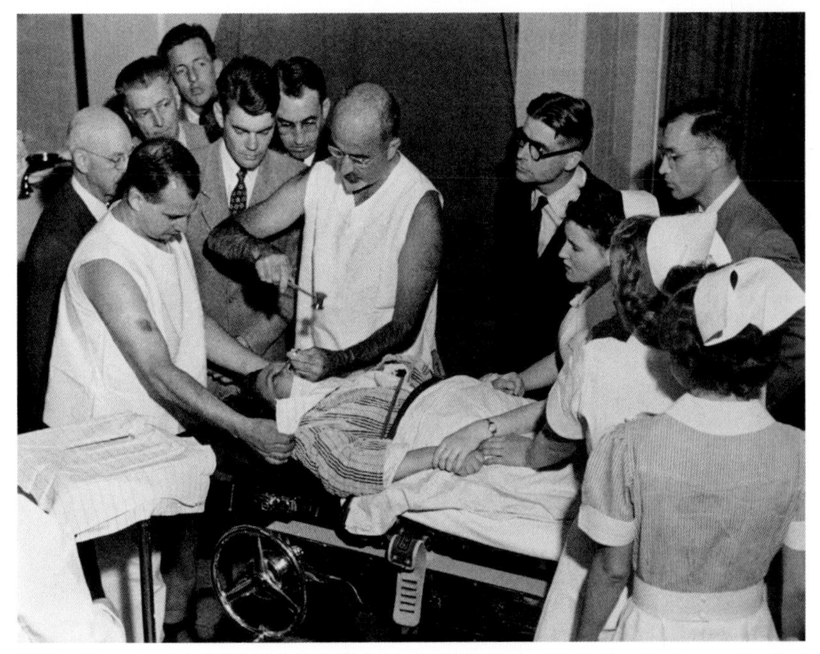

Walter Freeman, at work on one of the several thousand lobotomies he performed on patients across America during the mid-twentieth century. He used an ice pick to access his patients' brains through their eye sockets. Note the lack of mask, gown, and gloves.

Drawing by Cesare Lombroso dated 1888. The important and influential nineteenth-century Italian physiologist and criminologist developed a theory that criminality was inherited and that criminal instincts could be identified in features such as the slope of a forehead or the shape of an earlobe.

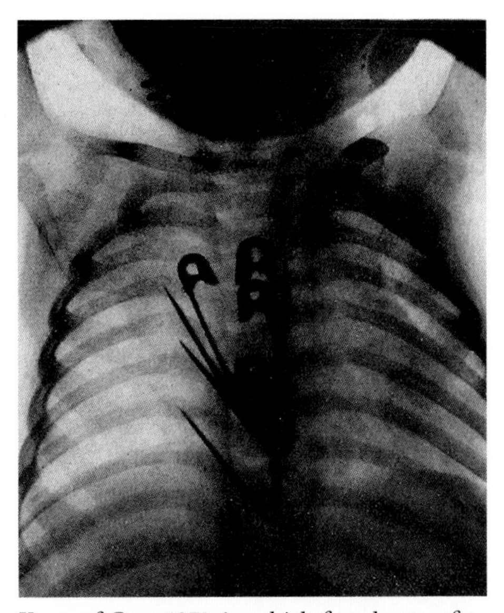

X-ray of Case 1071, in which four large safety pins were impacted in the esophagus of a nine-month-old child. Chevalier Quixote Jackson described this as his most difficult operation in a long career of removing swallowed objects and a reminder never to leave open safety pins within reach of small children—although in this case the baby's sister had fed the pins to her.

Illustration dated 1727 of the Reverend Stephen Hales supervising the insertion of a tube into an unfortunate horse's carotid artery in order to measure its blood pressure.

Werner Forssmann, who as a young doctor, out of curiosity and without any idea of the possible consequences, fed a catheter into an artery in his arm to see if he could reach his heart. He is photographed here twenty-seven years later, in 1956, the year he won a Nobel Prize for his revolutionary research.

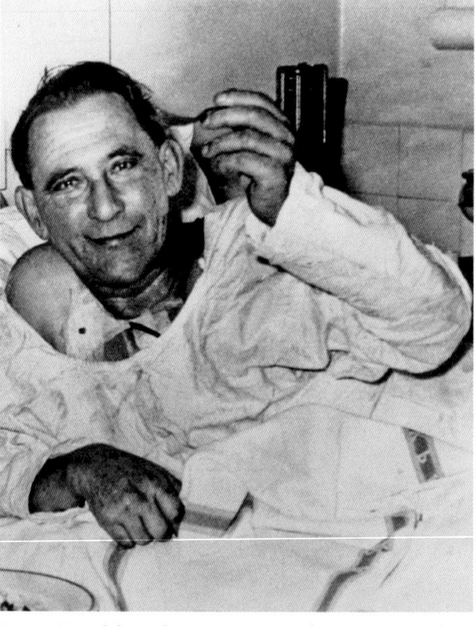

Louis Washkansky, recipient of the world's first heart transplant, in a hospital in Cape Town in 1967 soon after the procedure. While the operation was hailed as a breakthrough, he died eighteen days afterward.

William Harvey demonstrating to Charles I how blood circulates and the heart works. His theories were pretty much in line with our understanding now but were ridiculed at the time.

George Edward Bamberger and Charles Evan Watkins on their fifth birthday. The children were born in the same Chicago hospital in 1930, mislabeled, and sent home with the wrong parents; the error was not corrected until blood tests, at the time the height of technical sophistication, revealed who their true parents were.

Karl Landsteiner's research in Vienna at the start of the twentieth century marks the beginning of a modern understanding of blood; he established that it can be divided into different groups, which he labeled A, B, and O.

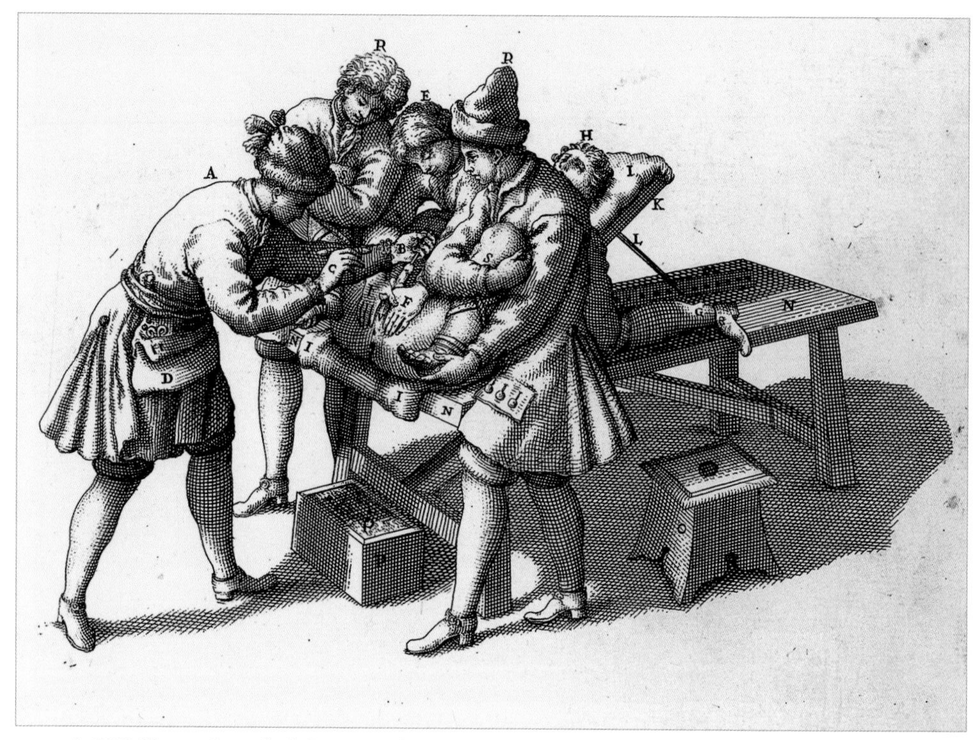

A 1707 illustration of a lithotomy, the procedure used for centuries to remove gallstones.

Charles-Édouard Brown-Séquard, who in the late 1880s, at the age of seventy-two, became famous for grinding up the testes of domestic animals in order to inject himself with the extract. He reported feeling "frisky as a forty-year-old," but it seriously damaged his scientific credibility with his peers.

Adolf Butenandt, the German biochemist and hormone expert, displaying the fencing scar of which he was so proud.

Canadian general practitioner Frederick Banting (right) and Toronto University laboratory assistant Charles Best, with whom Banting conducted his remarkably amateur but nevertheless successful trials on dogs in an attempt to cure diabetes. They are shown here in 1921 with one of the dogs from their lab.

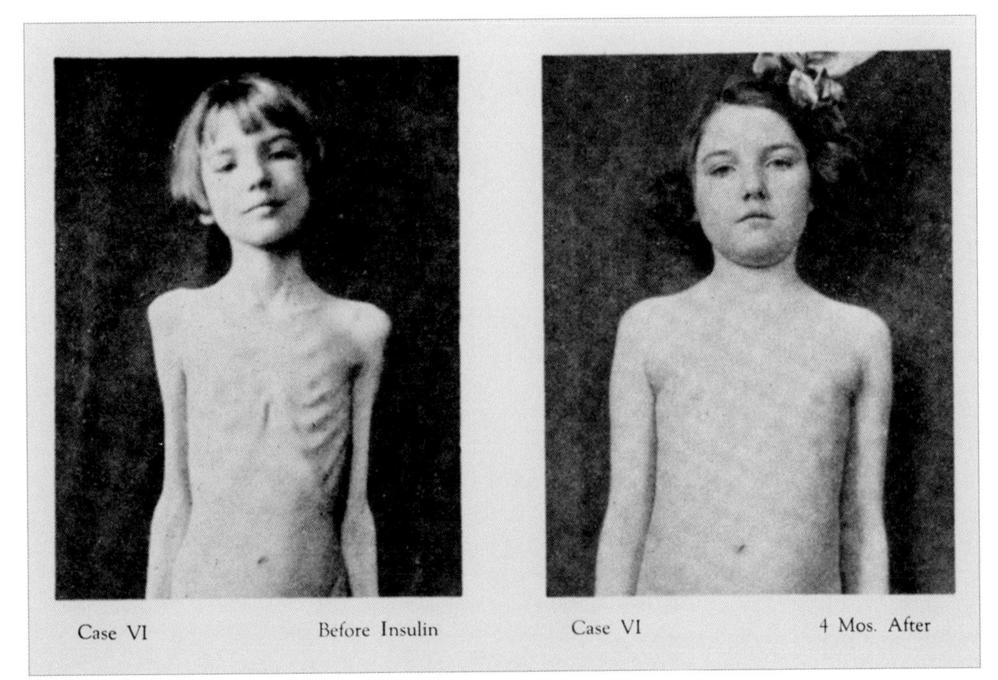

Photograph of Case VI: a young girl photographed before and after she was treated with insulin.

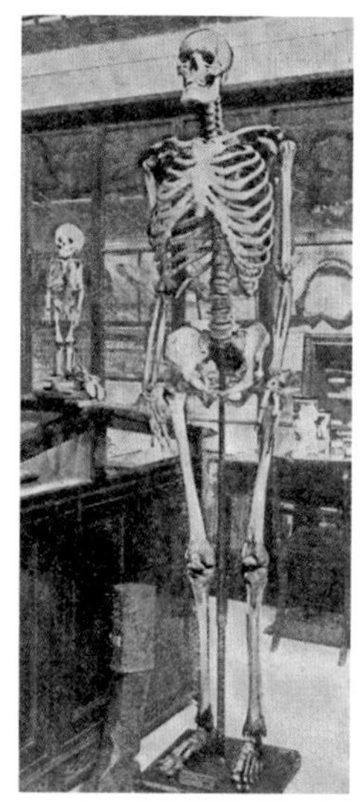

The skeletons of the celebrated "Irish giant" Charles Byrne, the tallest man in Europe when he died in 1783, and Caroline Crachami, known as the "Sicilian dwarf" (who died aged nine, measuring nineteen and a half inches, in 1824).

A page from *Gray's Anatomy,* first published in 1858. Henry Vandyke Carter's illustration shows the blood vessels of the neck.

The dissecting room of St. George's Hospital, photographed in 1860. Henry Gray, the author of *Gray's Anatomy,* sits next to the cadaver's feet, center left.

Walter Bradford Cannon, the "father of homeostasis"–our ability to maintain internal stability–in 1934: a genius whose stern expression belied a warm demeanor and a remarkable skill for persuading people to do uncomfortable things in the name of science.

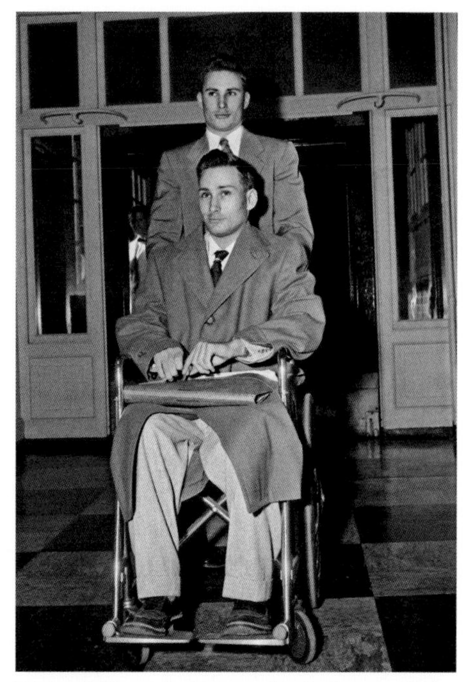

Richard Herrick being wheeled out of the hospital by his identical twin brother, Ronald, after the world's first kidney transplant in 1954.

British zoologist Peter Medawar in his laboratory at University College, London, having received the 1960 Nobel Prize for his pioneering studies of the immune system.

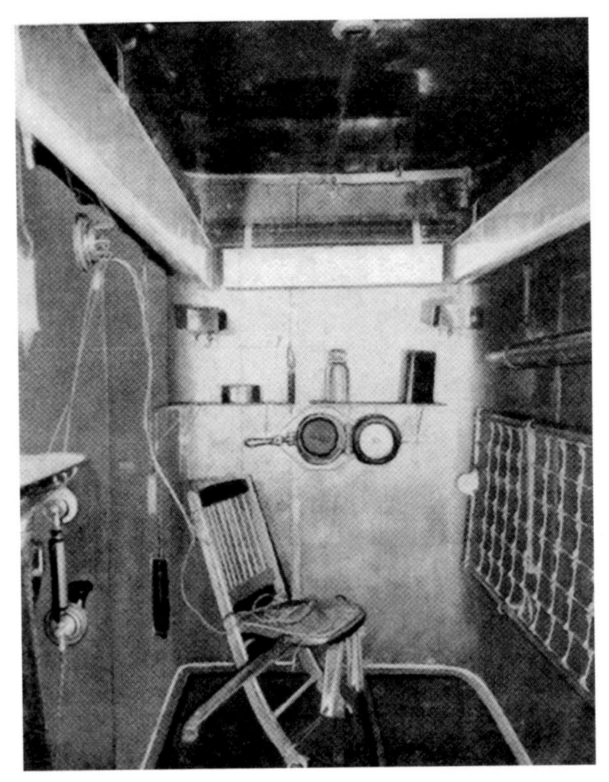

The interior of Wilbur Atwater's respiratory calorimeter, in which the subjects of his experiments would be confined for up five days at a time while Atwater and his assistants measured everything they ate, breathed, and excreted.

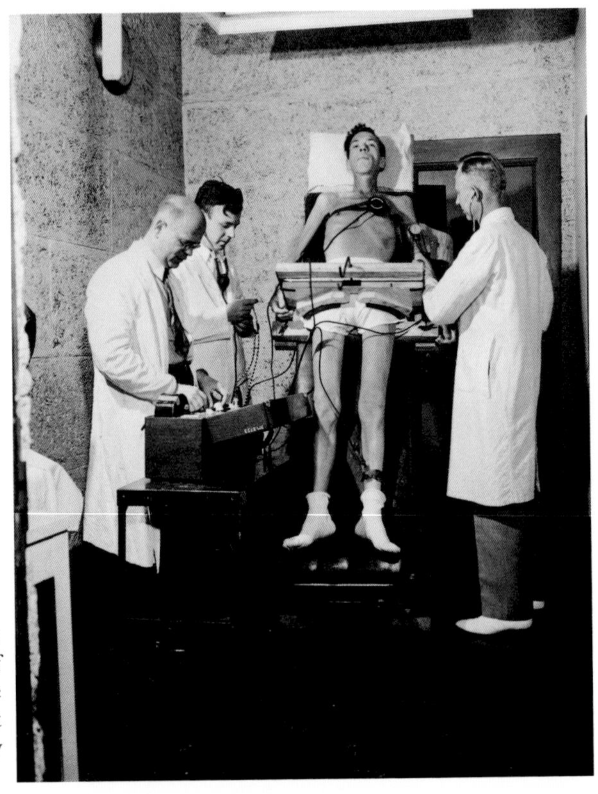

One of the thirty-six conscientious objectors who, toward the end of World War II, volunteered to be systematically starved for nutritionist Ancel Keys's study at the University of Minnesota.

William Beaumont, painted at the scene of one of the 238 experiments he conducted on Alexis St. Martin during the 1820s. Beaumont is shown holding part of the length of silk he has inserted through the open wound in St. Martin's stomach to examine the effects of his gastric juices.

French scientist Michel Siffre being hauled out of a cave deep inside a mountain in the Alps in 1962, after a self-imposed eight weeks spent isolated without daylight or any other clue to the passage of time.

Nettie Stevens, who, while studying the reproductive organs of mealworms in Pennsylvania in 1905, discovered the Y chromosome.

Ernst Gräfenberg, the German gynecologist who fled Nazi Germany for America, where he developed the intrauterine device first known as the Gräfenberg ring, and in 1944 identified an erogenous spot on the wall of the vagina: the Gräfenberg or G spot.

Early-nineteenth-century lithograph of a doctor examining his patient. For most of recorded history we have known shockingly little about how women are put together.

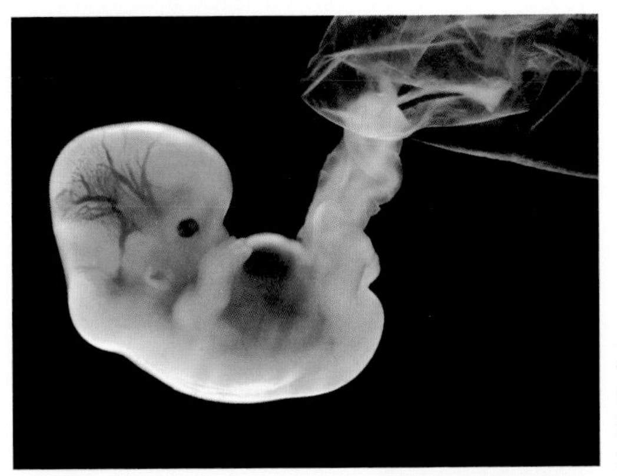

A six-week-old human embryo. It is about the size of a lentil and its heart is beating at one hundred beats per minute.

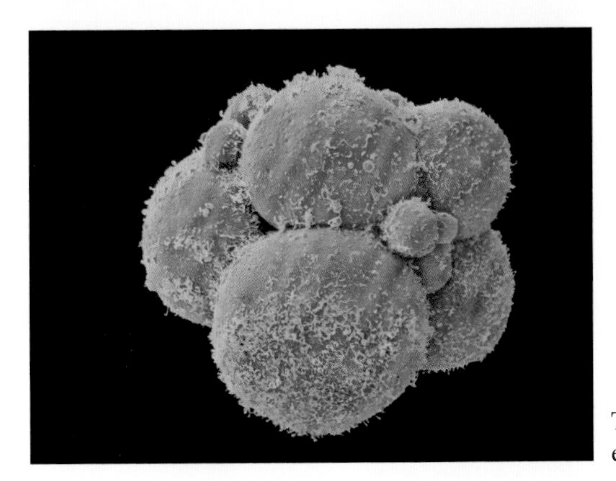

The human embryo at day three and eight-cell stage.

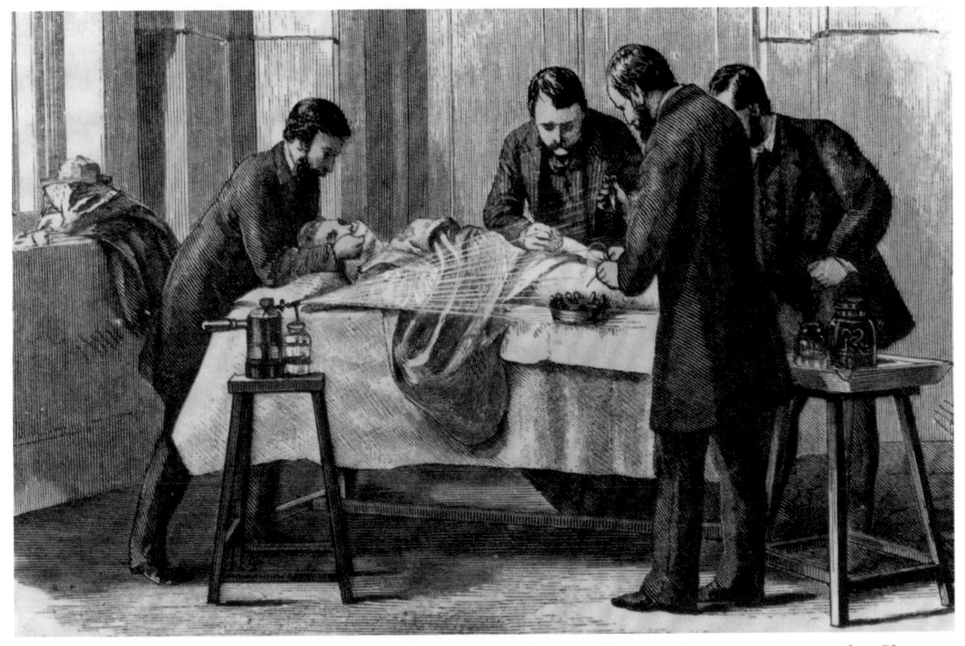

Joseph Lister, the pioneer in antiseptic surgery, using carbolic acid spray during surgery at the Glasgow Royal Infirmary.

The brilliant British scientist and *Boy's Own* hero Charles Scott Sherrington (right), to whom we owe much of our understanding of the central nervous system. He is photographed in 1938 with his former student Harvey Cushing.

London telephone operators perform a disinfectant mouthwash to fight the influenza epidemic, c. 1920.

A nurse at a sanatorium in the 1920s reads to tuberculosis patients taking fresh air while swaddled in blankets.

Dutch drawing of a mastectomy, seventeenth-century style: the breast is removed with a "tenaculum helvetianum," a type of forceps. Note the set of cautery irons smoldering in a pan on the left.

The brilliant American physicist Ernest Lawrence (bottom left) with a cyclotron, the particle accelerator he invented to energize protons, which doubled as a radiation gun that he used to cure his mother's cancer.

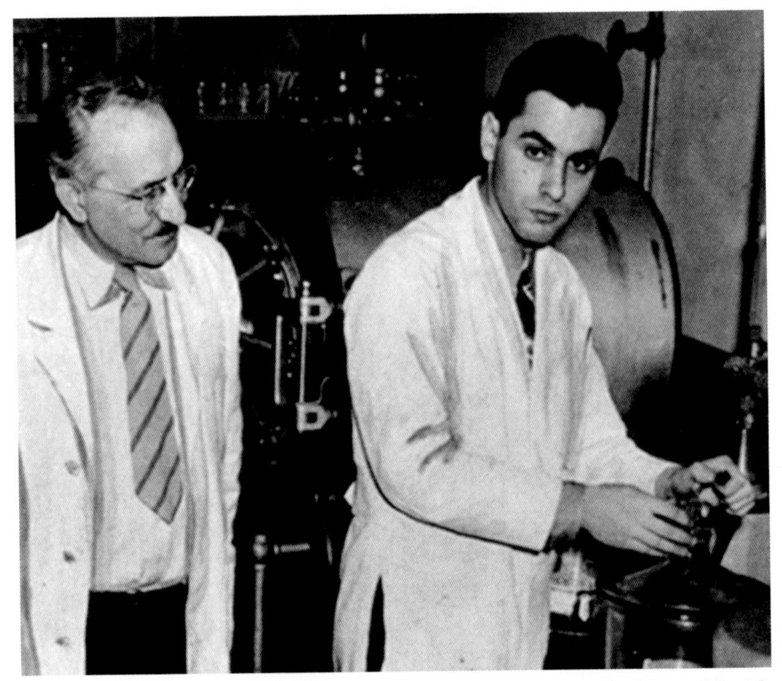

Albert Schatz, who discovered that soil microbes would provide the world with an additional antibiotic to penicillin, overseen by his supervisor, Selman Waksman, who took all the credit.

Alois Alzheimer, the Bavarian pathologist and psychiatrist whose 1906 report and lectures on pre-senile dementia in his patient Auguste Deter established the condition that became known as Alzheimer's disease.

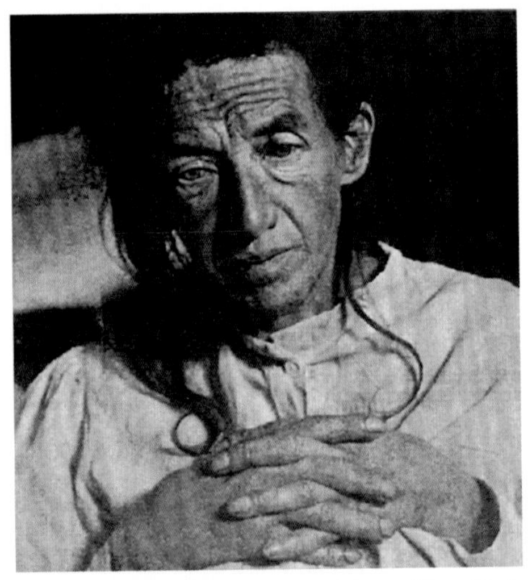

Auguste Deter first presented herself to Alois Alzheimer in 1901 at the age of fifty-one complaining of forgetfulness. When she died five years later, Alzheimer found her brain to be riddled with destroyed cells. She is the first person to have been diagnosed with Alzheimer's disease.

else in the world became the size of Americans, it would be equivalent to adding one billion people to the world's population.

Overweight is defined as a body mass index (BMI) of 25 to 30, and obesity as anything above that. BMI is a person's weight in kilograms divided by the square of their height in meters. The Centers for Disease Control has a very handy BMI calculator, which allows you to determine your BMI instantaneously by entering your height and weight. However, it must also be said that BMI is a crude measure for fatness because it doesn't distinguish between whether you are unusually muscular or just chubby. A bodybuilder and a couch potato could have identical BMI measures but entirely different health outlooks. But even if BMI is not faultless as a measure, you have only to look around to confirm that there is a lot of spare flesh about.

Perhaps no statistic to do with our increasing mass is more telling than that the average woman in the United States today weighs as much as the average man weighed in 1960. In that half a century or so, the average woman's weight has gone from 140 pounds to 166 pounds. The man's has gone from 166 to 196. The annual cost to the American economy in extra health care for overweight people has been put at $150 billion. What's worse, more than half of today's children are expected to be obese by age thirty-five, according to recent modeling at Harvard University. The current generation of young people is forecast to be the first in recorded history not to live as long as their parents because of weight-related health issues.

But the problem is hardly confined to America. People are getting fatter everywhere. In the rich countries of the OECD, the average rate for obesity is 19.5 percent, but it varies considerably across countries. The British are among the tubbiest after the United States, with about two-thirds of adults weighing more than they ought to and 27 percent of them registering as obese, up from 14 percent in 1990. Chile has the highest proportion of overweight citizens at 74.2 percent, closely followed by Mexico at 72.5 percent. Even in comparatively svelte France, 49 percent of adults are overweight and 15.3 percent weigh in as obese, compared with less than 6 percent just twenty-five years ago. The global figure for obesity is 13 percent.

There is no question that losing weight is hard. According to one calculation, you must walk thirty-five miles or jog for seven hours to lose just one pound. One big problem with exercise is that we don't track it very scrupulously. One study in America found that people overestimate the number of calories they burned in a workout by a factor of four. They also then consumed, on average, about twice as many calories as they had just burned off. As Lieberman noted in *The Story of the Human Body*, a worker on a factory floor will in a year expend about 175,000 more calories than a desk worker—equivalent to more than sixty marathons. That's pretty impressive, but here's a reasonable question: How many factory workers look as if they run a marathon every six days? To be cruelly blunt, not many. That's because most of them, like most of the rest of us, replace all those burned calories, and then some, when they are not working. The fact is, you can quickly undo a lot of exercise by eating a lot of food, and most of us do.

At the very least—and it really is the very least—you should get up and move around a little. According to one study, being a committed couch potato (defined as someone who sits for six hours or more per day) increases the mortality risk for men by nearly 20 percent and for women by almost double that. (Why sitting too much is so much more dangerous for women is unclear.) People who sit a lot are twice as likely to contract diabetes, twice as likely to have a fatal heart attack, and two and a half times as likely to suffer cardiovascular disease.

Amazingly, and alarmingly, it doesn't seem to matter how much you exercise the rest of the time. If you spend an evening on the seductive padding of your gluteus maximus, you may nullify any benefits you gained during an active day. As James Hamblin put it in *The Atlantic*, "You can't undo sitting." In fact, people with sedentary occupations and sedentary lifestyles—which is to say, most of us—can easily sit for fourteen or fifteen hours a day, and thus be completely and unhealthily immobile for all but a tiny part of their existence.

James Levine, an obesity expert from the Mayo Clinic and Arizona State University, coined the term "non-exercise activity thermogenesis," or NEAT, to describe the energy we expend from normal daily living. We actually burn a fair amount of calories just existing. The heart, brain, and kidneys burn about 400 calories a day each, the liver about 200. The process of eating and digesting food alone accounts for about one-tenth of the body's daily energy requirements. But we can do much more by simply getting up off our backsides. Even just standing burns an extra 107 calories an hour. Walking around burns 180. In one study, volunteers were instructed to watch television as normal through an evening, but to get up and walk around the room during every commercial break. That alone burned 65 extra calories an hour, about 240 calories over an evening.

Levine found that lean people tend to spend two and a half hours more a day on their feet than fat people, not consciously exercising, but just moving about, and it was this that kept them from accumulating fat. Then again, another study found that people in Japan and Norway are just as inactive as Americans, yet only half as likely to be obese, so exercise can only partly account for slimness.

In any case, a bit of extra weight may not be such a bad thing. A few years ago, *The Journal of the American Medical Association* caused a stir by reporting that people who are slightly overweight, particularly if middle-aged or older, may survive some serious illnesses better than those who are either lean or obese. The idea has become known as the obesity paradox, and it is hotly disputed by many scientists. Walter Willett, a researcher at Harvard, called it "a pile of rubbish" and said that "no one should waste their time reading it."

There's no doubt that exercise improves health, but it is hard to say by how much. A study of eighteen thousand runners in Denmark concluded that people who jog regularly can expect to live five to six years longer on average than non-joggers. But is that because jogging truly is that beneficial, or is it because people who jog tend to lead healthy, moderate lives anyway and are bound to have improved outcomes over us more slothful types, with or without sweatpants?

What is certain is that in a few tens of years at most you will close your eyes forever and cease to move at all. So it might not be a bad idea to take advantage of movement, for health and pleasure, while you still can.

11 EQUILIBRIUM

Life is an endless chemical reaction.

—STEVE JONES

THE SURFACE LAW is not something most of us ever have to think about, but it explains a lot about you. The law states simply that as the volume of an object grows, its relative surface area decreases. Think of a balloon. When a balloon is empty, it is mostly rubber with a trivial amount of air inside. But blow it up and it becomes mostly air with a comparatively small amount of rubber on the outside. The more you inflate it, the more its interior dominates the whole.

Heat is lost at the surface, so the more surface area you have relative to volume, the harder you must work to stay warm. That means that little creatures have to produce heat more rapidly than large creatures. They must therefore lead completely different lifestyles. An elephant's heart beats just thirty times a minute, a human's sixty, a cow's between fifty and eighty, but a mouse's beats six hundred times a minute—ten times a second. Every day, just to survive, the mouse must eat about 50 percent of its own body weight. We humans, by contrast, need to consume only about 2 percent of our body weight to supply our energy requirements. One area where animals are curiously—almost eerily—uniform is with the number of heartbeats they have in a lifetime. Despite the vast differences in heart rates, nearly all animals have about 800 million heartbeats in them if they live an average life. The exception is humans. We pass 800 million heartbeats after twenty-five years, and just keep on going for another fifty years and 1.6 billion heartbeats or so. It is tempting to attribute this exceptional vigor to some innate superiority on our part, but in fact it is only over the last ten or twelve generations that we have deviated from the standard mammalian pattern thanks to improvements in our life expectancy. For most of our history, 800 million beats per lifetime was about the human average, too.

We could reduce our energy needs considerably if we elected to be cold-blooded. A typical mammal uses about thirty times as much energy in a day as a typical reptile, which means that we must eat every day what a crocodile needs in a month. What we get from this is an ability to leap out of bed in the morning, rather than having to bask on a rock until the sun warms us, and to move about at night or in cold weather, and just to be generally more energetic and responsive than our reptilian counterparts.

We exist within extraordinarily fine tolerances. Although our body temperature varies slightly through the day (it is lowest in the morning, highest in the late afternoon or evening), it normally doesn't stray more than a degree or so from 98.6 degrees Fahrenheit. (That's in adults. Children tend to run about one degree higher.) To move more than a very few degrees in either direction is to invite a lot of trouble. A fall of just two degrees below normal, or a rise of four degrees above, can tip the brain into a crisis that can swiftly lead to irreversible damage or death. To avoid catastrophe, the brain has its trusty control center, the hypothalamus, which tells the body to cool itself by sweating or to warm itself by shivering and diverting blood flow away from the skin and into the more vulnerable organs.

That may not seem a terribly sophisticated way of dealing with such a critical matter, but the body does it remarkably well. In one well-known experiment cited by the British academic Steve Jones, a test subject ran a marathon on a treadmill while the room temperature was gradually raised from minus 49 degrees Fahrenheit to 131 degrees Fahrenheit—roughly the limits of human tolerance at both extremes. Despite the subject's exertions and the great range of temperatures, his core body temperature deviated by less than one degree over the course of the exercise.

That experiment largely recalled a series of experiments conducted more than two hundred years earlier for the Royal Society in London by Charles Blagden, a physician. Blagden built a heated chamber—essentially a walk-in oven—in which he and willing associates would stand for as long as they could bear it. Blagden managed ten minutes at a temperature of 198 degrees Fahrenheit. His friend the botanist Joseph Banks, freshly returned from circling the world with Captain James Cook and soon to become president of the Royal Society, managed 210 degrees Fahrenheit, but only for three minutes. "To prove that there was no fallacy in the degree of heat shewn by the thermometer," Blagden recorded, "we put some eggs and a beef-steak upon a tin frame, placed near the standard thermometer. . . . In about twenty minutes the eggs were taken out, roasted quite hard; and in forty-seven minutes the steak was not only dressed, but almost dry." The experimenters also measured the temperature of their urine immediately before and after the test and found that it was unchanged despite the heat. Blagden additionally deduced that perspiration had a central role in cooling the body—his most important insight, and indeed his only lasting contribution to scientific knowledge.

Occasionally, as we all know, our body temperature is elevated beyond normal in the condition known as a fever. Curiously, no one knows quite why this happens—whether fevers are an innate defense mechanism aimed at killing invading pathogens or simply a by-product of the body working hard to fight off infection. The question is important because if fever is a defense mechanism, then any effort to suppress or eliminate it may be counterproductive. Allowing a fever to run its course (within limits, needless to say) could be the wisest thing. An increase of only a degree or so in body temperature has been shown to slow the replication rate of viruses by a factor of two hundred—an astonishing increase in self-defense from only a very modest rise in warmth. The trouble is, we don't entirely understand what is going on with fevers. As Professor Mark S. Blumberg of the University of Iowa has put it, "If fever is such an ancient response to infection, one would think that the mechanism by which it benefits the host would be easy to determine. In fact, it has been difficult."

If elevating our temperature a degree or two is so helpful at fending off invading microbes, then why not raise it permanently? The answer is that it is just too costly. If we were to raise our body temperature permanently by only 3–4 degrees Fahrenheit, our energy requirements would shoot up by about 20 percent. The temperature we have is a reasonable compromise between utility and cost, as with most things, and actually even normal temperature is pretty good at keeping microbes in check. Just look at how swiftly they swarm in and devour you when you die. That's because your lifeless body falls to a delicious come-and-get-it temperature, like a pie left to cool on a windowsill.

The idea, incidentally, that we lose most of our heat through the top of our heads is, it seems, a myth. The top of your head accounts for no more than about 2 percent of your body surface area, and is, on most of us, pretty well insulated by hair, so the top of your head will never be a good radiator. On the other hand, it you are outdoors in cold weather and your head is the only part of you that is exposed, then it will play a disproportionate part in any heat loss, so listen to your mother when she tells you to put a hat on.

Maintaining equilibrium within the body is called homeostasis. The man who coined the term and is often referred to as the father of the discipline was the Harvard physiologist Walter Bradford Cannon (1871–1945). A stocky man whose grim and stiff gaze in photographs belied an apparently warm and genial manner in person, Cannon was undoubtedly a genius, and part of that genius seems to have been an ability to persuade others to do rash and uncomfortable things in the name of science. Curious to understand why our stomachs gurgle when we are hungry, he persuaded a student named Arthur L. Washburn to train himself to overcome the gag reflex in order to push a rubber tube down his throat and into his stomach, where a balloon on its end could be inflated to measure the contractions when he was deprived of food. Washburn would then spend the day going about his normal business–attending classes, working in the lab, running errands–while the balloon uncomfortably expanded and collapsed and people stared at him for being the source of strange noises and having a tube coming out of his mouth.

Cannon persuaded other of his students to consume food while being X-rayed so that he could watch as it proceeded from mouth to esophagus and onward into the digestive system. In so doing, he became the first person to observe the actions of peristalsis–that is, the muscular pushing of food through the digestive tract. These and other novel experiments became the basis of Cannon's classic text, *Bodily Changes in Pain, Hunger, Fear, and Rage,* which was the last word on physiology for years.

Cannon's interests seemed to know no bounds. He became the world authority on the autonomic nervous system–that is, all those things the body does automatically, like breathe, pump blood, and digest food–and on blood plasma. He did groundbreaking research on the amygdala and hypothalamus, deduced the role of adrenaline in survival response (he coined the term "fight or flight"), developed the first effective treatments for shock, and even found time to write an authoritative and respectful paper on the practice of voodoo. In his spare time, he was an enthusiastic outdoorsman. A mountain peak in Montana, in what is now Glacier National Park, was named Mount Cannon in honor of him and his wife after they were the first to scale it, on their honeymoon in 1901. At the outbreak of World War I, he enlisted as a volunteer for the Harvard Hospital Unit, even though he was forty-five years old and the father of five children. He spent two years in Europe as a field doctor. In 1932, Cannon distilled practically all of his knowledge and years of research into a popular book,

The Wisdom of the Body, outlining the body's extraordinary ability to regulate itself. A Swede named Ulf von Euler followed up on Cannon's studies into the fight-or-flight impulse in humans and won the Nobel Prize in Physiology or Medicine in 1970; Cannon himself was long dead by the time the importance of his work was fully appreciated, though he is now widely venerated retroactively.

One thing Cannon didn't understand—no one did yet—was what a staggering amount of energy the body requires at the cellular level in order to maintain itself. It took a very long time to figure that out, and when the answer came, it was provided not by some mighty research institute but by an eccentric Englishman working pretty much on his own in a pleasant country house in the west of England.

We now know that inside and outside the cell are charged particles called ions. Between them in the cell membrane is a kind of tiny air lock known as an ion channel. When the air lock is opened, the ions flow through, and that generates a little buzz of electricity—though "little" here is entirely a matter of perspective. Although each electrical twitch at the cellular level produces just one hundred millivolts of energy, that translates as thirty million volts per meter—about the same as in a bolt of lightning. Put another way, the amount of electricity going on within your cells is a thousand times greater than the electricity within your house. You are, in a very small way, exceedingly energetic.

It's all a matter of scale. Imagine, for purposes of demonstration, firing a bullet into my abdomen. It really hurts and it does a lot of damage. Now imagine firing the same bullet into a giant fifty miles tall. It doesn't even penetrate his skin. It's the same bullet and gun, just a different scale. That's more or less the situation with the electricity in your cells.

The stuff responsible for the energy in our cells is a chemical called adenosine triphosphate, or ATP, which may be the most important thing in your body you have never heard of. Every molecule of

ATP is like a tiny battery in that it stores up energy and then releases it to power all the activities required by your cells—and indeed by all cells, in plants as well as animals. The chemistry involved is magnificently complex. Here is one sentence from a chemistry textbook explaining a little of what it does: "Being polyanionic and featuring a potentially chelatable polyphosphate group, ATP binds metal cations with high affinity." For our purposes here it is enough to know that we are powerfully dependent on ATP to keep our cells humming. Every day you produce and consume your own body weight in ATP—some 200 trillion trillion molecules of it. From ATP's point of view, you are really just a machine for producing ATP. Everything else about you is by-product. Because ATP is consumed more or less instantaneously, you have only sixty grams—that is a little over two ounces—of it within you at any given moment.

It took a long time to figure any of this out, and when it came, almost no one at first believed it. The person who discovered the answer was a retiring, self-funded scientist named Peter Mitchell who in the early 1960s inherited a fortune from the Wimpey house-building company and used it to set up a research center in a stately home in Cornwall. Mitchell was something of an eccentric. He wore shoulder-length hair and an earring at a time when that was especially unusual among serious scientists. He was also famously forgetful. At his daughter's wedding, he approached another guest and confessed that she looked familiar, though he couldn't quite place her.

"I was your first wife," she answered.

Mitchell's ideas were universally dismissed, not altogether surprisingly. As one chronicler has noted, "At the time that Mitchell proposed his hypothesis there was not a shred of evidence in support of it." But he was eventually vindicated and in 1978 was awarded the Nobel Prize in Chemistry—an extraordinary accomplishment for someone who worked from a home lab. The eminent British biochemist Nick Lane has suggested that Mitchell should be as famous as Watson and Crick.

* * *

The surface law also dictates how big we can get. As the British scientist and writer J. B. S. Haldane observed almost a century ago in a famous essay, "On Being the Right Size," a human scaled up to the hundred-foot height of the giants of Brobdingnag in *Gulliver's Travels* would weigh 280 tons. That would make him forty-six hundred times heavier than a normal-sized human, but his bones would be just three hundred times thicker, not nearly robust enough to support such a load. In a word, we are the size we are because that is about the only size we can be.

Body size has a great deal to do with how we are affected by gravity. It will not have escaped your notice that a small bug that falls off a tabletop will land unharmed and continue on its way unperturbed. That is because its small size (strictly, its surface area-to-volume ratio) means that it is scarcely affected by gravity. What is less well known is that the same thing applies, albeit on a different scale, to small humans. A child half your height who falls and strikes her head will experience only one thirty-second the force of impact that a grown person would feel, which is part of the reason that children so often seem to be mercifully indestructible.

Adults are not nearly so fortunate. Few grown humans can normally survive a fall of much more than twenty-five or thirty feet, though there have been some notable exceptions—none more memorable perhaps than that of a British airman in World War II named Nicholas Alkemade.

In the late winter of 1944, while on a bombing run over Germany, Flight Sergeant Alkemade, the tail gunner on a British Lancaster bomber, found himself in a literally tight spot when his plane was hit by enemy flak and quickly filled with smoke and flames. Tail gunners on Lancasters couldn't wear parachutes because the space in which they operated was too confined, and by the time Alkemade managed to haul himself out of his turret and reach for his parachute, he found it was on fire and beyond salvation. He decided to leap from the plane anyway rather than perish horribly in flames, so he hauled open a hatch and tumbled out into the night.

He was three miles above the ground and falling at 120 miles per hour. "It was very quiet," Alkemade recalled years later, "the only sound being the drumming of aircraft engines in the distance, and no sensation of falling at all. I felt suspended in space." Rather to his surprise, he found himself to be strangely composed and at peace. He was sorry to die, of course, but accepted it philosophically, as something that happened to airmen sometimes. The experience was so surreal and dreamy that Alkemade was never certain afterward whether he lost consciousness, but he was certainly jerked back to reality when he crashed through the branches of some lofty pine trees and landed with a resounding thud in a snowbank, in a sitting position. He had somehow lost both his boots, and had a sore knee and some minor abrasions, but otherwise was quite unharmed.

Alkemade's survival adventures did not quite end there. After the war, he took a job in a chemical plant in Loughborough, in the English Midlands. While he was working with chlorine gas, his gas mask came loose, and he was instantly exposed to dangerously high levels of the gas. He lay unconscious for fifteen minutes before co-workers noticed his unconscious form and dragged him to safety. Miraculously, he survived. Some time after that, he was adjusting a pipe when it ruptured and sprayed him from head to foot with sulfuric acid. He suffered extensive burns but again survived. Shortly after he returned to work from that setback, a nine-foot-long metal pole fell on him from a height and very nearly killed him, but once again he recovered. This time, however, he decided to tempt fate no longer. He took a safer job as a furniture salesman and lived out the rest of his life without incident. He died peacefully, in bed, aged sixty-four in 1987.

Now, I am not suggesting that surviving a fall from the sky is something that anyone can count on, but it has happened more often than you might expect. In 1972, a flight attendant named Vesna Vulović survived a fall of 33,000 feet when the Yugoslav Airlines DC-9 on which she was flying broke up in midair over Czechoslovakia. And in 2007, an Ecuadorean-born window cleaner in Manhattan, Alcides Moreno, fell 472 feet when scaffolding he was standing on collapsed. His brother, working alongside, was killed instantly on impact, but Moreno miraculously survived. The human body, in short, can be a wonderfully resilient thing.

Indeed there is seemingly no challenge to human endurance that hasn't been overcome. Consider the case of little Erika Nordby, a toddler in Edmonton, Alberta, who woke up one night in the dead of winter and, wearing only diapers and a light top, walked out of her house through a back door that hadn't closed properly. When she was found, hours later, her heart had been stopped for at least two hours, but she was carefully warmed up at a local hospital and miraculously restored to life. She made a full recovery and became known, not surprisingly, as the Miracle Baby. Remarkably, just a couple of weeks later, a two-year-old boy on a farm in Wisconsin did almost exactly the same thing and was successfully revived and made a full recovery. Dying is, to coin a phrase, the last thing your body wants to do.

Children do much better with extreme cold than with extreme heat. Because their sweat glands aren't fully developed, they don't sweat freely as adults do. That is in large part why so many of them die so swiftly when left in cars in warm weather. In a sealed car with the temperature outside in the 80s, the inside can reach 130, and no child can cope with that for long. Between 1998 and August 2018, almost eight hundred children in the United States died when left unattended in hot cars. Half were under two years of age. Remarkably—indeed, I would say shockingly—more U.S. states have laws making it illegal to leave an animal unattended in a car than to leave a child unattended. The margin of difference is twenty-nine to twenty-one.

Because of our frailties, much of our own planet is off-limits to us. Earth may feel like a generally benign and kindly place, but a very large part of it is too cold or hot or arid or lofty for us to live successfully on it. Even with the advantage of clothing, shelter, and boundless ingenuity, humans can manage to live on only about 12 percent of Earth's land area and just 4 percent of the total surface area if you include the seas. It is a sobering thought that 96 percent of our planet is off-limits to us.

The thinness of the atmosphere puts a limit on how high we can live. The highest permanent settlements in the world are in the Andes in northern Chile on Mount Aucanquilcha, where miners live at 17,500 feet, but that appears to be absolutely at the limits of human tolerance. The miners themselves choose to trudge an additional 1,500 feet up the slopes to their workplace each day rather than sleep at 19,000 feet. For purposes of comparison, Mount Everest is about 29,000 feet.

At very high altitudes, any exertion becomes difficult and exhausting. Around 40 percent of people experience altitude sickness above thirteen thousand feet, and it is impossible to predict who the victims will be because it is not related to fitness. At extreme heights everyone struggles. Frances Ashcroft in *Life at the Extremes* notes how Tenzing Norgay and Raymond Lambert, on a climb of the South Col of Mount Everest in 1952, took five and a half hours to climb just 650 feet.

At sea level, about 40 percent of your blood volume is occupied by red blood cells, but that can increase by about half as much again with acclimatization to higher altitudes, though there is a price to be paid. The increase in red cells makes the blood thicker and more sluggish and puts extra pressure on the heart when pumping, and that can apply even to those who have lived their whole lives at great heights. Residents of lofty cities, such as La Paz in Bolivia (11,500 feet), sometimes suffer an illness called Monge's disease, which produces blue lips and clubbed fingers because their perpetually thickened blood is not flowing well. The problem goes away if they move to a lower altitude. Many sufferers are thus permanently exiled to the valleys, far from friends and families.

For reasons of economy, airlines normally keep cabins pressurized to an altitude equivalent of forty-nine hundred feet to seventy-nine hundred feet, which is why alcohol is more likely to go to your head while flying. It also accounts for why your ears pop during descent because the pressure changes as you reduce elevation. On an airliner flying at a normal cruising altitude of thirty-five thousand feet, if the cabin suddenly depressurized, passengers and crew could become confused and incompetent in as little as eight or ten seconds. Ashcroft notes the case of a pilot who passed out because he paused to put on his eyeglasses before his oxygen mask. Fortunately, the co-pilot was not incapacitated and took control of the plane.

One of the more infamous examples of oxygen deprivation—or hypoxia, as it is more formally known—was in October 1999 when the American professional golfer Payne Stewart, along with three business associates and two pilots, was on a chartered Learjet en route from Orlando to Dallas when the plane lost pressurization and all aboard blacked out. The plane's last contact was at 9:27 a.m., when the pilot acknowledged clearance to climb to thirty-nine thousand feet. Six minutes later, when a controller contacted the plane again, there was no response. Instead of turning west for Texas, the jet continued on a northwesterly track, on automatic pilot, across the central United States before eventually running out of fuel and crashing in a field in South Dakota. All six aboard were killed.

A disturbingly large amount of what we know about human survival abilities comes from experiments carried out on military prisoners, concentration camp inmates, and civilians during World War II. In Nazi Germany, healthy prisoners were subjected to amputations or experimental limb transplants and bone grafts in the hopes of finding better treatments for German casualties. Russian prisoners of war were plunged into ice water to determine how long a German pilot could survive a downing at sea. Others were kept outdoors naked in freezing weather for up to fourteen hours for similar purposes. Some experiments seem to have been driven by nothing more than morbid curiosity. In one, the subjects' eyes were injected with dyes to see if their eye color could be permanently changed. Many others were subjected to poisons and nerve gases of all types or infected with malaria, yellow fever, typhus, and smallpox. "Contrary to postwar apologies," write George J. Annas and Michael A. Grodin in *The Nazi Doctors and the Nuremberg Code*, "doctors were never forced to perform such experiments." They volunteered.*

Horrifying as the German experiments were, they were outdone, in scale if not cruelty, by the Japanese. Under a doctor named Shiro Ishii, the Japanese built an enormous complex of more than 150 buildings spread over almost 1,500 acres at Harbin in Manchuria with the avowed purpose of determining human physiological limitations through any means necessary. The facility was known as Unit 731.

In a typical experiment, Chinese prisoners were tied to stakes at staggered distances from a shrapnel bomb. The bomb was detonated and scientists then walked among them, carefully noting the nature and extent of the prisoners' injuries and how long it took them to die. Other prisoners were shot with flamethrowers for similar purposes, or starved, frozen, or poisoned. Some, for unfathomable reasons, were dissected while still conscious. Most of the victims were captured Chinese soldiers, but Unit 731 also experimented on selected Allied prisoners of war to make sure that toxins and nerve agents had the same effects on Westerners as on Asians. When pregnant women or young children were needed for experiments, they were randomly snatched from the streets of Harbin. Nobody knows how many people died in Unit 731, but one estimate has put the number as high as 250,000.

The outcome of all this was that Japan and Germany finished the war well ahead of the rest of the world in understanding microbiology, nutrition, frostbite, weapons injuries, and above all the effects of nerve

* The insensitivity in Nazi Germany could be breathtaking. In 1941, a psychiatric hospital at Hadamar, near Limburg, celebrated the putting to death of its ten thousandth cognitively deficient person with an official celebration with speeches and beer for the staff.

gases, toxins, and infectious diseases. Although many Germans were captured and tried for these war crimes, the Japanese almost entirely escaped punishment. Most were granted immunity from prosecution in return for sharing what they had learned with their American captors. Shiro Ishii, the physician who had conceived and run Unit 731, was extensively debriefed and then allowed to return to civilian life.

The existence of Unit 731 was a well-guarded secret, by Japanese and American officials alike, and would have remained unknown to the wider world forever except that in 1984 a student from Keio University in Tokyo came across a box of incriminating documents in a secondhand bookshop and brought them to the attention of others. By this time, it was far too late to bring to justice Shiro Ishii. He had died in 1959, peacefully in his sleep, at the age of sixty-seven after nearly a decade and a half of untroubled postwar life.

12 THE IMMUNE SYSTEM

The immune system is the most
interesting organ in the body.

—MICHAEL KINCH

I

THE IMMUNE SYSTEM is big and kind of messy and all over the place. It includes a lot of things that we don't usually think of in the context of immunity, like earwax, skin, and tears. Any invader that gets past these outer defenses—and comparatively few do—will quickly run into swarms of "proper" immune cells, which come pouring out of lymph nodes, bone marrow, the spleen, the thymus, and other corners of the body. There is a lot of chemistry involved. If you want to understand the immune system, you need to understand antibodies, lymphocytes, cytokines, chemokines, histamine, neutrophils, B cells, T cells, NK cells, macrophages, phagocytes, granulocytes, basophils, interferons, prostaglandins, pluripotent hematopoietic stem cells, and a great deal more—and I mean a great deal more.

Some of these overlap and some do multiple jobs. Interleukin-1, for instance, not only attacks pathogens but also plays a role in sleep, which may go some way to explaining why we are so often drowsy when unwell. By one calculation, we have some three hundred different types of immune cells at work within us, but Daniel Davis, professor of immunology at the University of Manchester in England, thinks the number is essentially incalculable. "A dendritic cell in the skin will be quite different from one in a lymph node, say, and so it all gets quite muddled to define specific types," he says.

On top of all that, every person's immune system is unique, making immune systems harder to generalize, harder to understand, and harder to treat when they go wrong. Moreover, the immune system doesn't just deal with germs. It has to respond to toxins, drugs, cancers, foreign objects, and even your own state of mind. If you are stressed or exhausted, you are much more likely to suffer an infection, for instance. Because protecting us from invasion is such a limitless challenge, the immune system sometimes makes mistakes and launches an attack on innocent cells. Given the number of inspections immune cells make day after day, the error rate is really low. It is a great irony nonetheless that a very high proportion of the suffering we do is inflicted on us by our own defenses in the form of autoimmune diseases like multiple sclerosis, lupus, rheumatoid arthritis, Crohn's disease, and many unappealing others. Altogether about 5 percent of us suffer from some form of autoimmune disease–a very high proportion for such an uncomfortable range of afflictions–and the numbers are growing faster than our abilities to treat them effectively.

"You could look at it and conclude that it's crazy that the immune system attacks itself," says Davis. "Alternatively, once you start to think about all that the immune system has to do, it's surprising that it doesn't happen all the time. Your immune system is constantly bombarded by things it has never seen before, things that may have only just come into existence–like new flu viruses, which are constantly mutating into new forms. So your immune system has to be able to identify and fight off a more or less infinite number of things."

Davis is a big but gentle man in his forties with a booming laugh and the happy air of someone who has found his niche in life. He studied physics at Manchester and Strathclyde Universities in Britain, but then moved to Harvard in the mid-1990s and decided that biology was where his real interests lay. By chance he ended up in the immunology lab at Harvard and became gripped by the elegant complexity of the immune system and the challenge of trying to unravel it all.

Despite the intricacy at the molecular level, all parts of the immune system contribute to a single task: to identify anything that is in the body that shouldn't be there and, if necessary, kill it. But the process is far from straightforward. A lot of things inside you are harmless or even beneficial, and it would be foolhardy or a waste of energy and resources to kill them. So the immune system has to be a bit like security people at airports watching stuff on a conveyor belt and only challenging those things that have nefarious intent.

At the heart of the system are five types of white blood cells: lymphocytes, monocytes, basophils, neutrophils, and eosinophils. They are all important, but lymphocytes are the ones that excite immunologists most. David Bainbridge calls lymphocytes "just about the cleverest little cells in the whole body" because of their ability to recognize almost any kind of unwanted invader and mobilize a swift and targeted response.

Lymphocytes are of two principal types: B cells and T cells. The *B* in B cells comes, a little oddly, from "bursa of Fabricius," an appendix-like organ in birds where B cells were first seen.* Humans and other

* The bursa of Fabricius is named for Hieronymus Fabricius (1537–1619), an Italian anatomist who thought it was connected to the production of eggs. Fabricius was wrong, but its actual purpose remained a mystery until 1955, when it was solved by a happy accident. Bruce Glick, then a graduate student at Ohio State University, removed bursas from chickens to see what effect it had on them in the hope of solving the mystery. But the removals had no discernible effect, so he gave up on the problem. The chickens were then passed on to another student, Tony Chang, who was studying antibodies. Chang discovered that the birds without bursas produced no antibodies. The two young researchers realized that the bursa of Fabricius was responsible for antibody production—a really big discovery in immunology. They submitted a paper to the journal *Science,* but it was returned as "uninteresting." Eventually, they got it published in *Poultry Science.* It has since become "one of the most cited papers in immunology," according to the British Society for Immunology. "Bursa," incidentally, comes from a Latin term for "bag" or "purse" and can describe various structures. Bursas

mammals don't have a bursa of Fabricius. Our B cells are made in the bone marrow, but it is entirely coincidental that that starts with a *b*, too. T cells are more faithful to their source. Though created in the bone marrow, they emerge from the thymus, a small organ in the chest just above the heart and between the lungs. For a very long time, the role of the thymus in the body was a complete mystery because it seemed to be just full of dead immune cells—"the place where cells went to die," as Davis put it in his superlative book *The Compatibility Gene*. In 1961, Jacques Miller, a young Franco Australian research scientist working in London, unraveled a mystery. What Miller established was that the thymus is a nursery for T cells. T cells are a kind of elite corps in the immune system, and the dead cells found in the thymus were lymphocytes that had failed to pass muster because they were either not very good at identifying and attacking foreign invaders or because they were too eager to attack the body's own healthy cells. They had, in short, failed to make the cut. It was an immensely significant discovery. As the medical journal *The Lancet* observed, it made Miller "the last person to identify the function of a human organ." Many people have wondered why he has never been honored with a Nobel Prize.

T cells subdivide into two further categories: helper T cells and killer T cells. Killer T cells, as the name suggests, kill cells that have been invaded by pathogens. Helper T cells help other immune cells act, including helping B cells produce antibodies. Memory T cells remember the details of earlier invaders and are therefore able to coordinate a swift response if the same pathogen shows up again—what is known as adaptive immunity.

Memory T cells are extraordinarily vigilant. I don't get mumps, because somewhere inside me are memory T cells that have been protecting me from a second attack for more than sixty years. When they identify an invader, they instruct B cells to produce proteins known

---

in humans (which are responsible for bursitis) are little sacs that help to cushion joints.

as antibodies, and these attack the invading organisms. Antibodies are clever things because they recognize and fight off previous invaders quickly if they dare come back. That's why so many diseases only make you sick once. It is also the process at the heart of vaccination. Vaccination is really a way of inducing the body to produce useful antibodies against a particular scourge without actually making oneself sick.

Microbes have developed various ways of fooling the immune system—by sending out confusing chemical signals, for instance, or by disguising themselves as benign or friendly bacteria. Some infectious agents, like *E. coli* and salmonella, can trick the immune system into attacking the wrong organisms. There are a lot of human pathogens out there, and much of their existence is devoted to evolving new and cunning ways to get inside us. The wonder isn't that we get sick sometimes but that we are not sick far more often. In addition, as well as killing invasive cells, the immune system must endeavor to kill our own cells when they misbehave, as when they turn cancerous.

Inflammation is essentially the heat of battle as the body defends itself from damage. Blood vessels in the vicinity of an injury dilate, allowing more blood to flow to the site, bringing with it white blood cells to fight off invaders. That causes the site to swell, increasing the pressure on surrounding nerves, resulting in tenderness. Unlike red blood cells, white blood cells can leave the circulatory system to pass through surrounding tissues, like an army patrol searching through jungle. When they encounter an invader, they fire off attack chemicals called cytokines, which is what makes you feel feverish and ill when your body is battling infection. It's not the infection that makes you feel dreadful, but your body defending itself. The pus that seeps from a wound is simply dead white cells that have given their lives in defense of you.

Inflammation is a tricky thing. Too much and it destroys neighboring tissues and can result in unnecessary pain, but too little and it fails to stop infection. Faulty inflammation has been linked to all kinds of maladies, from diabetes and Alzheimer's disease to heart attacks and strokes. "Sometimes," Michael Kinch, from Washington University in St. Louis, explained to me, "the immune system gets so ramped up that it brings out all its defenses and fires all its missiles in what is known as a cytokine storm. That's what kills you. Cytokine storms show up again and again in many pandemic diseases, but also in things like extreme allergic reactions to bee stings."

Much of what happens in the immune system at the cellular level is still very imperfectly understood. Quite a lot is not understood at all. During my visit to Manchester, Davis took me into his lab, where a team of postdoctoral scholars were hunched over computer screens studying images taken from very high-resolution microscopes. A postdoc named Jonathan Worboys showed me something they had only just discovered—rings made of protein scattered across the cell's surface, like portholes. No one outside this lab had ever seen these rings before.

"They're clearly formed for a reason," Davis said, "but we don't know yet what that reason is. It looks important, but it could be trivial. We just don't know. It may be four or five years before we really unravel it. It is the kind of thing that makes science exciting and difficult at the same time."

If the immune system has a patron saint, it is surely Peter Medawar, who was one of the very greatest of twentieth-century British scientists, as well as possibly the most exotic. The child of a Lebanese father and an English mother, he was born in 1915 in Brazil, where his father had business interests, though when Medawar was a boy the family moved to England. Medawar was tall, good-looking, and athletic. Max Perutz, a contemporary, described him as "vivacious, sociable, debonair, brilliant in conversation, approachable, restless, and intensely ambitious." Stephen Jay Gould called him "the cleverest man I have ever known." Although Medawar trained as a zoologist, it was his work with humans during World War II that brought him permanent fame.

In the summer of 1940, Medawar was sitting with his wife in

their garden in Oxford enjoying a sunny afternoon when they heard a plane sputtering overhead and looked up to see an RAF Spitfire falling from the sky. It crashed in flames just two hundred yards from their home. The pilot survived but suffered terrible burns. A day or so later, Medawar was presumably surprised to be asked by army doctors if he would come and have a look at the young pilot. Medawar was a zoologist, after all, but he was engaged in research on antibiotics, and there was a chance he might be able to help. It was the beginning of a wonderfully productive relationship that eventually culminated in a Nobel Prize.

The doctors were particularly troubled by the problem of getting skin grafts to take. Whenever skin was taken from one person and grafted onto another, it was accepted at first but then swiftly withered and died. Medawar was immediately gripped by the problem and couldn't understand why the body rejected something so clearly beneficial. "For all the clinical good-will and perhaps even mortal urgency that accompanies their transplantation, skin homografts are treated as if they were a disease of which their destruction is the cure," he wrote.

"People thought there was some problem with the surgery, that if surgeons could perfect their technique it would be all right," says Daniel Davis. But Medawar realized there was something more than that. Whenever he and his colleagues repeated a skin graft, it was always rejected even more quickly the second time. What Medawar subsequently found was that the immune system learns early in life not to attack its own normal, healthy cells. As Davis explained to me, "He discovered that if a mouse was exposed to skin from another mouse when it was very young, then when the mouse grew up, it would be able to accept a skin transplant from that second mouse. In other words, he discovered that at a young age the body learns what is self–what not to attack. You can get a skin transplant from one mouse to another as long as the recipient mouse has been trained in early life not to react to it." This was the insight that would, years later, win Medawar a Nobel Prize. As David Bainbridge has noted, "Although we take it for granted today, this sudden joining of transplantation and the immune system was a crucial point in medical science. It told us what immunity actually is."

II

TWO DAYS BEFORE Christmas in 1954, Richard Herrick of Marlborough, Massachusetts, was on the brink of death from kidney failure at the age of just twenty-three when he was given his life back by becoming the world's first kidney transplant recipient. Herrick was exceedingly lucky because he had an identical twin, Ronald, so had a donor with a perfect tissue match.

Even so, no one had ever attempted anything like this before, and his doctors weren't at all sure what the outcome would be. One distinct possibility was that both brothers might die. As Dr. Joseph Murray, the lead surgeon, explained years later, "None of us had ever asked a healthy person to accept this magnitude of risk solely for the sake of someone else." Happily, the outcome proved better than anyone had dared hope—indeed, had a certain fairy-tale quality to it. Richard Herrick not only survived the operation and regained his health but married his nurse and had two children with her. He lived eight years more before the original disease, glomerulonephritis, reasserted itself and killed him. His brother Ronald lived another fifty-six years with his one kidney. Herrick's surgeon, Dr. Joseph Murray, was awarded a Nobel Prize in Physiology or Medicine in 1990, though mostly for his later work on immunosuppression.

Problems with rejection, however, meant that most other attempts at transplants failed. Over the following decade, 211 people underwent kidney transplants, and most lived for no more than a few weeks, if that. Only six survived for as much as a year and in most of those cases because the donor was also a twin. It wasn't until the development of the miracle drug cyclosporine from a soil sample fortuitously collected on a Norwegian holiday (as you will recall from chapter 7) that transplants could start to become routine.

Advances in transplant surgery over the past few decades have been breathtaking. Today in the United States, of the 30,000 people who receive an organ transplant each year, over 95 percent are still alive twelve months later and 80 percent are alive five years later. The downside is that demand for replacement organs far outstrips supply. As of late 2018, 114,000 people were on transplant waiting lists in the United States. A new person joins the list every ten minutes, and twenty people a day die before a donated organ can be found. People on dialysis live an extra eight years on average, but that rises to twenty-three years with a transplant.

About a third of kidney transplants come from living donors (typically a close relative), but all other transplanted organs are from deceased donors, which is a real challenge. Anyone who needs an organ has to hope that someone dies in circumstances that leave a harvestable organ of the right size, that the victim isn't too far away, and that there are two teams of surgical specialists standing by—one to remove the necessary organ from the donor and another to reinstall it in the recipient. The median waiting time today for a kidney transplant in the United States is 3.6 years, up from 2.9 years in 2004, but many people can't wait that long. In the United States, seven thousand people a year on average die before they can receive a transplant.

One possible solution would be to use animal transplants. Organs taken from pigs could be grown to the right size, then harvested at will. Transplant surgeries could be scheduled instead of treated as emergencies. It is a wonderful solution in principle, but in practice it throws up two main problems. One is that organs from another animal species provoke a savage immune response—if there is one thing your immune system knows, it is that you shouldn't have a pig's liver inside you—and the second is that pigs are full of something called porcine endogenous retroviruses (or PERVs for short), which could infect any humans into which they are introduced. There are hopes that both problems can be overcome in the near future, which could transform prospects for thousands of people.

A separate and no less intractable problem is that immunosuppressive drugs are not ideal for several reasons. To begin with, they affect the entire immune system, not just the transplanted part, so the patient is left permanently vulnerable to infections and to cancers, which the immune system would normally tackle. The drugs can also be toxic.

Luckily, most of us will never need a transplant, but there are plenty of other things the immune system can do to us. Altogether humans are afflicted by some fifty types of autoimmune diseases, and the numbers are rising. Take Crohn's disease, the increasingly common inflammatory bowel disease. Before 1932, when Burrill Crohn, a New York physician, described it in a paper in *The Journal of the American Medical Association*, it wasn't even a recognized condition. At that time, Crohn's affected one person in 50,000. Then it became one in 10,000, then one in 5,000. Today the proportion is one in 250 and still rising.*

Why this has happened no one can say. Daniel Lieberman suggests that the overuse of antibiotics and the consequent depletion of our microbial reserves might have made us more susceptible to all autoimmune diseases, but acknowledges that the "causes remain elusive."

Equally bewildering is that autoimmune diseases are grossly sexist. Women are twice as likely as men to get multiple sclerosis, ten times more likely to get lupus, fifty times more likely to suffer a thyroid condition known as Hashimoto's thyroiditis. Altogether, 80 percent of all autoimmune diseases occur in women. Hormones are the presumed culprit, but how exactly female hormones trip up the immune system when male hormones don't is not at all clear.

* Crohn didn't use the term himself, preferring instead to call it regional ileitis, regional enteritis, or cicatrizing enterocolitis. It was later discovered that Thomas Kennedy Dalziel, a Glasgow surgeon, had described the same disease almost twenty years earlier. He called it chronic interstitial enteritis. Crohn obituary, *New York Times*, July 30, 1983; "Crohn of Crohn's Disease," *Gastroenterology*, May 1999.

The largest and in many ways most mystifying and intractable category of immune disorders is allergies. An allergy is simply an inappropriate response by the body to a normally harmless invader. Allergies are a surprisingly recent concept, too. The word's first appearance in English (spelled "allergie"), in *The Journal of the American Medical Association,* was only a little over a century ago. Yet allergies have become a bane of modern life. Roughly 50 percent of people claim to be allergic to at least one thing, and many claim to be allergic to lots of things (a condition known to medical science as atopy).

Allergy rates vary across the world from about 10 to 40 percent, with the rates closely following economic performance. The richer the country, the more allergies its citizens get. No one knows why being rich should be so bad for you. It may be that people of rich, urbanized nations are more exposed to pollutants—there is evidence that nitrogen oxides from diesel fuels correlate with higher incidences of allergies—or it may be that increased use of antibiotics in the rich nations has directly or indirectly affected our immune responses. Other contributory factors may be lack of exercise and increased obesity. Allergies are not specifically genetic as far as anyone can tell, but your genes can leave you more susceptible to getting certain allergies. If both your parents have a particular allergy, there is a 40 percent chance you will get it, too. So it's a greater likelihood but not a certainty.

Most allergies merely cause discomfort, but some can be life threatening. About seven hundred people a year die in America from anaphylaxis, the formal name for an extreme allergic reaction causing restriction of airways. These reactions are brought on most often by penicillin, foods, insect stings, and latex, in that order. Some people are extraordinarily sensitive to certain materials. Dr. Charles A. Pasternak in *The Molecules Within Us* notes how one child on an airplane had to be hospitalized for two days because a passenger two rows away ate peanuts. In 1999, just 0.5 percent of children had peanut allergies; today, less than twenty years later, the rate has increased fourfold.

In 2017, the National Institute of Allergy and Infectious Diseases declared that the best way to avoid or minimize peanut allergies was not to withhold peanuts from very young children, as had been believed for decades, but rather to give them small exposures as a way of hardening them to peanuts. Other authorities have suggested that leaving parents to, in effect, experiment on their own children is not a good idea and that any program of habituation should only be done under close, qualified supervision.

The most common explanation for soaring rates of allergies is the well-known hygiene hypothesis, which was first put forward in 1989 in a brief article in the *British Medical Journal* by an epidemiologist from the London School of Hygiene and Tropical Medicine named David Strachan, though he didn't use the term "hygiene hypothesis." That came later. The idea, very loosely, is that children in the developed world grow up in much cleaner environments than children of earlier ages did, and so don't develop resistance to infection as well as those who have a more intimate contact with dirt and parasites.

The hygiene hypothesis has some problems, however. One is that the big rise in allergies mostly dates from the 1980s, long after we began to get clean, so hygiene alone can't account for rising rates. A broader version of the hygiene hypothesis, known as the old friends hypothesis, has now largely supplanted the original theory. It postulates that our susceptibilities aren't based just on childhood exposures, but are a result of accumulated lifestyle changes dating back to the Neolithic period.

The bottom line in either case is that we don't know why allergies exist at all. Dying from ingesting a peanut is not something that confers any obvious evolutionary benefits, after all, so why this extreme sensitivity has been retained in some humans is, like so much else, a puzzle.

Disentangling the intricacies of the immune system is much more than just an intellectual exercise. Finding ways of using the body's own immune defenses to fight diseases—what is known as immunotherapy—has the promise of transforming whole areas of medicine. Two approaches in particular have attracted a good deal of attention in recent times. One is immune checkpoint therapy. Essentially, it is based on the idea that the immune system is programmed to fix a problem–kill an infection, say–and then withdraw. The immune system is a bit like a fire brigade in this respect. Once it has put out a fire, there's no point in it continuing to play water over the ashes, so it has built-in signals that tell it to pack up and go back to the firehouse to await the next crisis. Cancers have learned to exploit this by sending out stop signals of their own, fooling the immune system into retiring prematurely. Checkpoint therapy simply overrides the stop signals. The therapy works miraculously well with some cancers–some people with advanced melanomas who were near death have staged complete recoveries–but for reasons still not well understood, it only works sometimes. It also can have serious side effects.

The second type of therapy is called CAR T-cell therapy. CAR stands for "chimeric antigen receptor," and it is about as complicated and technical as it sounds, but essentially it involves genetically altering a cancer sufferer's T cells, then returning them to the body in a form that allows them to attack and kill cancer cells. The process works really well against some leukemias, but it kills healthy white blood cells as well as cancerous ones and therefore leaves the patient vulnerable to infections.

But the real problem with such therapies may be cost. CAR T-cell therapy, for instance, can cost the better part of $500,000 per patient. "What are we going to do," asks Daniel Davis, "cure a few rich people and tell everyone else that it is not available?" But that is, of course, another issue altogether.

13 DEEP BREATH: THE LUNGS AND BREATHING

> I am in the habit of going to sea whenever
> I begin to grow hazy about the eyes, and
> begin to be over conscious of my lungs.
>
> —HERMAN MELVILLE, *MOBY-DICK*

I

QUIETLY AND RHYTHMICALLY, awake or asleep, generally without thought, every day you breathe in and out about 20,000 times, diligently processing some 4,000 gallons (or 440 cubic feet) of air, depending on how big you are and how active. That's about 7.3 million breaths between birthdays, 550 million or so over the course of a lifetime.

In breathing, as in everything in life, the numbers are staggering—indeed fantastical. Every time you breathe, you exhale some 25 sextillion (that's 2.5×10^{22}) molecules of oxygen—so many that with a day's breathing you will in all likelihood inhale at least one molecule from the breaths of every person who has ever lived. And every person who lives from now until the sun burns out will from time to time breathe in a bit of you. At the atomic level, we are in a sense eternal.

For most of us, those molecules come pouring in through the nares, which is what anatomists call the nostrils (for no very compelling reason, it must be said). From there the air passes through the most mysterious space in your head, the sinus cavity. Proportionate to the rest of the head, the sinuses take up an enormous amount of space, and no one is at all sure why.

"Sinuses are strange," Ben Ollivere of the University of Nottingham and Queen's Medical Centre told me one day. "They are just cavernous spaces in your head. You would have room for a good deal more gray matter if you didn't have to devote so much of your head to the sinuses." The space isn't a complete void, but rather is riddled with a complex network of bones, which are thought to help with breathing efficiency, though no one can say quite how. Whether or not they have a function, the sinuses cause a lot of unhappiness. Thirty-five million Americans suffer sinusitis every year, and about 20 percent of all antibiotic prescriptions are for people with sinus conditions (even though sinus conditions are overwhelmingly viral and thus immune to antibiotics).

Incidentally, the reason your nose runs in chilly weather is the same reason your bathroom windows run with water in chilly weather. In the case of your nose, warm air from your lungs meets cold air coming into the nostrils and condenses, resulting in a drip.

The lungs are also wonderfully good at cleaning. According to one estimate, the average urban dweller inhales some twenty billion foreign particles every day—dust, industrial pollutants, pollen, fungal spores, whatever is adrift on the day's air. A lot of this stuff can make you very ill, but it doesn't, by and large, because your body is normally adept at challenging intruders. If an invading particle is big or especially irritating, you will almost certainly cough or sneeze it straight back out again (often in the process making it someone else's problem). If it is too small to provoke such a vehement response, it will in all likelihood be trapped in the mucus that lines your nasal passages or caught by the bronchi, or tubules, in your lungs. These tiny airways are lined with millions and millions of hairlike cilia that act like paddles (but beating furiously at sixteen times a second), and they swat the invaders back into the throat, where they are diverted to the stomach and dissolved by hydrochloric acid. If any invaders manage to get past these waving hordes, they will encounter little devouring machines called alveolar macrophages, which gobble them up. Despite all this, occasionally some pathogens get through and make you sick. That's the way life is, of course.

Only recently has it been discovered that sneezes are a much more drenching experience than anyone thought. A team led by Professor Lydia Bourouiba of MIT, as reported by *Nature,* studied sneezes more closely than anyone had ever chosen to before and found that sneeze droplets can travel up to eight meters and drift in suspension in the air for ten minutes before gently settling onto nearby surfaces. Through ultra-slow-motion filming, they also discovered that a sneeze isn't a bolus of droplets, as had always been thought, but more like a sheet—a kind of liquid Saran Wrap—that breaks over nearby surfaces, providing further evidence, if any were needed, that you don't want to be too close to a sneezing person. An interesting theory is that weather and temperature may influence how the droplets in a sneeze coalesce, which could explain why flu and colds are more common in cold weather, but that still doesn't explain why infectious droplets are more infectious to us when we pick them up by touch rather than when we breathe (or kiss) them in. The formal name for the act of sneezing, by the way, is sternutation, though some authorities in their lighter moments refer to a sneeze as an autosomal dominant compelling helio-ophthalmic outburst, which makes the acronym ACHOO (sort of).

Altogether the lungs weigh about 2.4 pounds, and they take up more space in your chest than you probably realize. They jut up as high as your neck and bottom out at about the breastbone. We tend to think of them as inflating and deflating independently, like bellows, but in fact they are greatly assisted by one of the least appreciated muscles in the body: the diaphragm. The diaphragm is a mammalian invention and it is a good one. By pulling down on the lungs from below, it helps them to work more powerfully. The increased respiratory efficiency that the diaphragm brings enables us to get more oxygen to our muscles, which helped us to become strong, and to our brains, which helped us to become smart. Efficiency is also assisted by a slight differential in air pressure between the outside world and the space around your lungs, known as the pleural cavity. Air pressure in the chest is less than atmospheric pressure, which helps to keep the lungs inflated. If air gets into the chest, because of a puncture wound, say, the differential vanishes and the lungs collapse to only about a third of their normal size.

Breathing is one of the few autonomic functions that you can control intentionally, though only up to a point. You can shut your eyes for as long as you wish, but you cannot shut off your breathing for long before the autonomic system reasserts itself and compels you to breathe. Interestingly, the discomfort you feel when you hold your breath for too long is caused not by the depletion of oxygen but by a buildup of carbon dioxide. That's why the first thing you do when you stop holding your breath is blow out. You would think that the most urgent need would be to get fresh air in rather than stale air out, but no. The body so abhors CO_2 that you must expel it before gulping in replenishment.

Humans are pretty poor at holding their breath—indeed, are inefficient breathers altogether. Our lungs can hold about six quarts of air, but normally we breathe in only about half a quart at a time, so there is plenty of scope for improvement. The very longest any human being has voluntarily held his breath was twenty-four minutes and three seconds by Aleix Segura Vendrell of Spain, who did it in a pool in Barcelona in February 2016, but that was after breathing pure oxygen for some time beforehand and then lying motionless in the water to reduce energy demand to a minimum. Compared with most aquatic mammals, this is really poor. Some seals can stay underwater for two hours. Most of us can't last much more than a minute, if that. Even the famous lady pearl divers of Japan, known as the ama, don't stay underwater for more than about two minutes normally (though they do make a hundred or more dives a day).

All in all, it takes a lot of lung to keep you going. If you are an averagely sized adult, you will have roughly twenty square feet of skin, but about a thousand square feet of lung tissue containing about fifteen hundred miles of airways. Packing such a lot of breathing apparatus into the modest space of your chest is a nifty solution to the very considerable problem of how to get a lot of oxygen efficiently to billions of cells. Without that intricate packaging, we might have to be like kelp—hundreds of feet long but with all the cells very near the surface to facilitate oxygen exchange.

Considering how complex an operation respiration is, it is not surprising that the lungs can cause us a lot of problems. What is perhaps surprising is how little we sometimes understand the causes of these problems, and of no condition is that more true than asthma.

II

IF YOU HAD to nominate someone to be a poster figure for asthma, you could do worse than the great French novelist Marcel Proust (1871–1922). But then you could nominate Proust as a poster figure for a great many medical conditions because he had a superabundance of them. He suffered from insomnia, indigestion, backaches, headaches, fatigue, dizziness, and crushing ennui. More than anything else, however, he was a slave to asthma. He had his first attack at nine and passed a wretched life thereafter. With his suffering came an acute germ phobia. Before opening his mail, he would have his assistant place it in a sealed box and expose it to formaldehyde vapors for two hours. Wherever he was in the world, he sent his mother detailed daily reports on his sleep, lung function, mental composure, and bowel movements. He was, as you will gather, somewhat preoccupied with his health.

Though some of his concerns were perhaps a touch hypochondriacal, the asthma was real enough. Desperate to find a cure, he submitted to countless (and pointless) enemas; took infusions of morphine, opium, caffeine, amyl, trional, valerian, and atropine; smoked medicated cigarettes; inhaled drafts of creosote and chloroform; underwent more than a hundred painful nasal cauterizations; adopted a milk diet; had the gas to his house cut off; and lived as much of his life as he could in the fresh air of spa towns and mountain resorts. Nothing worked. He died of pneumonia, his lungs worn out, in the autumn of 1922 aged just fifty-one.

In Proust's day, asthma was a rare disease and not well understood. Today it is common and still not understood. The second half of the twentieth century saw a rapid increase in asthma rates in most Western nations, and no one knows why. An estimated 300 million people in the world have asthma today, about 5 percent of adults and about 15 percent of children in those countries where it is measured carefully, though the proportions vary markedly from region to region and country to country, even from city to city. In China, the city of Guangzhou is highly polluted, while nearby Hong Kong, just an hour away by train, is comparatively clean as it has little industry and lots of fresh air because it is by the sea. Yet in clean Hong Kong asthma rates are 15 percent, while in heavily polluted Guangzhou they are just 3 percent, exactly the opposite of what one would expect. No one can account for any of this.

Globally, asthma is more common among boys than girls before puberty, but more common in girls than boys after puberty. It is more common in blacks than whites (generally but not everywhere) and in city people than rural people. In children, it is closely associated with both being obese and being underweight; obese children get it more often, but underweight children get it worse. The highest rate in the world is in the U.K., where 30 percent of children have shown asthma symptoms. The lowest rates are in China, Greece, Georgia, Romania, and Russia, with just 3 percent. All the English-speaking nations of the world have high rates, as do those of Latin America.

There is no cure, though in 75 percent of young people asthma resolves itself by the time they reach early adulthood. No one knows how or why that happens either, or why it doesn't happen for the unfortunate minority. Indeed, where asthma is concerned, no one knows much of anything.

Asthma (the word comes from a Greek term meaning "to gasp") has become not only more prevalent but more commonly lethal, and often quite suddenly. Among children who died between 1959 and 1966 in Great Britain, the proportion whose deaths were attributed to asthma leaped from 1 percent to 7.2 percent, and there were similar increases in Ireland, Norway, Australia, and New Zealand. These were linked to side effects of asthma medications that were heavily used in those countries at that time, and the death rate fell when the use of those medications was reduced. However, asthma remains the fourth leading cause of childhood death in Britain. In the United States, between 1980 and 2000 asthma rates doubled, but hospitalization rates tripled, suggesting that asthma is now not only more common but more severe. Similar rises have been found throughout much of the developed world—in Scandinavia, Australia, New Zealand, some of the richer parts of Asia—but not, curiously, everywhere. Japan, for instance, has not seen a great increase in asthma rates.

"You probably think asthma is caused by dust mites or cats or chemicals or cigarette smoke or air pollution," says Neil Pearce, professor of epidemiology and biostatistics at the London School of Hygiene and Tropical Medicine. "I have spent thirty years studying asthma, and the main thing I have achieved is to show that almost none of the things people think cause asthma actually do. They can provoke attacks if you have asthma already, but they don't cause it. We have very little idea what the primary causes are. We can do nothing to prevent it."

Pearce, who is from New Zealand originally, is one of the world's leading authorities on the spread of asthma but came to the field accidentally and quite late. "I had brucellosis"—a bacterial infection that leaves victims feeling as if they have flu permanently—"when I was in my early twenties, and that sidetracked me educationally," he says. "I'm from Wellington, and brucellosis isn't common in cities, so it took the doctors three years to diagnose it. Ironically, once they worked out what it was, it only took a two-week course of antibiotics to cure it." Though he had secured an honors degree in mathematics by then, he had missed his chance to go to medical school, so he gave up on higher education and worked for two years as a bus driver and in a factory.

It was only by chance, while looking for something more interesting to do, that he landed a job as a biostatistician at the Wellington Medical School. From there he became director of the Centre for Public Health Research at Massey University in Wellington. His interest in asthma epidemiology followed an outbreak of unexplained deaths among young asthmatics. Pearce was part of a team that traced the outbreak to an inhaled drug called fenoterol (no relation to the notorious opioid fentanyl). It was the beginning of a lasting association with asthma, though that is just one among many interests today. In 2010, he moved to England to take up a position at the venerable London School of Hygiene and Tropical Medicine in Bloomsbury.

"For a long time," he told me when we met, "the dogma was that asthma was a neurological disease—the nervous system sending the wrong signals to the lungs. Then, in the 1950s and '60s, the idea came along that it is an allergic reaction, and that has pretty much stuck. Even now textbooks say that the way people get asthma is by being exposed to allergens early in life. Basically everything in that theory is wrong. It's clear now that it is considerably more complicated than that. We now know that half the cases in the world involve allergies, but half are due to something else altogether—to nonallergic mechanisms. We don't know what those are."

For many sufferers, asthma can be triggered by cold air, stress, exercise, or other factors that have nothing to do with allergens or what is floating in the air. "More generally," Pearce added, "the dogma is that both allergic and nonallergic asthmas involve inflammation in the lungs, but with some asthmatics if you put their feet in a bucket of ice water, they begin to wheeze immediately. Now, that can't be due to inflammation, because it happens too fast. It has to be neurological. So now we are coming full circle for at least part of the answer."

Asthma is very different from other lung disorders in that it is normally present only some of the time. "If you test the lung function of asthmatics, most of the time for most of them it will be completely normal. It's only when they have an attack that problems with lung function become apparent and detectable. That's very unusual for a disease. Even when there are no symptoms present, the disease will nearly always be evident in blood or sputum tests. In asthma, in some cases, the disease just vanishes."

In an asthma attack, the airways narrow, and the sufferer struggles to get air in or out, especially out. In people with milder forms of asthma, steroids are nearly always effective at keeping attacks under control, but in people with more severe forms steroids rarely work.

"All we can really say about asthma is that it is primarily a Western disease," says Pearce. "There is something about having a Western lifestyle that sets up your immune system in a way that makes you more susceptible. We don't really understand why." One suggestion is the hygiene hypothesis—the idea that early exposures to infectious agents strengthen our resistance to asthma and allergies later in life. "It's a nice theory," says Pearce, "but it doesn't completely fit. There are countries like Brazil where asthma rates are high but so are infection rates."

The peak age for asthma onset is thirteen, but large numbers of people first experience it in adulthood. "Doctors will tell you that the first few years of life are crucial for asthma, but that's not exactly true," says Pearce. "It's the first few years of exposure. If you change jobs or change countries, you can still get asthma even as an adult."

Some years ago, Pearce made a curious discovery—that people who had had a cat early in life seemed to derive lifelong protection from getting asthma. "I like to joke that I've studied asthma for thirty years and I have never prevented a single case, but I have saved the lives of a lot of cats," he says.

In what way exactly Western lifestyles might provoke asthma isn't easy to say. Growing up on a farm seems to protect you, and moving to the city increases your risk, but once again we don't really know why. One intriguing theory, suggested by Thomas Platts-Mills of the University of Virginia, links asthma increases with children spending less time running around outdoors. As Platts-Mills has noted, children used to play outside after school. Now more often than not they go indoors and stay there. "We now have a population that sits around the house and sits still in ways that children have never sat still before," he told the journal *Nature*. Children who sit watching television not only are not exercising their lungs as they would if they were at play but even breathe differently from children who are not transfixed by a screen. Specifically, children who are reading take deeper breaths and sigh more often than children watching TV, and that slight difference in respiratory activity may be enough to increase TV watchers' susceptibility to asthma, according to this theory.

Other researchers have suggested that viruses may be responsible for asthma onset. A study at the University of British Columbia in 2015 suggested that an absence of four gut microbes (namely, *Lachnospira, Veillonella, Faecalibacterium,* and *Rothia*) in infants was closely associated with the development of asthma in the first years of life. But so far all these are just hypotheses. "The bottom line is that we just don't know yet," says Pearce.

III

ONE OTHER ALL-TOO-COMMON affliction of the lungs deserves a mention, not so much because of what it does to us as because of how extraordinarily long it took us to accept that it was doing it. I refer to smoking and lung cancer.

It would seem almost impossible to ignore a link between the two. A person who smokes cigarettes regularly (about a pack a day) is fifty times more likely than a nonsmoker to get cancer. In the thirty years between 1920 and 1950, which is when cigarette smoking took off in a big way in the world, the number of lung cancer cases soared. In America, they tripled. Similar increases were noted elsewhere. Yet it took forever to gain consensus that smoking caused lung cancer.

It seems crazy to us today, but it wasn't so crazy to people back then. The problem was that huge proportions of people smoked—80 percent of all men by the late 1940s—yet only some of them developed lung cancer. And some people who didn't smoke also developed lung cancer. So it was not especially straightforward to see a direct link between smoking and cancer. When lots of people are doing something and only some of them are dying from it, it makes it hard to impute blame to a single cause. Some people blamed air pollution for the rise of lung cancer. Others suspected the increased use of asphalt as a paving surface.

One leading skeptic was Evarts Ambrose Graham (1883–1957), a chest surgeon and professor at Washington University in St. Louis. Graham famously (but facetiously) maintained that we might as plausibly blame lung cancer on the development of nylon stockings because they had become popular at the same time as smoking. But when a student of his, the German-born Ernst Wynder, sought permission to conduct a study on smoking and cancer in the late 1940s, Graham gave his consent, mostly in the expectation that it would disprove the theory of a link between smoking and cancer once and for all. In fact, Wynder demonstrated conclusively that there was a link—so much so that Graham was persuaded by the evidence to change his mind. In 1950, the two men published a joint paper in *The Journal of the American Medical Association* outlining a clear statistical link between smoking and lung cancer. Soon afterward, the *British Medical Journal* ran a study with more or less identical findings by Richard Doll and A. Bradford Hill of the London School of Hygiene and Tropical Medicine.*

Although two of the world's most prestigious medical journals had now demonstrated a clear association between smoking and lung cancer, the findings had almost no effect. People just loved smoking too much to quit. Richard Doll in London and Evarts Graham in

* Hill had already made a signal contribution to medical science. Two years earlier, he had invented the randomized control trial, in a study of the effects of streptomycin.

St. Louis, both lifelong smokers, quit tobacco, but too late in the case of Graham. He died of lung cancer seven years after his own report. Elsewhere smoking just kept rising. The volume of smoking in the United States increased by 20 percent in the 1950s.

Spurred on by the tobacco industry, many commentators mocked the findings. Because Graham and Wynder could hardly train mice to smoke, they developed a machine that would extract tar from smoked cigarettes, which they then daubed on the skin of laboratory mice, causing tumors to erupt there. A writer from *Forbes* magazine wondered acidly (and, it must be said, a touch imbecilically), "How many men distill their tar from their tobacco and paint it on their backs?" Governments took little interest in the question. When Britain's minister of health, Iain Macleod, formally announced at a press conference that there was an unequivocal connection between smoking and lung cancer, he rather undercut his position by smoking conspicuously as he did so.

The Tobacco Industry Research Committee—a scientific panel paid for by cigarette manufacturers—argued that although cancer from tobacco had been induced in laboratory mice, it had never been demonstrated in humans. "No one has established that cigarette smoke, or any one of its known constituents, is cancer-causing to man," wrote the panel's scientific director in 1957, conveniently overlooking that there could never be an ethical way to experimentally induce cancer in a living person.

To further obviate concerns (and to make their products more appealing to women), cigarette manufacturers introduced filters in the early 1950s. Filters had the great effect that they could claim their cigarettes were now much safer. Most manufacturers charged a premium price for filtered cigarettes, even though the cost of filters was less than the tobacco they displaced. Moreover, most filters didn't filter out tars and nicotine any better than the tobacco itself had, and to compensate for a perceived loss of taste, the manufacturers started using stronger tobacco. The upshot was that by the late 1950s the average smoker was taking in more tar and nicotine than he had before filters were invented. By this point, the average American adult was smoking four thousand cigarettes a year. Interestingly, quite a lot of valuable cancer research in the 1950s was done by scientists funded by the cigarette industry who were urgently searching for causes of cancer other than cigarettes. As long as tobacco wasn't directly implicated, their research was often impeccable.

In 1964, the U.S. surgeon general announced an unequivocal link between smoking and lung cancer, but the announcement had little effect. The number of cigarettes smoked by the average American over the age of sixteen fell slightly from 4,340 a year before the announcement to 4,200 afterward, but then climbed back to about 4,500 and stayed there for years. Remarkably, the American Medical Association took fifteen years to endorse the surgeon general's finding. Throughout this period, one of the members of the board of the American Cancer Society was a tobacco magnate. As late as 1973, *Nature* ran an editorial backing women's smoking during pregnancy on the grounds that it calmed their stress.

How things have changed. Today just 18 percent of Americans smoke, and it is easy to think that we have pretty much solved the problem. But it's not quite as simple as that. Nearly one-third of people below the poverty line still smoke, and the habit continues to account for one-fifth of all deaths. It is a problem we are a long way from rectifying.

Finally, let us close on a common breathing affliction that is much less alarming (at least for most of us, most of the time), if no less mysterious: hiccups.

A hiccup is a sudden spasmodic contraction of the diaphragm, which essentially startles the larynx into closing abruptly, making the famous *hic* sound. No one knows why they happen. The world record for hiccups appears to have been held by a farmer in northwest Iowa, named Charles Osborne, who hiccuped continuously for sixty-eight years.* The hiccups began in 1922 when Osborne tried to lift a 350-pound hog for butchering and somehow triggered a hiccup response. At first he hiccuped about 40 times a minute. Eventually, that slowed to 20 times a minute. Altogether he was estimated to have hiccuped 430 million times over nearly seven decades. He never hiccuped when he was asleep. In the summer of 1990, a year before he died, Osborne's hiccups abruptly and mysteriously ceased.

If you do get hiccups and they don't go away spontaneously after a few minutes, medical science is at a more or less complete loss to help you. The best remedies any doctor can suggest are the same ones you've known about since you were small: startling the victims (by sneaking up and going, "Boo!" say), rubbing the back of their neck, having them take a bite of a lemon or a big sip of iced water or pulling on their tongues, and at least a dozen others. Whether any of these age-old remedies actually work is not a matter medical science has addressed. More significantly, no one appears to keep figures on how many people suffer from chronic or sustained hiccuping, but the problem seems not to be trivial. I was told by a surgeon that it happens fairly often after chest surgery–"more often than we like to admit," he added.

* Osborne was from the town of Anthon, Iowa. Although the town had a population of only about six hundred people, it was also the home of the tallest person in the world. Bernard Coyne stood over eight feet tall when he died at the age of twenty-three in 1921, just before Osborne began his hiccuping marathon.

14 FOOD, GLORIOUS FOOD

Tell me what you eat, and I will
tell you what you are.

—JEAN ANTHELME BRILLAT-SAVARIN,
THE PHYSIOLOGY OF TASTE

WE ALL KNOW that if we consume too much beer and cake and pizza and cheeseburgers and all the other things that make life frankly worth living, we will add pounds to our bodies because we have taken in too many calories. But what exactly are these little numerical oddments that are so keen to make us round and wobbly?

The calorie is a strange and complicated measure of food energy. Formally, it's a kilocalorie, and it is defined as the amount of energy required to heat one kilogram of water by one degree centigrade, but it seems safe to say that no one ever thinks of it in those terms when deciding what foods to eat. Just how many calories each of us needs is pretty much a personal matter. Until 1964, the official guidance in the United States was for thirty-two hundred calories per day for a moderately active man and twenty-three hundred for a similarly disposed woman. Today those inputs have been reduced to about twenty-six hundred calories for a moderately active man and two thousand for a moderately active woman. That's a big reduction. Over the course of a year, for a man that would be almost a quarter of a million fewer calories.

It probably won't come as a surprise to hear that in fact the inputs have gone in exactly the other direction. Americans today consume about 25 percent more calories than they did in 1970 (and, let's face it, we weren't exactly going without in 1970).

The father of caloric measurement–indeed of modern food science–was the American academic Wilbur Olin Atwater. A devout and kindly man with a walrus mustache and a stout frame that showed he was no stranger to the larder himself, Atwater was born in 1844 in upstate New York, the son of a traveling Methodist preacher, and studied agricultural chemistry at Wesleyan University in Connecticut. On a study trip to Germany, he was introduced to the exciting new concept of the calorie and returned to America with an evangelical urge to bring scientific rigor to the infant science of nutrition.* Taking a position as professor of chemistry at his alma mater, he embarked on a series of experiments to test every aspect of food science. Some of these experiments were a touch unorthodox, not to say risky. In one, he ate a fish poisoned with ptomaine to see what effect it would have on him. The effect was that it nearly killed him.

Atwater's most celebrated project was the building of a contraption he called a respiratory calorimeter. This was a sealed chamber, not much larger than a large cupboard, in which subjects were confined for up to five days while Atwater and his helpers minutely measured various facets of their metabolism–inputs of food and oxygen, outputs of carbon dioxide, urea, ammonia, feces, and so on–and so calculated caloric intake. The work was so exacting it took up to sixteen people to read all the dials and perform the calculations. Most of the subjects were students, though the lab janitor, Swede Osterberg, was also sometimes drafted in; quite how voluntarily is unknown. Wesleyan's

* There is a surprising lack of consensus on who actually invented the calorie with respect to diet. Some food historians say Nicolas Clément of France came up with the concept as far back as 1819. Others say it was a German, Julius von Mayer, in 1848, and still others credit two Frenchmen working together, P. A. Favre and J. T. Silbermann, in 1852. What is certain is that it was all the rage among European nutritionists by the 1860s, when Atwater first encountered it.

president was mystified by the point of the calorimeter–the calorie was an entirely new concept, after all–and especially appalled at the cost, and ordered Atwater to take a 50 percent pay cut or hire an assistant at his own expense. Atwater chose the latter and, undeterred, worked out the calories and nutritional values of practically all known foods–some four thousand in all. In 1896, he produced his magnum opus, *The Chemical Composition of American Food Materials,* which remained the last word on diet and nutrition for a generation. For a time he was one of the most famous scientists, of any type, in America.

Much of what Atwater concluded was ultimately wrong, but that wasn't really his fault. Nobody yet understood the concept of vitamins and minerals or even the need for a balanced diet. To Atwater and his contemporaries, all that made one food superior to another was how well it served as fuel. So he believed that fruits and vegetables provided comparatively little energy and needed to play no part in the average person's diet. Instead, he suggested that we should eat a lot of meat–two pounds every day, 730 pounds a year. The average American today eats 268 pounds of meat a year, about a third of Atwater's recommended amount, and most authorities say that is still too much.

Atwater's most unsettling discovery–to himself as much as to the world at large–was that alcohol was an especially rich source of calories, and thus an efficient fuel. As the son of a clergyman and a teetotaler himself, he was appalled to report it, but as a diligent scientist he felt his first duty was to the truth, however awkward. In consequence, he was swiftly disowned by his own, devoutly Methodist university and its already scornful president.

Before the controversy could be resolved, fate intervened. In 1904, Atwater suffered a massive stroke. He lingered for three years without recovering his faculties and died aged sixty-three, but his long efforts secured the calorie's place at the heart of nutrition science, evidently for all time.

* * *

As a measure of dietary intake, the calorie has a number of failings. For one thing, it gives no indication of whether a food is actually good for you or not. The concept of "empty" calories was quite unknown in the early twentieth century. Nor does conventional calorie measurement account for how foods are absorbed as they pass through the body. A great many nuts, for instance, are less completely digested than other foods, which means that they leave behind fewer calories than are consumed. You may eat 170 calories' worth of almonds, but keep only 130 of them. The other 40 sluice through without, as it were, touching the sides.

By whatever means you measure it, we are pretty good at extracting energy from food, not because we have an especially dynamic metabolism but because of a trick we learned a very long time ago: cooking. No one knows even approximately when humans first began cooking food. We have good evidence that our ancestors were utilizing fire 300,000 years ago, but Richard Wrangham of Harvard, who has devoted much of his career to studying the matter, believes that our ancestors mastered fire a million and a half years before that–which is to say long before we were properly human.

Cooking confers all kinds of benefits. It kills toxins, improves taste, makes tough substances chewable, greatly broadens the range of what we can eat, and above all vastly boosts the amount of calories humans can derive from what they eat. It is widely believed now that cooked food gave us the energy to grow big brains and the leisure to put them to use.

But in order to cook food, you also need to be able to gather and prepare it efficiently, and that is what Daniel Lieberman of Harvard believes is at the heart of our becoming modern. "You can't possibly have a large brain unless you've got the energy to fuel it," he told me when we met in the autumn of 2018. "And in order to fuel it, you need to master hunting and gathering. That's more challenging than people realize. It's not just a question of picking berries or digging up tubers; it is a matter of processing foods–making them easier to eat and digest, and safer to eat–and that involves toolmaking and communication and cooperation. That is the essence of what drove the shift from primitive to modern humans."

In nature, we actually starve pretty easily. We are incapable of deriving nutrition from most parts of most plants. In particular we cannot make use of cellulose, which is what plants primarily consist of. The few plants that we can eat are the ones we know as vegetables. Otherwise we are limited to eating a few botanical end products, such as seeds and fruits, and even many of those are poisonous to us. But we can benefit from a lot more foods by cooking them. A cooked potato, for instance, is about twenty times more digestible than a raw one.

Cooking frees up a lot of time for us. Other primates spend as many as seven hours a day just chewing. We don't need to eat constantly to ensure our survival. Our tragedy, of course, is that we eat more or less constantly anyway.

The fundamental components of the human diet–the macronutrients: water, carbohydrates, fat, and protein–were recognized nearly two hundred years ago by an English chemist named William Prout, but it was even then clear that some other, more elusive elements were needed to produce a fully healthy diet. No one knew for the longest time exactly what these elements were, but it was evident that in their absence people were likely to suffer a deficiency disease like beriberi or scurvy.

We now know them, of course, as vitamins and minerals. Vitamins are simply organic chemicals–that is, from things that are or were once alive, like plants and animals–while minerals are inorganic and come from soil or water. Altogether there are about forty of these little particles that we must get from our foods because we cannot manufacture them for ourselves.

Vitamins are a surprisingly recent concept. A little over four years after Wilbur Atwater died, a Polish émigré chemist in London, Casimir Funk, came up with the notion of vitamins, though he called them "vitamines," a contraction of "vital" and "amines" (amines being a type of organic compound). As it turned out, only some vitamins are amines, so the name was later shortened. (Other names were also tried, among them nutramines, food hormones, and accessory food factors, but failed to catch on.) Funk didn't discover vitamins but merely speculated, correctly, as to their existence. But because no one could produce these strange elements, many authorities refused to accept their reality. Sir James Barr, president of the British Medical Association, dismissed them as "a figment of the imagination."

The discovery and naming of vitamins didn't begin until almost the 1920s and has been a checkered affair, to put it mildly. In the beginning, vitamins were named in more or less strict alphabetical order–A, B, C, D, and so on–but then the system began to fall apart. Vitamin B was discovered to be not one vitamin but several, and these were renamed B_1, B_2, B_3, and so on up to B_{12}. Then it was decided that the B vitamins weren't so diverse after all, so some were eliminated and others reclassified, so that today we are left with six semi-sequential B vitamins: B_1, B_2, B_3, B_5, B_6, and B_{12}. Other vitamins came and went, so that the scientific literature is filled with a lot of what might be called ghost vitamins–M, P, PP, S, U, and several others. In 1935, a researcher in Copenhagen, Henrik Dam, discovered a vitamin that was central to blood coagulation and called it vitamin K (for the Danish *koagulere*). The next year, some other researchers came up with vitamin P (for "permeability"). The process hasn't entirely settled down yet. Biotin, for instance, was for a time called vitamin H, but then became B_7. Today it is mostly just called biotin.

Although Funk coined the term "vitamines," and is thus often given credit for their discovery, most of the real work of determining the chemical nature of vitamins was done by others, in particular Sir Frederick Hopkins, who was awarded the Nobel Prize for his work in 1929–a fact that left Funk permanently in one.

Even today vitamins are an ill-defined entity. The term describes thirteen chemical oddments that we need to function smoothly but are unable to manufacture for ourselves. Though we tend to think of them as closely related, they mostly have little in common apart from being useful to us. They are sometimes described as "hormones made outside the body," which is a pretty good definition except that it is only partly true. Vitamin D, one of the most vital of all vitamins, can both be made in the body (where it really is a hormone) or be ingested (which makes it a vitamin again).

A good deal of what we know about vitamins and their mineral cousins is surprisingly recent. Choline, for instance, is a micronutrient you have probably never heard of. It has a central role in making neurotransmitters and keeping your brain running smoothly, but that has only been known since 1998. It is abundant in foods that we don't generally eat a lot of–liver, Brussels sprouts, and lima beans, for instance–which doubtless explains why it is thought that some 90 percent of us have at least a moderate choline deficiency.

In the case of many micronutrients, scientists don't know quite how much you need or even what they do for you when you get them. Bromine, for instance, is found throughout the body, but nobody is sure if it is there because the body needs it or is just a kind of accidental passenger. Arsenic is an essential trace element for some animals, but we don't know if that includes humans. Chromium is definitely needed, but in such small amounts that it becomes toxic quite quickly. Chromium levels fall steadily as we age, but no one knows why they fall or what this indicates.

For nearly all vitamins and minerals, the risk of taking in too much is as great as the risk of getting too little. Vitamin A is needed for vision, for healthy skin, and for fighting infection, so it is vital to have it. Luckily, it is abundant in many common foods, like eggs and dairy products, so it's easy to get more than enough. But there's the rub. The recommended daily level is seven hundred micrograms for women and nine hundred for men; the upper limit for both is about three thousand micrograms, and exceeding that regularly can become risky. How many of us could begin to guess even roughly how close we are to getting the balance right? Iron similarly is vital for healthy red blood cells. Too little iron and you become anemic, but too much is toxic, and there are some authorities who believe that quite a number of people may be getting too much of it. Curiously, too much or too little iron both provide the same symptom, lethargy. "Too much iron in the form of supplements can accumulate in our tissues causing our organs literally to rust," Leo Zacharski of Dartmouth-Hitchcock Medical Center in New Hampshire told *New Scientist* in 2014. "It's a far stronger risk factor than smoking for all sorts of clinical disorders," he added.

In 2013, an editorial in the highly respected *Annals of Internal Medicine,* based on a study led by researchers at Johns Hopkins University, said that nearly everyone in high-income countries was sufficiently well nourished not to require vitamins or other health supplements and that we should stop wasting our money on them. The report came in for some swift and withering criticism, however.

Professor Meir Stampfer of the Harvard Medical School said it was regrettable that "such a poorly done paper would be published in a prominent journal." According to the Centers for Disease Control, far from having plenty in our diet, some 90 percent of American adults don't get the recommended daily dose of vitamins D and E and about half don't get sufficient vitamin A. No less than 97 percent, according to the CDC, don't get enough potassium, a vital electrolyte, which is particularly alarming because potassium helps to keep your heart beating smoothly and your blood pressure within tolerable limits. Having said that, there is often disagreement over what precisely we do need. In America, the daily recommended dose of vitamin E is fifteen milligrams, for instance, but in the U.K. it is three to four milligrams—a very considerable difference.

What can be said with some confidence is that many people have a faith in health supplements that goes some way beyond the fully rational. Americans can choose from among a truly staggering eighty-seven thousand different dietary supplements and we spend a no less impressive $40 billion a year on them.

The greatest of vitamin controversies was stirred up by the American chemist Linus Pauling (1901–94), who had the distinction of winning not one but two Nobel Prizes (for chemistry in 1954 and for peace eight years later). Pauling believed that massive doses of vitamin C were effective against colds, flu, and even some cancers. He took up to forty thousand milligrams of vitamin C daily (the recommended daily dose is sixty milligrams) and maintained that his large intake of vitamin C had kept his prostate cancer at bay for twenty years. He had no evidence for any of his claims, and all have been pretty well discredited by subsequent studies. Thanks to Pauling, to this day many people believe that taking a lot of vitamin C will help to get rid of a cold. It won't.

Of all the many things we take in with our foods (salts, water, minerals, and so on), just three need to be altered as they proceed through the digestive tract: proteins, carbohydrates, and fats. Let's look at them in turn.

PROTEINS

PROTEINS ARE COMPLICATED molecules. About a fifth of our body weight is made up of them. In simplest terms, a protein is a chain of amino acids. About a million different proteins have been identified so far, and nobody knows how many more are to be found. They are all made from just twenty amino acids, even though hundreds of amino acids exist in nature that could do the job just as well. Why evolution has wedded us to such a small number of amino acids is one of the great mysteries of biology. For all their importance, proteins are surprisingly ill-defined. Although all proteins are made from amino acids, there is no accepted definition as to how many amino acids you need in a chain to qualify as a protein. All that can be said is that a small but unspecified number of amino acids strung together is a peptide. Ten or twelve strung together is a polypeptide. When a polypeptide begins to get bigger than that, it becomes, at some ineffable point, a protein.

It is a slightly strange fact that we break down all the proteins we consume in order to reassemble them into new proteins, rather as if they were Lego toys. Eight of the twenty amino acids cannot be made in the body and must be consumed in the diet.* If they are missing from the foods we eat, then certain vital proteins cannot be made. Protein deficiency is almost never a problem for people who eat meat, but it can be for vegetarians because not all plants provide all the necessary amino acids. It is interesting that most traditional diets in the world are based around combinations of plant products that do provide all the necessary amino acids. So people in Asia eat a lot of rice and soybeans, while indigenous Americans have long combined corn with black or pinto beans. This isn't just a matter of taste, it seems, but an instinctive recognition of the need for a rounded diet.

CARBOHYDRATES

CARBOHYDRATES ARE COMPOUNDS of carbon, hydrogen, and oxygen, which are bound together to form a variety of sugars—glucose, galactose, fructose, maltose, sucrose, deoxyribose (the stuff found in DNA), and so on. Some of these are chemically complex and known as polysaccharides, some are simple and known as monosaccharides, and some are in between and known as disaccharides. Although all are sugars, not all are sweet. Some, like the starches found in pasta and potatoes, are too big to activate the tongue's sweet detectors. Virtually all carbohydrates in the diet come from plants, with one conspicuous exception: lactose, from milk.

We eat a lot of carbohydrates, but we use them up quickly, so the total amount in your body at any given time is modest—usually less than a pound. The main thing to bear in mind is that carbohydrates, upon being digested, are just more sugar—often quite a lot more.

* The eight are isoleucine, leucine, lysine, methionine, phenylalanine, tryptophan, threonine, and valine. The bacterium *E. coli* is unusual among living things in its ability to utilize a twenty-first amino acid, called selenocysteine.

That means that a 150-gram serving of white rice or a small bowl of cornflakes will have the same effect on your blood glucose levels as nine teaspoons of sugar.

FATS

THE THIRD MEMBER of the trio, fats, are also made up of carbon, hydrogen, and oxygen, but in different proportions. This has the effect of making fat easier to store. When fats are broken down in the body, they are teamed up with cholesterol and proteins in a new molecule called lipoproteins, which travel through the body via the bloodstream. Lipoproteins come in two principal types: high density and low density. Low-density lipoproteins are the ones frequently referred to as "bad cholesterol" because they tend to form plaque deposits on the walls of blood vessels. Cholesterol is not as fundamentally evil as we tend to think it. Indeed, it is vital to a healthy life. Most of the cholesterol in your body is locked up in your cells, where it is doing useful work. Just a small part–about 7 percent–floats about in the bloodstream. Of that 7 percent, one-third is "good" cholesterol and two-thirds is "bad."

So the trick with cholesterol is not to eliminate it but to maintain it at a healthy level. One way to do so is to eat a lot of fiber, or roughage. Fiber is the material in fruits, vegetables, and other plant foods that the body cannot fully break down. It contains no calories and no vitamins, but it helps to lower cholesterol and slows the rate at which sugar gets into the bloodstream and is then turned into fat by the liver, among many other benefits.

Carbohydrates and fats are the principal fuel reserves of the body, but they are stored and used in different ways. When the body needs fuel, it tends to burn up the available carbohydrates and store any spare fat. The main point to bear in mind–and you are no doubt well aware of it each time you take your shirt off–is that the human body likes to hold on to its fat. It burns some of the fat we consume for energy, but a good deal of the rest is sent off to tens of billions of tiny storage terminals called adipocytes, which exist all over the body. The upshot of all this is that the human body is designed to take in fuel, use what it needs, and store the rest to call on later as required. That makes it possible for us to be active for hours at a time without eating. Your body below the neck doesn't do a lot of complicated thinking, and it is only too happy to hold on to any surplus fat you give it. It even rewards you for overeating with a lovely feeling of well-being.

Depending on where the fat ends up, it is known as subcutaneous (beneath the skin) or visceral (around the belly). For complex chemical reasons, visceral fat is much worse for you than the subcutaneous kind.

Fat comes in several varieties. "Saturated fat" sounds greasy and unhealthy, but in fact it is a technical description of carbon-hydrogen bonds rather than how much of it runs down your chin when you bite into it. As a rule, animal fats tend to be saturated and vegetable fats to be unsaturated, but there are many exceptions, and you can't tell by looking whether a food is high in saturated fat or not. Who would guess, for instance, that an avocado has five times as much saturated fat as a small bag of potato chips? Or that a large latte has more than almost any pastry? Or that coconut oil is almost nothing but saturated fat?

Even more invidious are trans fats, an artificial form of fat made from vegetable oils. Invented by a German chemist in 1902, they were long thought of as a healthy alternative to butter or animal fat, but we now know the opposite to be true. Also known as hydrogenated oils, trans fats are much worse for your heart than any other kind of fat. They raise levels of bad cholesterol, lower levels of good cholesterol, and damage the liver. As Daniel Lieberman has rather chillingly put it, "Trans fats are essentially a form of slow-acting poison."

As early as the mid-1950s, Fred A. Kummerow, a biochemist at the University of Illinois, reported clear evidence of a link between high intake of trans fats and clogged coronary arteries, but his findings were widely dismissed, particularly with the influence of lobbying by the food processing industry. Not until 2004 did the American Heart Association finally accept that Kummerow was right, and not until

2015—almost sixty years after Kummerow first reported the dangers—did the Food and Drug Administration finally decree trans fats unsafe to eat. Despite their known dangers, it remained legal to add them to foods in America until July 2018.

Finally, we should say a word or two about the most vital of our macronutrients: water. We consume about two and a half quarts of water a day, though we are not generally aware of it because about half is contained within our foods. The conviction that we should all drink eight glasses of water a day is the most enduring of dietary misunderstandings. The idea has been traced to a 1945 paper from the U.S. Food and Nutrition Board, which noted that that was the amount that the average person consumed in a day. "What happened," Dr. Stanley Goldfarb of the University of Pennsylvania told the BBC radio program *More or Less* in 2017, "was that people sort of confused the idea that this was the required intake. And the other confusion that occurred was then people said that it is not so much that you should take in eight ounces eight times a day, but that you should consume that in addition to whatever fluid you consume in association with your diet and your meals. And there was never any evidence for that."

One other enduring myth concerning water intake is the belief that caffeinated drinks are diuretics and make you pee out more than you have taken in. They may not be the most wholesome of options for liquid refreshment, but they do make a net contribution to your personal water balance. Thirst, curiously, is not a reliable indication of how much water you need. People allowed to drink all the water they want after getting very thirsty usually report feeling slaked after drinking only one-fifth the amount they have lost through perspiration.

Drinking too much water can actually be dangerous. Normally, your body manages fluid balance very well, but occasionally people take in so much water that the kidneys cannot get rid of it fast enough and they end up dangerously diluting the sodium levels in their blood, setting off a condition known as hyponatremia. In 2007, a young woman in California named Jennifer Strange died after drinking six quarts of water in three hours in a clearly ill-judged water-drinking competition held by a local radio station. Similarly in 2014, a high school football player in Georgia, complaining of cramps after practice, downed two gallons of water and two of Gatorade and soon afterward fell into a coma and died.

Over a lifetime, we eat about sixty tons of food, which is equivalent, notes Carl Zimmer in *Microcosm,* to eating sixty small cars. In 1915, the average American spent half his weekly income on food. Today it's just 6 percent. We live in a paradoxical situation. For centuries, people ate unhealthily out of economic necessity. Now we do it out of choice. We are in the historically extraordinary position that far more people on Earth suffer from obesity than from hunger. To be fair, it doesn't take much to put on weight. One chocolate chip cookie a week, in the absence of any offsetting extra exercise, will translate into about two pounds of extra weight a year.

It took a surprisingly long time to realize that a lot of the things we eat can make you seriously unhealthy. The person most responsible for our enlightenment was a nutritionist from the University of Minnesota named Ancel Keys.

Keys was born in 1904 into a moderately distinguished family in California (his uncle was the movie star Lon Chaney, to whom he bore a striking resemblance). He was a bright but undermotivated child. Professor Lewis Terman of Stanford, who studied intelligence in youngsters (he was responsible for putting "Stanford" in the Stanford-Binet IQ test), declared the young Keys a potential genius, but Keys chose not to fulfill his potential. Instead, he dropped out of school at fifteen and worked at a variety of exotic jobs, from sailor in the merchant navy to a shoveler of bat guano in Arizona. Only then did he belatedly embark on an academic career, but he made up for lost time in a big way, rapidly acquiring degrees in biology and economics from the University of California at Berkeley, a PhD in oceanography from the Scripps Institution in La Jolla, California, and a second PhD, in physiology, from Cambridge University in England. After settling briefly at Harvard, where he became a world authority on high altitude physiology, he was lured to the University of Minnesota to become the founding director of its Laboratory of Physiological Hygiene. There he rapidly became an expert on human nutrition. When America joined the Second World War, the War Department commissioned Keys to devise a lightweight food pack for paratroopers. The result was the imperishable army food known as K rations. The K stood for Keys.

In 1944, as much of Europe faced the prospect of starvation because of the disruptions and privations of war, Keys embarked on what became known as the Minnesota Starvation Experiment. He recruited thirty-six healthy male volunteers–all conscientious objectors–and for six months allowed them just two meager meals a day (one on Sundays) for a total daily intake of about 1,500 calories. Over the six months, the men's average weight dropped from 152 pounds to 115. The idea of the experiment was to establish how well people could cope with the experience of chronic hunger and how well they would recover afterward. Essentially, it just confirmed what anyone could have guessed at the outset–that chronic hunger made the volunteers irritable, lethargic, and depressed, and left them more susceptible to illness. On the plus side, when their normal diet was resumed, they quickly recovered their lost weight and missing vitality. On the basis of the study, Keys produced a two-volume work, *The Biology of Human Starvation,* which was highly regarded, though not particularly timely. By the time it came out, in 1950, nearly everyone in Europe was well fed again and starvation was not an issue.

Soon afterward, Keys embarked on the study that would permanently seal his fame. The Seven Countries Study compared the dietary habits and health outcomes of 12,000 men in seven nations: Italy, Greece, the Netherlands, Yugoslavia, Finland, Japan, and the United States. Keys found a direct correlation between levels of dietary fat and heart disease–a conclusion that is hardly surprising now but was revolutionary then. In 1957, with his wife, Margaret, Keys produced a popular book called *Eat Well and Stay Well,* which promoted what we now know as the Mediterranean diet. The book infuriated the dairy and meat industries, but it made Keys rich and universally famous, and it marked a milestone in the history of dietary science. Before Keys, nutritional studies had been directed almost entirely at combating deficiency diseases. Now, people realized that too much nutrition could be as dangerous as too little.

Keys's findings have come in for some sharp criticism over the years. One commonly heard complaint is that Keys focused on countries that supported his thesis and ignored those that did not. The French, for example, eat more cheese and drink more wine than almost anybody else on Earth and yet have some of the lowest rates of heart disease. This "French paradox," as it is known, led Keys to exclude France from the study because it didn't fit with his findings, critics have claimed. "When Keys didn't like data," says Lieberman, "he just eliminated them. By today's standards he would have been accused and fired for scientific misconduct."

Keys's defenders have argued, however, that the French dietary anomaly wasn't widely noted outside France until 1981, so Keys wouldn't have known to include it. Whatever else anyone concludes, Keys surely deserves credit for drawing attention to the role of diet in maintaining heart health. And it must be said it did him no harm. Keys devoted himself to a Mediterranean-style diet long before anyone had heard of the term and lived to be a hundred. (He died in 2004.)

Keys's findings have had a lasting effect on dietary recommendations. The official guidance in most countries is that fats should account for no more than 30 percent of a person's daily diet, and saturated fats no more than 10 percent. The American Heart Association puts it even lower at 7 percent.

Now, however, we are not quite so sure how solid that advice is. In 2010, two large studies (in *The American Journal of Clinical Nutrition* and the *Annals of Internal Medicine*) involving almost a million people in eighteen countries concluded that there was no clear evidence that avoiding saturated fat reduced the risk of heart disease. A similar and more recent study in the British medical journal *The Lancet* in 2017 found that fat was "not associated with cardiovascular diseases, myocardial infarction, or cardiovascular disease mortality" and that dietary guidelines consequently needed to be readdressed. Both conclusions have been heatedly disputed by some academics.

The problem with all dietary studies is that people eat foods that have oils, fats, good and bad cholesterol, sugars, salts, and chemicals of every description all mixed together in ways that make it impossible to attribute any particular outcome to any one input, and that is not to mention all the other factors that affect health: exercise, drinking habits, where you carry fat on your body, genetics, and much more. According to another, oft-quoted study, a forty-year-old man who eats a hamburger every day will knock a year off his life expectancy. The trouble is that people who eat a lot of hamburgers also tend to do things like smoke, drink, and fail to get adequate exercise that are just as likely to cause an early checkout. Eating a lot of hamburgers is not good for you, but it doesn't come with a timeline.

These days the most frequently cited culprit for dietary concern is sugar. It has been linked to a lot of horrible diseases, notably diabetes, and there is no question that most of us take in way more sugar than we need. The average American consumes twenty-two teaspoons of added sugar a day. For young American men, it's closer to forty. The World Health Organization recommends a maximum of five.

It doesn't take much to go over the limit. A single standard-sized can of soda pop contains about 50 percent more sugar than the daily recommended maximum for an adult. One-fifth of all young people in America consume five hundred calories or more a day from soft drinks, which is all the more arresting when you realize that sugar isn't actually very high in calories—just sixteen per teaspoon. You have to take in a lot of sugar to get a lot of calories. The problem is that we do take in a lot, more or less all the time, often when we don't even know it. For one thing, nearly all processed foods include added sugar.

By one estimate, about half the sugar we consume is lurking in foods where we are not even aware of it—in breads, salad dressings, spaghetti sauces, ketchup, and other processed foods that don't normally strike us as sugary. Altogether about 80 percent of the processed foods we eat contain added sugars. Heinz ketchup is almost one-quarter sugar. It has more sugar per unit of volume than Coca-Cola.

Complicating matters is that there is also a lot of sugar in the good stuff we eat. Your liver doesn't know whether the sugar you consume comes from an apple or a candy bar. A sixteen-ounce bottle of Pepsi has about thirteen teaspoons of added sugar and no nutritive value at all. Three apples would give you just as much sugar but compensate by also giving you vitamins, minerals, and fiber, not to mention a greater feeling of satiation. That said, even the apples are a lot sweeter than they really need to be. As Lieberman has noted, modern fruits have been selectively bred to be vastly more sugary than they once were. The fruits that Shakespeare ate were, for the most part, probably no sweeter than the modern carrot.

Many of our fruits and vegetables are nutritionally less good for us than they were even in the fairly recent past. Donald Davis, a biochemist at the University of Texas, in 2011 compared the nutritive values of various foods in 1950 with those of our own era and found substantial drops in almost every type. Modern fruits, for instance, are almost 50 percent poorer in iron than they were in the early 1950s, and about 12 percent down in calcium and 15 percent in vitamin A. Modern agricultural practices, it turns out, focus on high yields and rapid growth at the expense of quality.

In the United States, we are left in the bizarre and paradoxical situation that we are essentially the world's most overfed nation but also one of its most nutritionally deficient ones. Comparisons with the past are a bit difficult to make because in 1970 Congress canceled the only comprehensive federal nutrition survey ever attempted after the preliminary results proved embarrassing. "A significant proportion of the population surveyed is malnourished or at a high risk of developing nutritional problems," the survey reported, just before it was axed.

It is hard to know what to make of any of this. According to the *Statistical Abstract of the United States*, the amount of vegetables eaten by the average American between 2000 and 2010 dropped by thirty pounds. That seems an alarming decline until you realize that the most popular vegetable in America by a very wide margin is the French fry. (It accounts for a quarter of our entire vegetable intake.) These days, eating thirty pounds less "vegetables" may well be a sign of an improved diet.

A striking marker of just how confused nutrition advice can be was a finding by an advisory committee for the American Heart Association that 37 percent of American nutritionists rate coconut oil—which is essentially nothing but saturated fat in liquid form—as a "healthy food." Coconut oil may be tasty, but it is no better for you than a big scoop of deep-fried butter.

"It is," says Lieberman, "a reflection of how deficient dietary education can be. People just aren't always getting the facts. It's possible for doctors to go through medical school without being taught nutrition. That's crazy."

Perhaps nothing is more emblematic of the unsettled state of knowledge on the modern diet than the long and unresolved controversy over salt. Salt is vital to us. There is no question of that. We would die without it. That's why we have taste buds devoted exclusively to it. Lack of salt is nearly as dangerous to us as lack of water. Because our bodies cannot produce salt, we must consume it in our diets. The problem is in determining how much is the right amount. Take too little and you grow lethargic and weak, and eventually you die. Take too much and your blood pressure soars and you run the risk of heart failure and stroke.

The average American consumes about 3,400 milligrams of sodium a day. It is very difficult not to. A lightish lunch of soup and a sandwich, none of it conspicuously salty, can easily push you over the limit. Many authorities believe that 3,400 milligrams is way too much and that it vastly increases the risk of heart attack and stroke. The World Health Organization suggests that we consume no more than 2,000 milligrams of sodium a day. But other authorities say that reducing sodium intake to that level has no proven health benefit and may actually be harmful.

One study in Britain estimated that as many as 30,000 people a year died in the U.K. from consuming too much salt over too long a period, but another study concluded that salt did no harm to anyone except for those with elevated blood pressure, and yet another concluded that people who ate a lot of salt actually lived longer. A meta-analysis at McMaster University in Canada of 133,000 people in four dozen countries found a link between high salt intake and heart problems only for those with existing hypertension, while low salt intake (less than three thousand milligrams a day) had an increased risk of heart problems for people from both groups. In other words, according to the McMaster study, too little salt is at least as risky as too much.

A central reason for the lack of consensus, it turns out, is that both sides indulge in what statisticians call confirmation bias. Simply put, they don't listen to each other. A 2016 study in the *International Journal of Epidemiology* found that researchers on both sides of the argument overwhelmingly cite papers that support their own views and ignore or dismiss those that do not. "We found that the published literature bears little imprint of an ongoing controversy, but rather contains two almost distinct and disparate lines of scholarship," the study's authors wrote.

To try to find an answer, I met Christopher Gardner, director of nutrition studies and professor of medicine at Stanford University in Palo Alto, California. He is a friendly man, with a ready laugh and relaxed manner. Though nearing sixty, he looks at least fifteen years younger. (This seems to be true of most people in Palo Alto.) We met at a restaurant in a neighborhood shopping center. He arrived, almost inevitably, on a bicycle.

Gardner is a vegetarian. I asked him if that was for health or ethical reasons. "Well, actually originally it was to impress a girl," he said, grinning. "That was in the 1980s. But then I decided I liked it." In fact, he liked it so well he decided to start a vegetarian restaurant but felt he needed to understand the science better, so he did a PhD in nutrition science and got sidetracked into academia. He is refreshingly reasonable about what we should and shouldn't eat.

"In principle, it's really pretty simple," he says. "We should eat less added sugar, less refined grain, and more vegetables. It's essentially a question of trying to eat mostly good things and avoiding mostly bad things. You don't need a PhD for that."

In practice, however, things are not so straightforward. We are all habituated, at an almost subliminal level, to go for the bad stuff. Gardner's students demonstrated that with a beautifully simple experiment in one of the university's cafeterias. Each day they gave the cooked carrots a different label. The carrots were always the same and the labels always truthful, but they just emphasized a different quality each day. So one day the carrots were labeled as plain carrots, then the next day as low-sodium carrots, then as high-fiber carrots, and finally as twisted glaze carrots. "The students took 25 percent more of the sugary-sounding glazed carrots," Gardner says with another broad smile. "These are smart kids. They are aware of all the issues about weight and health and all that, and yet they took the bad option anyway. It's a reflex. We had the same results with asparagus and broccoli. It's not easy to overcome the dictates of your subconscious."

It is a frailty food manufacturers are very good at manipulating, Gardner says. "Lots of food products are advertised as low in salt, fat, or sugar, but nearly always when manufacturers reduce one of the three, they boost the other two to compensate. Or they put some omega 3 in a brownie, and emphasize that in large letters on the packaging as if it is a health product. But it's still a brownie! The problem for society is that we eat a lot of crappy foods. Even food banks mostly give out processed foods. We just have to change people's habits."

Gardner thinks that's happening, albeit slowly. "I'm really confident that the ground is moving," he says. "But you don't change habits overnight."

As we part, he adds an afterthought. "There's a really easy way to do food shopping in supermarkets," he says. "Just stick to the outside aisles. The aisles in between are almost entirely filled with processed foods. If you stick to the outside, you will automatically have a healthier diet."

It is easy to make risk sound scary. It is often written that eating a daily helping of processed meat increases your risk of colorectal cancer by 18 percent, which is doubtless true. But as Julia Belluz of *Vox* has pointed out, "A person's lifetime risk of colorectal cancer is about 5 percent, and eating processed meat every day appears to boost a person's absolute risk of cancer by 1 percentage point, to 6 percent (that's 18 percent of the 5 percent lifetime risk)." So, put another way, if a hundred people eat a hot dog or bacon sandwich every day, over the course of a lifetime one of them will get colorectal cancer (in addition to the five who would have gotten it anyway). That's not a risk you may want to take, but it's not a death sentence.

It is important to distinguish between probability and destiny. Just because you are obese or a smoker or couch potato doesn't mean you are doomed to die before your time, or that if you follow an ascetic regime you will avoid peril. Roughly 40 percent of people with diabetes, chronic hypertension, or cardiovascular disease were fit as a fiddle before they got ill, and roughly 20 percent of people who are severely overweight live to a ripe old age without ever doing anything about it. Just because you exercise regularly and eat a lot of salad doesn't mean you have bought yourself a better life span. What you have bought is a better chance of having a better life span.

So many variables have been implicated in heart health—exercise and lifestyle, consumption of salt, alcohol, sugar, cholesterol, trans fats, saturated fats, unsaturated fats, and so on—that it is almost certainly a mistake to pin the blame decisively on any one component. A heart attack, as one doctor has put it, is "50 percent genetic and 50 percent cheeseburger." That exaggerates matters, of course, but the underlying point is valid.

The most prudent option, it seems, is to have a balanced and moderate diet. A sensible approach is, in short, the sensible approach.

15 THE GUTS

> Happiness is a good bank account, a good cook and a good digestion.
>
> —JEAN-JACQUES ROUSSEAU

INSIDE, YOU ARE enormous. Your alimentary canal is about forty feet long if you are an average-sized man, a bit less if you are a woman. The surface area of all that tubing is about half an acre.

Bowel transit time, as it is known in the trade, is a very personal thing and varies widely between individuals, and in fact within individuals depending on how active they are on a given day and what and how much they have been eating. Men and women evince a surprising amount of difference in this regard. For a man, the average journey time from mouth to anus is fifty-five hours. For a woman, typically, it is more like seventy-two. Food lingers inside a woman for nearly a full day longer, with what consequences, if any, we do not know.

Roughly speaking, however, each meal you eat spends about four to six hours in the stomach, a further six to eight hours in the small intestine, where all that is nutritious (or fattening) is stripped away and dispatched to the rest of the body to be used or, alas, stored, and up to three days in the colon, which is essentially a large fermentation tank where billions and billions of bacteria pick over whatever the rest of the intestines couldn't manage—fiber mostly. That's why you are constantly told to eat more fiber: because it keeps your gut microbes happy and at the same time, for reasons not well understood, reduces the risk of heart disease, diabetes, bowel cancer, and indeed death of all types.

Nearly everyone equates the location of the stomach with the belly, but in fact it is much higher up and markedly off center to the left. It is about ten inches long and shaped like a boxing glove. The wrist end, where the food enters, is called the pylorus, and the fist part is the fundus. The stomach is less vital than you might think. We give it way too much credit in popular consciousness. It contributes a bit to digestion both chemically and physically, by squeezing its contents with muscular contractions and bathing them in hydrochloric acid, but its contribution to digestion is helpful rather than vital. Many people have had their stomachs removed without serious consequence. The real digestion and absorption—the feeding of the body—happens further down.

The stomach holds about one and a half quarts, which is not very much compared with other animals. The stomach of a big dog will hold up to twice as much food as yours does. When food reaches the consistency of pea soup, it is known as chyme (pronounced "kime"). The rumblings of your gut, incidentally, come mostly from the large intestine, not the stomach. The technical term for gut rumblings is "borborygmi."

One thing the stomach does do is kill off many microbes, by soaking them in hydrochloric acid. "Without your stomach, a lot more of what you ate would make you ill," Katie Rollins, a general surgeon and lecturer at the University of Nottingham, told me one day in the dissecting room there.

It is a miracle that any microbes get through, but some do, as we all know to our cost. Part of the problem is that we bombard ourselves with a lot of tainted stuff. An investigation by the Food and Drug Administration in 2016 found that 84 percent of chicken breasts, nearly 70 percent of ground beef, and getting on for half of pork chops contained intestinal *E. coli,* which is not good news for anything but the *coli.**

Foodborne illness is America's secret epidemic. Every year three thousand people, the equivalent of a small town, die of food poisoning in the United States, and around 130,000 are hospitalized. It can be a decidedly horrible way to die. In December 1992, Lauren Beth Rudolph had a cheeseburger at a Jack in the Box restaurant in Carlsbad, California. Five days later, she was taken to the hospital suffering excruciating abdominal cramps and bloody diarrhea, and her condition was rapidly deteriorating. In the hospital, she suffered three massive cardiac arrests and died. She was six years old.

Over the next few weeks, seven hundred customers who had visited seventy-three Jack in the Box restaurants in four states grew ill. Three of them died. Others suffered permanent organ failure. The source was *E. coli* in undercooked meat. According to *Food Safety News,* the Jack in the Box company knew that its hamburgers were being undercooked "but had decided that cooking them to the required 155 degrees made them too tough."

Equally pernicious is salmonella, which has been called "the most ubiquitous pathogen in nature." According to a USDA study, about a quarter of all chicken pieces sold in stores are contaminated with salmonella. Some 40,000 cases of salmonella infection are reported in the United States each year, but the real number is thought to be much higher. By one estimate, for every reported case a further 28 go unreported. That works out to 1,120,000 cases a year. There is no treatment for salmonella poisoning.

* *E. coli* is a strange organism in that most strains do us no harm and some are positively beneficial–so long as they don't end up in the wrong place. *E. coli* in your colon, for instance, produces vitamin K for you–and most welcome that is. We are talking here about strains of *E. coli* that hurt you or end up where they shouldn't be.

Salmonella has nothing to do with spawning fish. It is named for Daniel Elmer Salmon, a U.S. Department of Agriculture scientist, though it was actually discovered by his assistant Theobald Smith, yet another of those forgotten heroes of medical history. Smith, born in 1859, was the son of German immigrants (the family name was Schmitt) in upstate New York and grew up speaking German, so was able to follow and appreciate the experiments of Robert Koch more quickly than most of his American contemporaries. He taught himself Koch's methods for culturing bacteria and was thus able to isolate salmonella in 1885, long before any other American could do so. Daniel Salmon was head of the Bureau of Animal Husbandry at the U.S. Department of Agriculture and was primarily an administrator, but the convention of the day was to list the bureau head as lead author on the department's papers, and that was the name that got attached to the microbe. Smith was also robbed of credit for the discovery of the infectious protozoa *Babesia,* which is wrongly named for a Romanian bacteriologist, Victor Babeş. In a long and distinguished career, Smith also did important work on yellow fever, diphtheria, African sleeping sickness, and fecal contamination of drinking water, and showed that tuberculosis in humans and in livestock was caused by different microorganisms, proving Koch wrong on two vital points. Koch also believed that TB could not jump from animals to humans, and Smith showed that that was wrong, too. It was thanks to this discovery that pasteurization of milk became a standard practice. Smith was, in short, the most important American bacteriologist during what was the golden age of bacteriology and yet is almost completely forgotten now.

Responsibility for food safety is split among a raft of federal agencies in America in a way that rather defies logic. The Food and Drug Administration is responsible for the skin of sausages, but the Food Safety and Inspection Service is responsible for what goes inside them. Cheese pizzas are looked after by one agency, but meat pizzas by another. And so it goes through a whole range of foodstuffs. Altogether fifteen agencies have a regulatory role in some aspect or other of American food safety. No one agency has overall control.

Incidentally, most nausea-inducing microbes need time to proliferate inside you before they make you sick. A few, like *Staphylococcus aureus,* can make you ill in as little as an hour, but most take at least twenty-four hours. As Dr. Deborah Fisher of Duke University told *The New York Times,* "People tend to blame the last thing they ate, but it's probably the thing before the last thing they ate." Actually, many infestations take far longer than that to manifest. Listeriosis, which kills about three hundred people a year in America, can take up to seventy days to show symptoms, which makes tracking down a source of infection a nightmare. In 2011, thirty-three people died of listeriosis before the source–cantaloupe from Colorado–was identified.

The largest source of foodborne illness is not meat or eggs or mayonnaise, as commonly supposed, but green leafy vegetables. They account for one in five of all food illnesses.

For a very long time, nearly everything we knew about the stomach was thanks to an unfortunate accident in 1822. In the summer of that year, on Mackinac Island in Lake Huron in upper Michigan, a customer was examining a rifle in the island general store when it suddenly went off. A young Canadian fur trapper named Alexis St. Martin had the misfortune to be standing just three feet away and directly in the line of fire.* The shot tore a hole in his chest just below the left breast and gave him something he really didn't want: the most famous stomach in medical history. St. Martin miraculously survived, but the wound never entirely healed. His doctor, a U.S. Army surgeon named William Beaumont, realized that the inch-wide hole gave him an unusual window into the trapper's interior and direct access to his

* St. Martin lived for a time in Cavendish, Vermont, site of the accident that drove an iron bar through the skull of another hapless laborer, Phineas Gage, and also the birthplace of Nettie Stevens, discoverer of the Y chromosome. None of the three were in Cavendish at the same time, however.

stomach. He took St. Martin into his home and looked after him, but with the understanding (sealed with a formal contract) that Beaumont would be free to perform experiments on his guest. For Beaumont, this was a peerless opportunity. In 1822, no one knew quite what happened to food once it disappeared down one's throat. St. Martin had the only stomach on Earth that could be studied directly.

Beaumont's experiments principally consisted in suspending different foods on lengths of silken thread into St. Martin's stomach, leaving them for a measured interval, then pulling them out to see what had happened. Sometimes, in the interests of science, he tasted the contents to judge their tartness and acidity, and by so doing deduced that the principal digestive agent of the stomach is hydrochloric acid. This was a breakthrough that caused great excitement in gastric circles and made Beaumont famous.

St. Martin was not the most cooperative of subjects. Often he disappeared, once for four years before Beaumont was able to track him down. Despite these interruptions, Beaumont eventually published a landmark book, *Experiments and Observations on the Gastric Juice and the Physiology of Digestion*. For about a century, almost all medical knowledge of the process of digestion was thanks to St. Martin's stomach.

Ironically, St. Martin outlived Beaumont by twenty-seven years. After drifting around for a few years, he returned to his hometown of St. Thomas, Quebec, married, raised a family of six children, and died aged eighty-six in 1880, nearly sixty years after the accident that made him famous.

The heart of the digestive tract is the small intestine, twenty-five feet or so of coiled tubing where most of the body's digestion takes place. The small intestine is traditionally divided into three sections: the duodenum (meaning "twelve," because that is how many finger widths of space it was deemed to take up in the average man in ancient Rome);

the jejunum (meaning "without food" because the jejunum was often found to be empty in corpses); and the ileum (meaning "groin" on account of its proximity to same). In fact, however, the divisions are entirely notional. If you took your intestines out and laid them on the ground, you wouldn't be able to tell where one part began and the other ended.

The small intestine is lined with tiny hairlike projections called villi, which add enormously to its surface area. Food is passed along by a process of contraction known as peristalsis—a kind of stadium wave for the gut. It advances at the rate of about one inch per minute. A natural question is, Why don't all our ferocious digestive juices eat through our own gut lining? The answer is that the alimentary canal is lined with a single layer of protective cells called the epithelium. These vigilant cells, and the gooey mucus they produce, are all that stand between you and digesting your own flesh. If that tissue is breached and the gut contents get into another part of the body, death without immediate medical treatment inevitably follows. Yet that only rarely happens. So battered is this front line of cells that each is replaced after just three or four days, just about the highest turnover rate for the whole body.

Wrapped around the outside of it, like a wall around a garden, is six feet of broader-gauge plumbing known variously as the large intestine, bowel, or colon. Where the small and large intestines join (just above the belt line on the right-hand side of your body), there is a pouch called the cecum, which is important in herbivores but of no particular consequence in humans, and jutting off the cecum is the fingerlike protrusion known as the appendix, which has no certain purpose but kills about 80,000 people around the world every year when it ruptures or grows infected. In the United States, nearly 400,000 people are hospitalized with appendicitis annually and about 300 die, according to the U.S. National Library of Medicine.

The appendix is strictly the vermiform appendix, in recognition of its wormlike shape. For a long time, all that could be said about the appendix was that you could remove it and not miss it, which strongly suggested that it had no purpose at all. Now the best thinking is that the appendix serves as a reservoir for gut bacteria. About one person in every sixteen in the developed world will suffer appendicitis at some point, enough to make it the most common cause of emergency surgery. Without surgery, many appendicitis victims would die. Once it was quite a common cause of death. The incidence of acute appendicitis in the rich world is about half today what it was in the 1970s, and no one is quite sure why. It remains more common in wealthy countries than in developing ones, though the rates in developing countries have been rising rapidly, presumably because of changing dietary habits, but again no one knows for sure.

The most extraordinary story of appendectomy survival that I know of occurred aboard the U.S. submarine *Seadragon* in Japanese-controlled waters in the South China Sea during World War II when a sailor named Dean Rector from Kansas developed an acute and obvious case of appendicitis. With no qualified medical personnel on board, the commander ordered the ship's pharmacist's assistant, one Wheeler Bryson Lipes (of no known relation to the present author), to perform the surgery. Lipes protested that he had no medical training, did not know what an appendix looked like or where it was to be found, and had no surgical equipment to work with. The commander instructed him to do what he could anyway as the senior medical person aboard.

Lipes's bedside manner was not perhaps the most reassuring. His pep talk to Rector was this: "Look, Dean, I never did anything like this before, but you don't have much chance to pull through anyhow, so what do you say?"

Lipes succeeded in anesthetizing Rector—in itself an achievement because he had no instructions on the dosage to give—then, wearing a tea strainer lined with gauze as a surgical mask and guided by little more than a first aid manual, he cut into Rector with a galley knife and somehow managed to find and remove the inflamed appendix and to sew up the wound. Rector miraculously survived and enjoyed a full and healthy recovery. Unfortunately, he did not get to enjoy a full and healthy life. Three years after his appendectomy, he was killed in action on another submarine in nearly the same location. Lipes served in the navy until 1962 and lived to the ripe age of eighty-four but never performed surgery again, which is of course just as well.

The small intestine empties into the large intestine via a connection called the ileocecal sphincter. The large intestine really is a kind of fermentation tank, home of feces, flatus, and all our microbial flora, and a place where nothing much happens in a hurry. In the early twentieth century, Sir William Arbuthnot Lane, an otherwise distinguished British surgeon, became convinced that all that sluggish muck promoted a buildup of morbid toxins, leading to a condition he called autointoxication. He identified an abnormality that became known as Lane's kinks and began surgically excising lengths of large intestine from sufferers. Gradually, he extended the practice until he was performing total colectomies—a procedure that was entirely unnecessary. People flocked to him from all around the world to be parted from their bowels. After his death, it was shown that Lane's kinks were entirely imaginary.

In America, Henry Cotton, superintendent of Trenton State Hospital in New Jersey, also took an unfortunate interest in the large intestine. Cotton became convinced that psychiatric disorders were due not to disturbances in the brain but to congenitally misshapen bowels and embarked on a program of surgery for which he had no apparent aptitude. He managed to kill 30 percent of his patients and cure none—but then none had any conditions that needed curing. Cotton also became an enthusiast for tooth extraction, taking out almost sixty-five hundred teeth (an average of ten per patient) in a single year, 1921, without the use of anesthetic.

The large intestine is in fact engaged in a lot of important work. It reabsorbs large volumes of water, which it returns to the body. It also provides a warm home for vast colonies of microbes that chew away at whatever the smaller intestine hasn't taken already, in the process capturing lots of useful vitamins like B_1, B_2, B_6, B_{12}, and K, which are also returned to the body. What's left is dispatched for evacuation as feces.

Adults in the West produce about 200 grams of feces a day–a little under half a pound, about 180 pounds a year, 14,000 pounds in a lifetime. Stools consist in large part of dead bacteria, undigested fiber, sloughed-off intestinal cells, and the residues of dead red blood cells. Every gram of feces you produce contains 40 billion bacteria and 100 million archaea. Analysis of stool samples also finds many fungi, amoebas, bacteriophages, alveolates, ascomycetes, basidiomycetes, and a great deal else, though whether some of these things are permanently resident or just passing through is rarely certain. Stool samples taken two days apart can give strikingly different results. Even samples taken from two ends of the same stool can seem to come from two different people. We are a long way from understanding it all.

Nearly all cancer that occurs in the gut is found in the large intestine and almost never in the small intestine. Although no one knows why for sure, many researchers think that it is because of the abundance of bacteria in the former. Professor Hans Clevers of the University of Utrecht in the Netherlands thinks it is related to diet. "Mice get cancer in the small intestine but not in the colon," he says. "But if you give them a Western-style diet, that reverses. It is the same for Japanese people when they move to the West and adopt a Western lifestyle. They get less stomach cancer, but more colon cancer."

The first person in modern times to take a close scientific interest in stools was Theodor Escherich (1857–1911), a young pediatric researcher in Munich who began microscopically examining babies' stools in the late nineteenth century. He found nineteen different kinds of microorganisms there, which was considerably more than he expected to find because the only obvious sources of input were their mothers' milk and the air they breathed. The most abundant of these is called *Escherichia coli* in his honor. (Escherich himself called it *Bacteria coli commune.*)

E. coli has become the most studied microbe on the planet. It has spawned literally hundreds of thousands of papers, according to Carl Zimmer, who has written a fascinating book, *Microcosm,* on this single extraordinary bacillus. Two strains of *E. coli* have more genetic variability than all the mammals on Earth put together. Poor Theodor Escherich never knew any of this. *E. coli* wasn't named for him until 1918, seven years after his death, and the name wasn't officially accepted until 1958.

Finally, a word or two about flatus, the well-bred term for a fart. Flatus consists primarily of carbon dioxide (up to 50 percent), hydrogen (up to 40 percent), and nitrogen (up to 20 percent), though the exact proportions will vary from person to person and indeed from day to day. About a third of people produce methane, a notorious greenhouse gas, while two-thirds produce none at all. (Or at least none on the occasions on which they have been tested; flatus testing is not the most exacting of disciplines.) The smell of a fart is composed largely of hydrogen sulfide, even though hydrogen sulfide accounts for only about one to three parts per million of what is expelled. Hydrogen sulfide in concentrated form—as in sewage gas—can be highly lethal, but why we are so sensitive to it in trace exposures is a question science has yet to answer. Curiously, we don't smell it at all when it rises to lethal levels. As Mary Roach put it in her splendid study of all things alimentary, *Gulp,* "The olfactory nerves become paralyzed."

All the gases of flatus can make a pretty explosive combination, as was tragically demonstrated in Nancy, France, in 1978 when surgeons stuck an electrically heated wire up the rectum of a sixty-nine-year-old man to cauterize a polyp and caused an explosion that literally tore the patient apart. According to the journal *Gastroenterology,* this was just one of "many recorded examples of explosion of colonic gas during anal surgery." Nowadays, most patients undergo laparoscopic, or keyhole, surgery, which involves being insufflated with–pumped full of–carbon dioxide, and that not only reduces discomfort and scarring but eliminates the risk of explosive mishaps.

16 SLEEP

O sleep, O gentle sleep,
Nature's soft nurse.

—WILLIAM SHAKESPEARE, *HENRY IV, PART 2*

I

SLEEPING IS THE most mysterious thing we do. We know that it is vital; we just don't know exactly why. We can't say with any certainty what sleep is for, what is the right amount for maximum health and happiness, or why some people fall into it with ease while others struggle perpetually to attain it. We lose a third of our lives to it. I am sixty-six years old as I write this. I have in effect slept through the whole of the twenty-first century.

There isn't any part of the body that does not benefit from sleep or suffer from its absence. If you are deprived of it for long enough, you will die—though what exactly it is that kills you from lack of sleep is a mystery, too. In 1989, in an experiment unlikely to be repeated on grounds of cruelty, researchers from the University of Chicago kept ten rats awake until they died and discovered that it took between eleven and thirty-two days for exhaustion to fatally overcome them. Postmortems showed the rats had no abnormalities that could explain their deaths. Their bodies just gave up on them.

Sleep has been tied to a great many biological processes—

consolidating memories, restoring hormonal balance, emptying the brain of accumulated neurotoxins, resetting the immune system. People with early signs of hypertension who slept for one hour more per night than previously showed a significant improvement in their blood pressure readings. It would seem to be, in short, a kind of nightly tune-up for the body. As Professor Loren Frank of the University of California at San Francisco told the journal *Nature* in 2013, "The story that everyone tells is that sleep is important for transferring memories to the rest of the brain. But the problem is there's basically no direct evidence for this idea." But why we should be required to so utterly give up consciousness for this is a question yet to be answered. It isn't just that we are disengaged from the outside world when slumbering, but for much of the time are actually paralyzed.

Sleep is clearly about more than just resting. One curious fact is that animals that are hibernating also have periods of sleep. It comes as a surprise to most of us, but hibernation and sleep are not the same thing at all, at least not from a neurological and metabolic perspective. Hibernating is more like being concussed or anesthetized: the subject is unconscious but not actually asleep. So a hibernating animal needs to get a few hours of conventional sleep each day within the larger unconsciousness. A further surprise to most of us is that bears, the most famous of wintry slumberers, don't actually hibernate. Real hibernation involves profound unconsciousness and a dramatic fall in body temperature—often to around 32 degrees Fahrenheit. By this definition, bears don't hibernate, because their body temperature stays near normal and they are easily roused. Their winter slumbers are more accurately called a state of torpor.

Whatever sleep gives us, it is more than just a period of recuperative inactivity. Something must make us crave it deeply to leave ourselves so vulnerable to attack by brigands or predators, yet as far as can be told, sleep does nothing for us that couldn't equally be done while we were awake but resting. We also do not know why we pass much of the night experiencing the surreal and often unsettling hallucinations that we call dreams. Being chased by zombies or finding yourself unaccountably naked at a bus stop doesn't seem, on the face of it, a terribly restorative way to while away the hours of darkness.

And yet it is universally assumed that sleep must answer some deep elemental need. As the eminent sleep researcher Allan Rechtschaffen observed many years ago, "If sleep does not serve an absolutely vital function, then it is the biggest mistake the evolutionary process has ever made." Nonetheless, as far as we know, all sleep does is (in the word of another researcher) "make us fit to be awake."

All animals seem to sleep. Even quite simple creatures like nematodes and fruit flies have periods of quiescence. The amount of sleep needed varies markedly among animals. Elephants and horses get by on just two or three hours a night. Why they need so little is unknown. Most other mammals require a great deal more. The animal that used to be thought the mammalian sleep champion, the three-toed sloth, is still often said to sleep for up to twenty hours a day, but that number came from studying captive sloths, who have no predators and not a lot to do. Wild sloths slumber for more like ten hours a day—not a huge amount more than we do. Extraordinarily, some birds and marine mammals are able to switch off one half of their brain at a time, so that one half remains alert while the other is snoozing.

Our modern understanding of sleep may be said to date from a December night in 1951 when a young sleep researcher at the University of Chicago named Eugene Aserinsky tried out a machine for measuring brain waves that his lab had acquired. Aserinsky's volunteer subject for the first night's test was his eight-year-old son, Armond.

Ninety minutes after young Armond had settled down into what was normally a peaceful night's sleep, Aserinsky was surprised to see the monitor's unspooling graph paper jerk to life and begin the kinds of jagged tracings associated with an active, wakeful mind. But when Aserinsky went next door, he found Armond still fast asleep. His eyes, however, were moving visibly beneath his lids. Aserinsky had just discovered rapid eye movement sleep, the most interesting and mysterious of the multiple phases of our nightly sleep cycle. Aserinsky didn't exactly rush the news into print. Almost two years passed before a small report on the discovery appeared in the journal *Science*.*

We now know that a normal night's sleep consists of a series of cycles, each involving four or five phases (depending on whose categorization methods you favor). First comes the business of relinquishing consciousness, which for most of us takes between five and fifteen minutes to achieve fully. This is followed by a period in which we slumber lightly but restoratively, as in a nap, for about twenty minutes. Sleep is so shallow in these first two stages that you may be asleep but think you are awake. Then comes a deeper sleep, lasting about an hour, from which it is much harder to rouse a sleeper. (Some authorities divide this period into two stages, giving the sleep cycle five distinct periods rather than four.) Finally comes the rapid eye movement (or REM) phase, when we do most of our dreaming.

During the REM part of the cycle, the sleeper becomes mostly paralyzed, but the eyes dart about beneath closed lids as if witnessing some urgent melodrama, and the brain grows as lively as at any time during wakefulness. In fact, some parts of the forebrain are livelier during REM sleep than when we are fully conscious and moving around.

Why the eyes move during REM sleep is uncertain. One obvious idea is that we are "watching" our dreams. Not all of you is paralyzed during the REM phase. Your heart and lungs continue to function, for obvious reasons, and clearly your eyes are free to swivel, but the muscles that control bodily movement are all restrained. The explanation most often proposed is that immobilization stops us from

* Aserinsky was an interesting, if restless, fellow. Before coming to the University of Chicago in 1949 at the age of twenty-seven, he had attended two colleges and majored successively in sociology, pre-med, Spanish, and dentistry without completing his studies in any of them. In 1943, he was drafted into the army and, despite being blind in one eye, passed the war as a bomb disposal expert.

harming ourselves by thrashing about or trying to flee from attack when caught up in a bad dream. A very few people suffer from a condition called REM sleep behavior disorder in which the limbs don't become paralyzed, and they do indeed sometimes hurt themselves or their partners with their thrashing. For others, paralysis doesn't immediately abate upon awaking and the victim finds himself awake but unable to move—a deeply unnerving experience, it seems, but one that mercifully tends to last only for a few moments.

REM sleep accounts for up to two hours of every night's sleep, roughly a quarter of the total. As the night passes, the periods of REM sleep tend to lengthen, so that your most dreamy spells are usually in the final hours before waking.

The cycles of sleep are repeated four or five times a night. Each cycle lasts about ninety minutes, but can vary. REM sleep is seemingly important for development. Newborn babies spend at least 50 percent of their sleep time (which is most of their time anyway) in the REM phase. For fetuses it may be as much as 80 percent. For a long time, it was thought that we did all our dreaming during REM sleep, but a 2017 study at the University of Wisconsin found that 71 percent of people dreamed during non-REM sleep (as compared with 95 percent during REM sleep). Most men have erections during REM sleep. Women likewise experience increased blood flow to the genitals. No one knows why, but it seems not to be overtly associated with erotic impulses. Typically, a man will be erect for two hours or so a night.

We are more restless at night than most of us realize. The average person turns over or significantly changes position between thirty and forty times in the course of a night. We also wake up far more than you might think. Arousals and brief awakenings in the night can add up to thirty minutes without being noticed. On a visit to a sleep clinic for his 1995 book, *Night,* the writer A. Alvarez thought he had experienced an unbroken night's sleep but discovered when his chart was reviewed in the morning that he had woken up twenty-three times. He also had five dreaming periods of which he had no recollection.

As well as normal overnight sleep, we also commonly indulge

in snatches of wakeful-hours sleep in a state known as hypnagogia, a netherworld between waking and unconsciousness, often without being aware of it. Alarmingly, when a dozen airline pilots on long-haul flights were studied by sleep scientists, almost all were found to have been asleep, or all but asleep, at various times during the flight without realizing it.

The relationship between the sleeping person and the outside world is often a curious one. Most of us have experienced that abrupt feeling of falling while asleep known as a hypnic or myoclonic jerk. No one knows why we have this sensation. One theory is that it goes back to the days when we slept in trees and had to take care not to fall off. The jerk may be a kind of fire drill. That may seem far-fetched, but it is a curious fact, when you think about it, that no matter how profoundly unconscious we get, or how restless, we almost never fall out of bed, even unfamiliar beds in hotels and the like. We may be dead to the world, but some sentry within us keeps track of where the bed's edge is and won't let us roll over it (except in unusually drunk or fevered circumstances). Some part of us, it seems, pays heed to the outside world, even for the heaviest sleepers. Studies at Oxford University, related by Paul Martin in his book *Counting Sheep,* found that EEG readings for test subjects twitched whenever their own names were read aloud as they slept but didn't react when other, unknown names were recited. Tests have also shown that people are pretty good at waking themselves at a predetermined time without an alarm clock, which means that some part of the sleeping mind must be tracking the real world outside the skull.

Dreaming may simply be a by-product of this nightly cerebral housecleaning. As the brain clears wastes and consolidates memories, neural circuits fire randomly, briefly throwing up fragmentary images, a bit like someone jumping between television channels when looking for something to watch. Confronted with this incoherent flow of memories, anxieties, fantasies, suppressed emotions, and the like, the brain possibly tries to make a sensible narrative out of it all, or possibly, because it is itself resting, doesn't try at all, and just lets the incoherent pulses flow past. That may explain why we generally don't remember dreams much despite their intensity—because they are not actually meaningful or important.

II

IN 1999, AFTER ten years of careful work, a researcher at Imperial College in London named Russell Foster proved something that seemed so unlikely that most people refused to believe it. Foster found that our eyes contain a third photoreceptor cell type in addition to the well-known rods and cones. These additional receptors, known as photosensitive retinal ganglion cells, have nothing to do with vision but exist simply to detect brightness—to know when it is daytime and when night. They pass this information on to two tiny bundles of neurons within the brain, roughly the size of a pinhead, embedded in the hypothalamus and known as suprachiasmatic nuclei. These two bundles (one in each hemisphere) control our circadian rhythms. They are the body's alarm clocks. They tell us when to rise and shine and when to call it a day.

All that may seem eminently sensible and good to know, but when Foster announced his discovery, it caused the most enormous outcry in the ophthalmological world. Almost no one could believe that such a fundamental thing as an ocular cell type could have been overlooked for so long. One member of an audience shouted, "Bullshit!" at one of Foster's presentations and stalked out.

"They struggled to accept that something they had been studying for 150 years—namely, the human eye—had a type of cell whose function they had completely overlooked," he says. In fact, Foster was right and has since been completely vindicated. "They're much more gracious about it now," he jokes. Today Foster is professor of circadian neuroscience and head of the Nuffield Laboratory of Ophthalmology at Oxford University.

"What's really interesting about these third receptors," Foster told me when we met in his office at Brasenose College, just off the High

Street, "is that they function completely independently of sight. As an experiment, we asked a lady who was completely blind—she had lost her rods and cones as a result of a genetic disease—to tell us when she thought the lights in the room were switched on or off. She told us not to be ridiculous because she couldn't see anything, but we asked her to try anyway. It turned out she was right every time. Even though she had no vision—no way of 'seeing' the light—her brain detected it with perfect fidelity at a subliminal level. She was astonished. We all were."

Since Foster's discovery, scientists have found that we have body clocks not just in the brain but all over—in our pancreas, liver, heart, kidneys, fatty tissue, muscle, virtually everywhere—and these operate to their own timetables, dictating when hormones are released or organs are busiest or most relaxed. Your reflexes, for instance, are at their sharpest in mid-afternoon, while blood pressure peaks toward evening. Men tend to pump more testosterone early in the morning than later in the day. If any of these systems get too out of sync, problems can result. Disturbances to the daily rhythms of the body are thought to contribute to (and in some cases may directly account for) diabetes, heart disease, depression, and serious weight gain.*

The suprachiasmatic nuclei work closely with a nearby and long mysterious pea-sized structure, the pineal gland, which is more or less in the middle of the head. Because of its central location and its solitary nature—most structures in the brain come in pairs, but the pineal stands alone—the philosopher René Descartes concluded that the pineal is where the soul resides. Its actual function, to produce melatonin, a hormone that helps the brain track day length, wasn't discovered until the 1950s, making it the last of the main endocrine glands to be decoded. How exactly melatonin relates to sleep is still not understood. Melatonin levels within us rise as evening falls and peak

* Even our teeth mark the passing of time by acquiring daily microscopic accretions, not unlike tree rings, until they stop growing at about the age of twenty. Scientists count the rings in ancient teeth to work out how long it took children to grow up in the very distant past.

in the middle of the night, so it would seem logical to associate them with drowsiness, but in fact melatonin production also rises at night in nocturnal animals when they are most active, so it is not promoting sleepiness. The pineal, in any case, tracks not just day/night rhythms but also seasonal changes, which are really important for animals that hibernate or breed seasonally. They are consequential for humans, too, but in ways that we mostly don't notice. Your hair grows faster in the summertime, for instance.

As David Bainbridge has neatly put it, "The pineal is not our soul, it is our calendar." But it is also a very curious fact that several of our fellow mammals—elephants and dugongs to name just two—don't have pineals and don't seem to suffer for it. In humans, the seasonal role of melatonin is not entirely clear. Melatonin is a more or less universal molecule; it is found in bacteria, jellyfish, plants, and almost anything else that is subject to circadian rhythms.* In humans, production falls significantly as we age. A seventy-year-old produces only a quarter as much melatonin as a twenty-year-old. Why this should be, and what effect it has on us, remain to be determined.

What is certain is that the circadian system can get seriously confused if its normal daily rhythms are disturbed. In a famous experiment in 1962, a French scientist named Michel Siffre isolated himself for about eight weeks deep inside a mountain in the Alps. Without daylight, clocks, or other clues to the passage of time, Siffre had to guess when twenty-four hours had elapsed and discovered to his astonishment that when he had calculated thirty-seven days to have passed, it was actually fifty-eight. He became hopeless at gauging even short increments of time. When asked to estimate the passage of two minutes, he waited more than five.

* In the United States, melatonin is commonly taken as a treatment for jet lag or insomnia. It is, as James Hamblin has written, "one of the very few hormones that you can purchase in the United States without a prescription. It is considered a dietary supplement and therefore held to essentially no premarket standards of quality, safety, or efficacy."

In recent years, Foster and his colleagues have come to realize that we have more seasonal rhythms than formerly thought. "We've been finding rhythms," he says, "in lots of unexpected areas—self-harm, suicide, child abuse. We know it is not just coincidental that these things have seasonal peaks and troughs because the patterns are six-month-shifted from the Northern Hemisphere to the Southern." Whatever people do in a northern spring—like commit suicide in greater numbers—they do six months later in the southern spring.

Circadian rhythms may also make a big difference to the effectiveness of the medications we take. As the Manchester University immunologist Daniel Davis has noted, fifty-six of the one hundred bestselling drugs in use today target parts of the body that are time sensitive. "Around half of these bestselling drugs stay active in the body for only a short time after being taken," he writes in *The Beautiful Cure*. Take them at the wrong time and they may well be less effective, or possibly not effective at all.

We are really at the beginning of our understanding of the importance of circadian rhythms for all living things, but as far as we can tell, all organisms, even bacteria, have internal clocks. "It may be," as Russell Foster says, "a signature of life."

The suprachiasmatic nuclei don't entirely account for why we get sleepy and want to go to bed. We are also subject to a natural sleep pressure—a profound and eventually irresistible urge to nod off—governed by something called sleep homeostats. The pressure to sleep grows more intense the longer we stay awake. This is in large part a consequence of an accumulation of chemicals in the brain as the day goes by, in particular one called adenosine, which is a by-product of the output of ATP (or adenosine triphosphate), the little molecule of intense energy that powers our cells. The more adenosine you accumulate, the drowsier you feel. Caffeine slightly counteracts its effects, which is why a cup of coffee perks you up. Normally, the two systems operate in synchronicity, but occasionally they deviate, as when we cross several time zones on a long-distance plane flight and we experience jet lag.

Exactly how much sleep you need appears to be a personal matter, but nearly all of us fall somewhere in the range of a nightly requirement of seven to nine hours. Much depends on age, health, and what you have been up to lately. We sleep less as we get older. Newborns may sleep for nineteen hours a day, preschoolers for up to fourteen, young children for eleven or twelve, teens and young adults for ten or so—though they, like most adults, may not get all that they need because of staying up too late and having to rise too early. The problem is particularly acute for teenagers because their circadian cycles can be up to two hours adrift from those of their elders, turning them into comparative night owls. When a teenager struggles to get up in the morning, that isn't laziness; it's biology. Matters are compounded in America by what *The New York Times* in an editorial called "a dangerous tradition: starting high school abnormally early." According to the *Times*, 86 percent of U.S. high schools start their day before 8:30 a.m., and 10 percent start before 7:30. Later start times have been shown to produce better attendance, better test results, fewer car accidents, and even less depression and self-harm.

Nearly all authorities agree that we are sleeping less than we used to at all age levels. According to the journal *Baylor University Medical Center Proceedings*, the average amount of sleep people get on a night before work has fallen from eight and a half hours fifty years ago to under seven now. Another study found a similar decline among schoolchildren. The cost to the U.S. economy of all this tossing and turning has been estimated at more than $60 billion from absenteeism and diminished performance.

Between 10 and 20 percent of adults in the world suffer from insomnia, according to various studies. Insomnia has been linked to diabetes, cancer, hypertension, stroke, heart disease, and (not surprisingly) depression. A study in Denmark, noted in *Nature*, found that women who regularly worked night shifts showed a 50 percent greater risk of developing breast cancer than their counterparts who worked by day.

"There's also now good data to show that sleep-deprived individuals have higher levels of beta amyloid [a protein associated with Alzheimer's disease] than those who have slept normally," Foster told me. "I wouldn't say that sleep disruption causes Alzheimer's, but it is probably a contributing factor and may well speed the decline."

For many people, the principal cause of insomnia is the snoring of a partner. It is a very common problem. About half of us snore at least sometimes. Snoring is the rattling of the soft tissues in the pharynx when one is unconscious and relaxed. The more relaxed, the greater the snoring, which is why drunken people snore particularly robustly. The best way to reduce snoring is to lose weight, sleep on your side, and not drink alcohol before retiring. Sleep apnea (from a Greek word meaning "breathless") is when the airways become obstructed and victims either stop breathing or nearly stop breathing while asleep, and it is more common than generally appreciated. About 50 percent of people who snore have some degree of sleep apnea.

The most extreme and horrifying form of insomnia is a very rare condition known as fatal familial insomnia, which was first medically described as recently as 1986. It is an inherited disorder (hence, familial) that is known to affect only about three dozen families in the world. Sufferers simply lose the ability to fall asleep and slowly die of exhaustion and multiple organ failure. The disease is always fatal. The destructive agent is a type of corrupted protein called a prion (short for proteinaceous infectious particle). Prions are rogue proteins. They are the wicked little particles behind Creutzfeldt-Jakob disease and mad cow disease (bovine spongiform encephalopathy) and some other horrible neurological illnesses, like Gerstmann-Sträussler-Scheinker disease, that most of us have never heard of because they are mercifully rare (but without exception very bad news for coordination and cognition). Some authorities think prions may also have a role in Alzheimer's and Parkinson's diseases.* In fatal familial insomnia, prions attack the thalamus, the walnut-sized body deep in the brain that controls our autonomic responses—blood pressure, heart rate, the release of hormones, and so on. How exactly prion disruption interferes with sleep is unknown, but it is a wretched way to go.

Another disorder that disrupts sleep is narcolepsy. It is commonly associated with extreme drowsiness at inappropriate times, but many with the condition have as much trouble staying asleep as staying awake. The condition affects four million people around the world. It is caused by a lack of a chemical in the brain called hypocretin, which exists in such tiny amounts that it was only discovered in 1998. Hypocretins are neurotransmitters that keep us wakeful. Without them, sufferers may abruptly nod off in the middle of a conversation or while eating, or slip into a kind of twilight state that is closer to hallucination than to consciousness. Conversely, they may be quite exhausted but unable to sleep at all. It can be a miserable condition, and has no cure, but mercifully it is quite rare, affecting just one person in twenty-five hundred in the Western world.

More common sleep disorders, collectively known as parasomnias, include sleepwalking, confusional arousal (when the victim appears to be awake but is profoundly muddled), nightmares, and night terrors. The last two are not easily distinguished except that night terrors are more intense and tend to leave the victim more shaken, though

* Prions were discovered by Dr. Stanley Prusiner of the University of California at San Francisco. In 1972, while still training as a neurologist, he examined a sixty-year-old woman who was suffering from a sudden onset dementia so severe that she couldn't manage even the simplest and most familiar tasks, like how to put a key in a door. Prusiner became convinced that the cause was a misshapen infectious protein which he called a prion. His theory was widely derided for years, but Prusiner was eventually vindicated and was awarded a Nobel Prize in 1997. The death of neurons leaves the brain pocked with cavities, like a sponge—hence the term "spongiform."

curiously victims of night terrors very often have no recollection of the experience the following morning. Most parasomnias are much more common in young children than in adults and tend to disappear around puberty, if not before.

The longest anyone has intentionally gone without sleep was in December 1963 when a seventeen-year-old high school student in San Diego named Randy Gardner managed to stay awake for 264.4 hours (11 days and 24 minutes) as part of a school science project. The first few days were comparatively easy for him, but gradually he became irritable and confused until his entire existence was a kind of hallucinatory blur. When he finished the project, Gardner fell into bed and slept for 14 hours. "I remember when I woke up, I was groggy, but not any groggier than a normal person," he told an NPR interviewer in 2017. His sleep patterns returned to normal, and he suffered no noticeable ill effects. Later in life, however, he experienced terrible insomnia, which he believed was "karmic payback" for his youthful adventure.*

Finally, we should say a word about that mysterious but universal harbinger of weariness, the yawn. No one understands why we yawn. Babies yawn in the womb. (They hiccup, too.) People in comas yawn. It is a ubiquitous part of life, and yet what exactly it does for us is unknown. One suggestion is that it is somehow connected with shedding excess carbon dioxide, though no one has ever explained in what way. Another is that it brings a rush of cooler air into the head, thus slightly banishing drowsiness, though I have yet to meet anyone who felt refreshed and energized after yawning. More to the point, no scientific study has ever shown a relationship between yawning and energy levels. Yawning doesn't even correlate reliably with how tired

* There have been surprisingly few challenges to the record. In 2004, ten people competed to stay awake the longest for a television series called *Shattered* on Channel 4 in Britain. The winner, Clare Southern, lasted 178 hours, more than three days less than Randy Gardner.

you are. Indeed, when we yawn the most is often in the first couple of minutes after rising from a good night's sleep when we are at our most rested.

Perhaps the least explicable aspect of yawning is its extreme infectiousness. Not only do we more or less have to yawn when we see others do so, but just hearing or thinking about yawning causes us to yawn. You will almost certainly want to yawn now. And frankly there is nothing wrong with that.

17 INTO THE NETHER REGIONS

On a Presidential visit to a farm, Mrs. Coolidge asked her guide how many times the rooster copulated daily. "Dozens of times" was the reply. "Please tell that to the President," Mrs. Coolidge requested. When the President passed the pens and was told about the rooster, he asked: "Same hen every time?" "Oh no, Mr. President, a different one each time." The President nodded slowly, then said: "Tell that to Mrs. Coolidge."

—*LONDON REVIEW OF BOOKS*, JANUARY 25, 1990

I

IT IS A slightly startling fact that for the longest time we didn't know why some people are born male and some female. Although chromosomes had been discovered in the 1880s by the very busy and lushly named German Heinrich Wilhelm Gottfried von Waldeyer-Hartz, their importance wasn't understood or appreciated.* (He called them chromosomes because of how well they absorbed chemical dyes under the microscope.) We now know, of course, that females have two X chromosomes and males have one X and a Y, which is what accounts

* For most of his career, he was just plain Wilhelm Waldeyer. The more effusive title came in 1916, near the end of his life, when he was ennobled by the German state.

for their sexual differences, but that knowledge was a long time in coming. Even in the late nineteenth century, scientists commonly thought that sex was determined not by chemistry but by external factors like diet or air temperature or even a woman's mood during the early stages of pregnancy.

The first step in solving the problem came in 1891 when a young zoologist at the University of Göttingen in central Germany, Hermann Henking, noticed an odd thing while studying the testicles of a genus of fire wasp called *Pyrrhocoris*. In all the specimens he studied, one chromosome always remained aloof from the others. Henking dubbed it "X" because it was mysterious, not because of its shape, as is nearly always assumed. His finding generated a ripple of interest among other biologists but seems not to have captivated Henking himself. He took a job soon afterward with the German Fisheries Association, where he spent the rest of his life surveying North Sea fish stocks, and, as far as can be told, never looked at another insect testicle again.

Fourteen years after Henking's accidental discovery, on the other side of the Atlantic, came the real breakthrough. A scientist at Bryn Mawr College in Pennsylvania named Nettie Stevens was doing similar work with the reproductive apparatus of mealworms when she discovered another aloof chromosome and—her crucial insight—realized that it seemed to have a role in determining sex. She called it the Y chromosome to continue the alphabetical sequence begun by Henking.

Nettie Stevens deserves to be better known. Born in 1861 in Cavendish, Vermont (the place coincidentally where Phineas Gage had an iron bar shot through his skull while building a railway there thirteen years earlier), Stevens grew up in modest circumstances, and it took her a very long time to fulfill her dream of attaining higher education. She worked for several years as a teacher and librarian before finally entering Stanford University in 1896 at the advanced age of thirty-five, and she was forty-two and tragically near the end of her short life when she finally earned her PhD. Taking a position as a junior researcher at Bryn Mawr, she embarked on a blizzard of activity, publishing thirty-eight papers as well as discovering the Y chromosome.

Had the importance of her discovery been more widely appreciated, Stevens would almost certainly have won a Nobel Prize. Instead, for many years the credit was usually given to Edmund Beecher Wilson, who had independently made the same discovery at almost the same time (exactly who was first has long been a matter of contention) but without fully appreciating its significance. Stevens doubtless would have achieved greater things still, but she contracted breast cancer and died aged just fifty in 1912 after only eight years as a qualified scientist.

In illustrations, the X and Y chromosomes are always portrayed as being roughly in the shape of an X or a Y, but in fact most of the time they don't look like any letters of the alphabet. During cell division, the X chromosome does indeed briefly assume an X shape, but then so do all the non-sex chromosomes. The Y chromosome only superficially resembles a Y. It is just an extraordinary coincidence that they bear a passing or occasional resemblance to the letters for which they are named.

Historically, chromosomes were not at all easy to study. They spend most of their existence balled up in an indistinguishable mass in the cell nucleus. The only way to count them was to get fresh samples from living cells at the moment of cell division, and that was a tall order. Cell biologists, according to one report, "literally waited at the foot of the gallows in order to fix the testis of an executed criminal immediately after death before the chromosomes could clump." Even then the chromosomes tended to overlap and blur, making anything but a rough count difficult. But in 1921, a cytologist at the University of Texas named Theophilus Painter announced that he had secured good images and declared with reassuring confidence that he had counted twenty-four pairs of chromosomes. That number stuck, universally unquestioned, for thirty-five years until a closer examination in 1956 showed that in fact we have just twenty-three pairs—a fact that had been clearly evident in photographs for years (including in at least one popular textbook illustration) had anyone taken the trouble to count.

As to what precisely makes some of us males and some females, that knowledge is even more recent. It wasn't until 1990 that two teams in London, at the National Institute for Medical Research and at the Imperial Cancer Research Fund, identified a sex-determining region on the Y chromosome that they dubbed the SRY gene, for "sex-determining region Y." After countless generations of making little boys and little girls, humans finally knew how they did it.

The Y chromosome is a curious and runty thing. It has only about seventy genes; other chromosomes have as many as two thousand. The Y chromosome has been shrinking for 160 million years. At its current rate of deterioration, it has been estimated, it could vanish altogether in another 4.6 million years. That doesn't mean, happily, that males will cease to exist in 4.6 million years. The genes that determine gender traits would probably just move across to another chromosome. Moreover, our ability to manipulate the reproductive process is likely to be rather more refined in 4.6 million years, so this is probably not something we should lose sleep over.*

Interestingly, sex isn't actually necessary. Quite a number of organisms have abandoned it. Geckos, the little green lizards that are often encountered clinging like suckered bath toys to walls in the tropics, have done away with males altogether. It is a slightly unsettling thought if you are a man, but what we bring to the procreative party is easily dispensed with. Geckos produce eggs, which are clones of the mother, and these grow into a new generation of geckos. From the mother's point of view, this is an excellent arrangement because it means that 100 percent of her genes are inherited. With conventional sex, each partner passes on just half its genes, and that number is relentlessly thinned with each succeeding generation. Your grandchildren have only a quarter of your genes, your great-grandchildren only an eighth,

* Other geneticists, it is worth noting, have suggested that the extinction could happen in as little as 125,000 years or as much as 10 million.

your great-great-grandchildren a mere sixteenth, and so shrinkingly on it goes. If genetic immortality is your ambition, then sex is a very poor way of achieving it. As Siddhartha Mukherjee observed in *The Gene: An Intimate History*, humans don't actually reproduce at all. Geckos reproduce; we recombine.

Sex may dilute our personal contribution to posterity, but it is great for the species. By mixing and matching genes, we get variety and that gives us safety and resilience. It makes it harder for diseases to sweep through whole populations. It also means that we can evolve. We can hold on to beneficial genes and discard ones that impede our collective happiness. Cloning gives you the same thing over and over. Sex gives you Einstein and Rembrandt–and a lot of dorks, too, of course.

Probably no area of human existence has generated less certainty, or been more inhibitive to open discussion, than sex. Perhaps nothing says more about our delicacy toward matters genital than that the word "pudendum"–meaning the external genitals, particularly those of a woman–comes from the Latin for "to be ashamed." It is next to impossible to get reliable figures about almost anything to do with sex as a pastime. How many people are unfaithful to their partners at some point in a relationship? Somewhere between 20 percent and 70 percent, depending on which of many studies you consult.

One problem, which should surprise no one, is that survey respondents are inclined to embrace alternative realities when they think their answers cannot be checked. In one study, the number of sexual partners women were prepared to recall increased by 30 percent when they thought they were hooked up to a lie detector. Remarkably, for a 1995 survey called the Social Organization of Sexuality: Sexual Practices in the United States, conducted jointly by the University of Chicago and the National Opinion Research Center, respondents were permitted to have someone else, usually a child or current sexual partner, present when they were interviewed, which is hardly likely to have resulted in fully candid responses. Indeed, it was shown afterward that the proportion of people answering that they had had sex with more than one person in the previous year fell from 17 percent to 5 percent when another person was present.

The survey was criticized for lots of other deficiencies. Because of funding problems, only 3,432 people were interviewed instead of the 20,000 originally intended, and because all the respondents were aged eighteen or older, it offered no conclusions on teenage pregnancies or birth control practices, or much else of crucial importance to public policy. Moreover, the survey focused only on households, so it excluded people in institutions—college students, prisoners, and members of the armed forces most notably. All of these made the report's findings questionable if not entirely useless.

Another problem with sex surveys—and there is no delicate way of putting this—is that people are sometimes just stupid. In another analysis, reported by Cambridge University's David Spiegelhalter in the wonderful *Sex by Numbers: The Statistics of Sexual Behaviour,* when asked to state what in their view constituted full sex, some 2 percent of male respondents said that penetrative intercourse did not qualify, leaving Spiegelhalter to wonder what exactly they might be waiting for "before they feel they have gone all the way."

Because of the difficulties, the field of sex studies has a long history of providing dubious statistics. In his 1948 work, *Sexual Behavior in the Human Male,* Alfred Kinsey of Indiana University reported that nearly 40 percent of men had had a homosexual experience resulting in orgasm and that nearly a fifth of young men brought up on farms had had sex with livestock. Both figures are now thought highly unlikely. Even more dubious were the 1976 *Hite Report on Female Sexuality* and the companion *Hite Report on Male Sexuality* published soon afterward. The author, Shere Hite, used questionnaires and had a very low, nonrandom, highly selective response rate. Nonetheless, Hite confidently declared that 84 percent of women were dissatisfied with their male partners and that 70 percent of women married for more than five years were in an adulterous relationship. The findings were heavily criticized at the time, but the books were huge best sellers. (A more scientific, and more recent, U.S. National Health and Social Life Survey found that 15 percent of married women and 25 percent of married men said they had been unfaithful at some time.)

On top of all that, the subject of sex is full of statements and statistics that are often repeated but based on nothing. Two durable ones are these: "Men think of sex every seven seconds" and "The average amount of time spent kissing in a lifetime is 20,160 minutes" (that is 336 hours). In fact, according to genuine studies, men of college age think about sex nineteen times a day, roughly once every waking hour, which is about the same frequency as they think about food. College women think about food more often than they think about sex, but they don't think about either terribly often. No one does anything at all every seven seconds other than perhaps respire and blink. Similarly, no one knows how much of an average lifetime is devoted to kissing or where that weirdly precise and durable figure of 20,160 minutes comes from.

On a more positive note, we can say with some confidence that the median time for sex (in Britain at least) is nine minutes, though the whole act, including foreplay and undressing, is more like twenty-five minutes. According to Spiegelhalter, energy use on average per sexual session is about a hundred calories for men and about seventy for women. A meta-analysis showed that for older people the risk of a heart attack was raised for up to three hours after sex, but it was similarly raised for shoveling snow, and sex is more fun than shoveling snow.

II

IT IS SOMETIMES said there are more genetic differences between men and women than there are between humans and chimpanzees. Well, perhaps. It all depends on how you measure genetic differences. But the statement is in any case clearly meaningless in any practical sense. A chimpanzee and a human may have as much as 98.8 percent of genes in common (depending on how they are measured), but that doesn't mean they are just 1.2 percent different as beings. Chimpanzees cannot hold a conversation, cook dinner, or outwit a human four-year-old. Clearly, it is a question of not what genes you have but how they are expressed—how they are put to use.

That said, men and women unquestionably are very unalike in many important ways. Women (and we are talking here about healthy, fit women) carry about 50 percent more fat on their frames than fit, healthy men. This not only makes the woman more agreeably soft and shapely to suitors but also gives her reserves of fat she can call upon for milk production during times of hardship. Women's bones wear out sooner, particularly after menopause, so they suffer more breaks in later life. Women get Alzheimer's twice as often (partly because they also live longer) and experience higher rates of autoimmune diseases. They metabolize alcohol differently, which means they get intoxicated more easily and succumb to alcohol-related diseases like cirrhosis faster than men do.

Women even tend to carry bags differently than men do. It is thought that their wider hips necessitate a less perpendicular carrying angle for their forearm so that their swinging arms aren't constantly banging against their legs. That's why women generally carry bags with their palms facing forward (allowing their arms to be slightly splayed) while men carry them with palms facing back. Far more significantly, women and men have heart attacks in quite different ways. A woman suffering a heart attack is more likely to experience abdominal pain and nausea than a man, which makes it more probable that it will be misdiagnosed. In a thousand ways large and small, they are quite different beings.

Men have their own differences. They get Parkinson's disease more often and commit suicide more, even though they suffer less from clinical depression. They are more vulnerable to infection than females (and not just humans but across nearly all species). That may indicate some hormonal or chromosomal difference that hasn't yet been determined, or it may simply be that males on the whole lead riskier, more infection-prone lives. Men are also more likely to die from their infections and from physical injury, though again whether that is because we are hormonally compromised or just too proud and foolish to seek medical attention promptly (or both) is an unanswerable question.

All this is important because until recently drug trials very often excluded women, largely because it was feared their menstrual cycles could skew results. As Judith Mank of University College London told the BBC program *Inside Science* in 2017, "People had been assuming that women are just 20 percent smaller than men, but otherwise are much the same." We now know that there is much more to it than that. In 2007, the journal *Pain* reviewed all of its published findings over the previous decade and found that almost 80 percent had come from male-only tests. A similar gender bias, based on hundreds of clinical studies, was reported for cancer trials in the journal *Cancer* in 2009. Such findings are seriously consequential because women and men can respond to drugs in very different ways—ways often overlooked by clinical trials. The drug phenylpropanolamine was commonly used in over-the-counter medications for colds and coughs for years until it was discovered that it appreciably increased the risk of hemorrhagic stroke in women but not in men. Similarly, an antihistamine called Hismanal and an appetite suppressant called Pondimin were withdrawn after they were shown to pose serious risks to women, but only after the first had been on the market for eleven years and the second for twenty-four. Ambien, a popular sleep medication in America, had its recommended dosage for women cut in half in 2013 when it was found that a high proportion of female users were suffering impaired performance if they tried to drive the next morning. Men did not suffer in anything like the same way.

Women are anatomically different in one other very significant way: they are the sacred keepers of human mitochondria—the vital little powerhouses of our cells. Sperm pass on none of their mitochondria during conception, so all mitochondrial information is transferred from generation to generation through mothers alone. Such a system means that there will be many extinctions along the way. A woman endows all her children with her mitochondria, but only her daughters have the mechanism to pass it onward to future generations. So if a woman has only sons or no children at all—and that happens quite often, of course—her personal mitochondrial line will die with her. All her descendants will still have mitochondria, but it will come from other mothers on other genetic lines. In consequence, the human mitochondrial pool shrinks a little with every generation because of these localized extinctions. Over time, the mitochondrial pool for humans has shrunk so much that, almost unbelievably but rather wonderfully, we are all now descended from a single mitochondrial ancestor—a woman who lived in Africa about 200,000 years ago. You might have heard her referred to as Mitochondrial Eve. She is, in a sense, mother of us all.

For most of recorded history, we have known shockingly little about women and how they are put together. As Mary Roach notes in her delightfully irreverent book *Bonk*, "Vaginal secretions [were] the only bodily fluid about which virtually nothing was known" despite their importance to conception and a woman's general sense of well-being.

Matters specific to women—menstruation above all—were almost totally a mystery to medical science. Menopause, clearly another milestone event in a woman's life, didn't attract formal notice until 1858, when the word is first recorded in English, in the *Virginia Medical Journal*. Abdominal examinations were conducted rarely, vaginal examinations almost never, and any investigations below the neck usually involved the doctor feeling blindly under the bedclothes while gazing fixedly at the ceiling. Many doctors kept dummies so that a woman could point to the affected part without having to reveal or even mention it by name. When René Laënnec invented the stethoscope in 1816 in Paris, the greatest benefit wasn't that it improved sound transmission (an ear to the chest was actually about as good) but that it allowed a physician to check a woman's heart and other inner workings without directly touching her flesh.

Even now, there is a huge amount concerning female anatomy about which we are uncertain. Consider the G spot. It is named for Ernst Gräfenberg, a German gynecologist and scientist who fled Nazi Germany for America and there developed the intrauterine contraceptive device, which was originally called the Gräfenberg ring. In 1944, he wrote an article for the *Western Journal of Surgery, Obstetrics and Gynecology* in which he identified an erogenous spot on the wall of the vagina. The *Western Journal of Surgery, Obstetrics and Gynecology* did not normally attract a great deal of attention, but this article got passed around. Thanks to it, the newly identified erogenous location became known as the Gräfenberg spot, subsequently shortened to G spot. But whether or not women actually possess a G spot is a matter of continuing, and sometimes heated, debate. Imagine the amount of research funding that would follow if someone were to suggest that males have an erogenous spot that they have not been fully utilizing. In 2001, the *American Journal of Obstetrics and Gynecology* declared the G spot a "modern gynecologic myth," but other studies have shown that a majority of women, in America at least, believe they have one.

Male ignorance of female anatomy is quite arresting, it appears, particularly when you consider how keen men are to get to know it in other respects. A survey of a thousand men, conducted in conjunction with a campaign called Gynecological Cancer Awareness Month, found that the majority could not reliably define or identify most of a female's private parts—vulva, clitoris, labia, and so on. Half could not even find the vagina on a diagram. So perhaps a brief rundown is in order here.

The vulva is the complete genital package—vaginal opening, labia, clitoris, and so on. The fleshy mound above the vulva is called the mons pubis. At the top of the vulva itself is the clitoris (probably from a Greek word for "hillock," but there are other candidates), which is packed with some eight thousand nerve endings—more per unit of area than any other part of the female anatomy—and exists, as far as can be told, only to give pleasure. Most people, including females, are unaware that the visible part of the clitoris, called the glans, is literally only the tip of it. The rest of the clitoris plunges into the interior and extends down both sides of the vagina for about five inches. Until the early twentieth century, "clitoris" seems generally to have been pronounced "kly-to-rus."

The vagina (Latin for "scabbard") is the channel connecting the vulva to the cervix and uterus beyond. The cervix is a doughnut-shaped valve that stands between the vagina and the uterus. "Cervix" in Latin means "neck of the womb," which is precisely what it is. It serves as a gatekeeper, deciding when to let substances (like sperm) in and when to let others (like blood during menstruation and babies during birth) out. Depending on the size of a man's organ, the cervix is sometimes hit during sex, which some women find pleasurable and others find uncomfortable or painful.

The uterus is simply a more formal name for the womb, where babies grow. The uterus normally weighs two ounces, but at the end of a pregnancy it may weigh two pounds. Flanking the uterus are the ovaries, where eggs are stored, but they are also where hormones like estrogen and testosterone are produced. (Women produce testosterone, too, just not as much as men do.) The ovaries are connected to the uterus by Fallopian tubes (properly called oviducts). These are named for Gabriele Falloppio (sometimes spelled "Fallopio"), the Italian anatomist who first described them in 1561. Eggs are usually fertilized in the tube and then pushed outward into the uterus.

And there you have, very briefly, the principal pieces of sexual anatomy that are unique to women.

Male reproductive anatomy is considerably more straightforward. It consists essentially of three external parts—penis, testicles, and scrotum—with which nearly everyone is familiar, at least conceptually. For the record, however, I will note that the testicles are factories for producing sperm and some hormones; the scrotum is the sac in which they are housed; and the penis is the delivery device for sperm (the active part of semen), as well as outlet for urine. But behind the scenes in supporting roles are other structures, known as accessory sex organs, that are much less familiar but nonetheless vital. Most men, I daresay, have never heard of their epididymis and would be a little surprised to learn that they have twelve meters of it—that's forty feet, the length of a Greyhound bus—tucked inside their scrotal sacs. The epididymis is fine tubing, neatly coiled, in which sperm mature. The word is from the Greek for "testicles" and, a touch surprisingly, was first used in English by Ben Jonson in his play *The Alchemist* in 1610. He was presumably showing off because no one in the audience was likely to know what he meant by it.

Similarly obscure but no less important are the other accessory sex organs: bulbourethral glands, which produce a lubricating fluid, and are sometimes also known as Cowper's glands after their seventeenth-century discoverer; seminal vesicles, where semen is in large part produced; and the prostate, which everyone has at least heard of, though I have yet to meet a layman under fifty who knows quite what it does. The prostate, it might be said, produces seminal fluid throughout a man's adulthood and anxiety in his later years. We shall discuss this latter attribute in a later chapter.

One perennial mystery of male reproductive anatomy is why the testicles are on the outside, where they are exposed to trauma. It is usually said that it is because testicles function better in cooler air, but this overlooks that many mammals get along perfectly well with their testicles on the inside: elephants, anteaters, whales, sloths, and sea lions, to name but a few. Temperature regulation may indeed be a factor in testicular efficiency, but the human body is perfectly capable of dealing with that without leaving the testicles so disconcertingly vulnerable to harm. Ovaries, after all, are kept safely hidden away.

There is also a great deal of uncertainty over what is normal in terms of penis size. In the 1950s, the Kinsey Institute for Sex Research recorded the average length of the erect penis at 5 to 7 inches. By 1997, a sample of over a thousand men put it at 4.5 to 5.75 inches, a fairly notable demotion. Either men are shrinking, or there is a great deal more variability in penis size than has traditionally been allowed. The bottom line is that we don't know.

Sperm appears to have enjoyed (if that is the word) more careful clinical study, almost certainly because of concerns about fertility. Authorities seem to be universally agreed that the average quantity of semen released at orgasm is 3 to 3.5 milliliters (about a teaspoonful), with an average spurt distance of seven to eight inches, though according to the British scientist and writer Desmond Morris a launch of three feet has been scientifically recorded. (He does not specify the circumstances.)

The most interesting experiment involving sperm was almost certainly that undertaken by Robert Klark Graham (1906–97), a California businessman who made a fortune manufacturing shatterproof lenses for eyeglasses and then in 1980 founded the Repository for Germinal Choice, a sperm bank that promised to stock the sperm only of Nobel laureates and others of exceptional intellectual stature. (Graham modestly included himself among the select worthies.) The idea was to help women produce babies of genius by giving them the very best sperm modern science could provide. Some two hundred children were born as a result of the bank's efforts, though none, it seems, proved to be an outstanding genius or even an accomplished eyeglass engineer. The bank closed in 1999, two years after the death of its founder, and, all in all, does not seem to have been greatly missed.

18 IN THE BEGINNING: CONCEPTION AND BIRTH

To begin my life with the beginning of
my life, I record that I was born.

—CHARLES DICKENS, *DAVID COPPERFIELD*

IT'S A LITTLE hard to know what to make of sperm. On the one hand, they are heroic: the astronauts of human biology, the only cells designed to leave our bodies and explore other worlds.

But on the other hand, they are blundering idiots. Shoot them into a womb and they seem curiously ill-prepared for the one task evolution has given them. They are terrible swimmers and appear to have almost no sense of direction. Unaided, it could take a sperm ten minutes to swim across a space the width of one of the words on this page. That's why a male orgasm is such a vigorous endeavor. What seems to the man purely a burst of pleasure really is a kind of rocket launch. Once the sperm are expelled, it isn't known whether they move about randomly until one strikes lucky or whether they are drawn to the waiting egg by some chemical signal.*

* From a Greek word meaning "to sow," the term "sperm" is first recorded in English in *The Canterbury Tales*. In those days, and at least until the time of Shakespeare, it was generally pronounced "sparm." Spermatozoa, the more formal designation, dates only from 1836, in a British anatomical guide.

In either case, overwhelmingly they fail. The chances of a successful fertilization from a single randomly timed act of sex have been calculated to be only about 3 percent. And matters seem to be getting worse across the Western world. About one in seven couples now seek help in conceiving.

Several studies have reported serious falls in sperm counts in recent decades. A meta-analysis in the journal *Human Reproduction Update,* based on 185 studies over nearly forty years, concluded that sperm counts in Western nations fell by more than 50 percent between 1973 and 2011.

Among the suggested causes have been diet, lifestyle, environmental factors, frequency of ejaculation, and even (seriously) wearing tight underpants, but no one knows. In a *New York Times* article titled "Are Your Sperm in Trouble?," the columnist Nicholas Kristof concluded that, yes, they probably are and attributed it to "a common class of chemical called endocrine disruptors, found in plastics, cosmetics, couches, pesticides and countless other products." He suggested that the average young man's sperm in the United States is about 90 percent faulty. Other studies in Denmark, Lithuania, Finland, Germany, and elsewhere have reported sharp falls in sperm counts.

Richard Bribiescas, a professor of anthropology, ecology, and evolutionary biology at Yale, believes that many of the reported counts are dubious and that even if correct there is no reason to suppose that there has been a decline in overall fertility. Diet and lifestyle, body temperature at the time of testing, and frequency of ejaculation are all likely to influence sperm counts, and the totals may vary widely over time in the same person. "Even if modest declines in sperm count have indeed occurred, there is no reason to believe that male fecundity has been compromised," Bribiescas wrote in *Men: Evolutionary and Life History*.

The fact is, it is really hard to say because there is such enormous variability in sperm production among healthy men anyway. The number of spermatozoa produced by the average man in the prime of life varies from 1 million to 120 million per milliliter, with an average of about 25 million per milliliter. The average ejaculation is about three milliliters, which means that a typical sex act produces enough sperm to repopulate a medium-sized country at the very least. Why there is such a broad range of wriggling potentiality, and indeed why such an extravagance of production even at the lower end when only one sperm is required for conception, are questions that science has yet to answer.

Women likewise are endowed with a massive surplus of reproductive possibility. It is a curious fact that every woman is born with her lifetime's supply of eggs already inside her. They are formed when she is still in the womb and sit in the ovaries for years and years before being called into play. As noted in the previous chapter, the idea of women being born with a full load of eggs–the formal name is ova–was first suggested by the great and busy German anatomist Heinrich von Waldeyer-Hartz, but even he would have been astonished at just how quickly and abundantly eggs are formed within the growing child. A twenty-week-old fetus will weigh no more than three or four ounces but will already have 6 million eggs inside her. That number falls to 1 million by the time of birth and continues to fall, though at a slower rate, through life. As she enters her childbearing years, a woman will have about 180,000 eggs primed and ready to go. Why she loses so many eggs along the way and yet enters her childbearing years with vastly more than she will ever need are two of life's many imponderables.

The bottom line is that as a woman ages, the number and quality of her eggs diminishes, and that can be a problem for those who postpone childbirth to the later stages of their productive years, which is exactly what is happening throughout the developed world. In six nations–Italy, Ireland, Japan, Luxembourg, Singapore, and Switzerland–the average age of women at their first birth is now over thirty, and in six others–Denmark, Germany, Greece, Hong Kong, the Netherlands, and Sweden–it is just under. (The United States is an outlier here. The average age of women at first birth there is 26.4, the lowest among rich nations.) Buried within these national averages are even greater variations within social or economic groupings. In Britain, for instance, the average age for women at first birth is 28.5 years, but for university graduates it is 35. As Carl Djerassi, the father of the contraceptive pill, noted in an essay in *The New York Review of Books*, by the age of thirty-five a woman's stock of eggs is 95 percent exhausted and those that remain are more liable to produce faults or surprises, like multiple births. Once women pass thirty, they are much more likely to have twins. The one certainty of procreation is that the older both parties get, the more difficulty they are likely to have conceiving, and the more problems they may encounter if they do conceive.

One intriguing paradox of reproduction is that women are having babies later but preparing for it earlier. The age of first menstruation for women has fallen from fifteen in the late nineteenth century to just twelve and a half today, at least in the West. That is almost certainly because of improved nutrition. But what cannot be explained is that the rate has accelerated even further in more recent years. Just since 1980, the age of puberty has fallen in America by eighteen months. About 15 percent of girls now begin puberty by age seven. That could be a reason for alarm. According to the *Baylor University Medical Center Proceedings*, evidence suggests that the prolonged exposure to estrogen substantially increases the risk of breast and uterine cancer in later life.

But let us suppose, for the sake of a happy narrative, that one hardy or lucky spermatozoon has reached the waiting egg. The egg is one hundred times larger than the sperm it pairs with. Fortunately, the sperm doesn't have to force its way in, but is welcomed like a long-lost if curiously diminutive friend. The sperm passes through an outer barrier called the zona pellucida and, all being well, fuses with the egg, which immediately activates a kind of electrical force field around itself to stop other sperm from getting through. The DNA from sperm and egg are combined into a new entity called a zygote. A new life has begun.

Success from this point is by no means assured. Perhaps as many as half of all conceptions are lost without being noticed. Without this, the rate of birth defects would be 12 percent instead of 2 percent. About 1 percent of implanted eggs end up stuck in the Fallopian tube, or somewhere else other than the womb, in what is known as an ectopic pregnancy (from a Greek word meaning "wrong place"). This can be very dangerous even now. Once it was a death sentence.

But if all goes well, within a week the zygote has produced ten or so cells known as pluripotent stem cells. These are the master cells of the body and one of the great miracles of biology. They determine the nature and organization of all the billions of cells that transform a little ball of possibility (known formally as a blastocyst) into a functioning and adorable little human (known as a baby). This moment of transition, when cells begin differentiating, is called gastrulation and has been described many times as the most important event of your life.

The system isn't perfect, however, and occasionally a fertilized egg splits to form identical (or monozygotic) twins. Identical twins are clones: they share the same genes and normally are very alike in appearance. They are in contrast to fraternal (or dizygotic) twins, which result when two eggs are produced in the same ovulation and both are fertilized by separate spermatozoa. In that case, the two babies develop side by side in the womb and are born together but are no more alike than any other two siblings. About one in 100 natural births result in fraternal twins, one in 250 in identical twins, one in 6,000 in triplets, and one in 500,000 in quadruplets, but fertility treatments greatly increase the likelihood of multiple births. Twins and other multiples are about twice as common today as they were in 1980. Women who have had twins already are ten times more likely to produce a second set than women who have not.*

* Doctors also sometimes use the terms "binovular" for fraternal twins and "uniovular" for a matched set.

* * *

Now things speed up considerably. After three weeks, the budding embryo has a beating heart. After 102 days, it has eyes that can blink. In 280 days, you have a new child. Along the way, at about eight weeks, the developing infant stops being called an embryo (from Greek and Latin words meaning "swollen") and starts being a fetus (from the Latin for "fruitful"). Altogether it takes just forty-one cycles of cell divisions to get from conception to a fully formed little human.

For much of this early period, the mother is likely to suffer from morning sickness, which, as almost any pregnant woman can tell you, doesn't happen just in the morning. About 80 percent of mothers-to-be suffer nausea, especially during the first three months, though for an unlucky few the condition can last for the whole nine months. Sometimes it becomes so severe that it gets a medical name: hyperemesis gravidarum. In such cases, it may require hospitalization. The most common theory for why women suffer morning sickness is that it encourages them to eat cautiously during the early stages of pregnancy, though that fails to explain why morning sickness then usually stops after a few weeks, when women should still probably be conservative in their food choices, or why women who eat a safe and bland diet get sick anyway. A big part of the reason that there are no cures for morning sickness is that the tragic experience in the 1960s of thalidomide, which was designed to combat morning sickness, left pharmaceutical companies permanently reluctant to try to make drugs of any type for pregnant women.

The business of pregnancy and birth has never been easy. However tedious and painful childbirth is now, it was much worse in the past. Until the modern era, levels of care and expertise were often pretty appalling. Just determining whether a woman was pregnant was a long-standing challenge for medical men. "We have known a practitioner of thirty years' standing blister the abdomen in the ninth month under the idea that he was treating a morbid growth," wrote one authority as late as 1873. The only truly reliable test, one doctor noted drily, was to wait nine months and see if a baby emerged. Medical students in England weren't required to study any part of obstetrics until 1886.

Women who suffered from morning sickness and were rash enough to declare it were likely to be bled, given enemas, or dosed with opiates. Women were sometimes bled even if they had no symptoms at all, as a precaution. They were also encouraged to loosen their corsets and to abjure "conjugal enjoyments."

Almost anything to do with reproduction was considered suspect—pleasure above all. In a popular book of 1899, *What a Young Woman Ought to Know*, Mary Wood-Allen, an American doctor and social reformer, told women that they could engage in conjugal relations within marriage so long as it was done "without a particle of sexual desire." In the same period, surgeons developed a new procedure called an oophorectomy—the surgical removal of the ovaries. For a decade or so, it was the operation of choice for well-off women with menstrual cramps, back pain, vomiting, headaches, even chronic coughing. In 1906, an estimated 150,000 American women underwent oophorectomies. It more or less goes without saying that it was an entirely pointless procedure.

Even with the best care, the long process of creating life and giving birth was agonizing and dangerous. Pain was considered a more or less necessary correlate of the process because of the biblical injunction "in sorrow thou shalt bring forth children." Death for mother or baby or both was not uncommon. "Maternity is another word for eternity" was a common saying.

For 250 years, the great fear was puerperal fever, or childbed fever as it was more commonly known. Like so many other diseases, it seemed to leap into ugly existence from out of nowhere. It was first recorded in Leipzig, Germany, in 1652 and then swept through Europe. It came on suddenly, often after a successful delivery when the new mother was feeling quite well, and left the victims fevered and delirious, and all too often dead. In some outbreaks, 90 percent of those infected died. Women often begged not to be taken to the hospital to give birth.

In 1847, a medical instructor in Vienna named Ignaz Semmelweis realized that if doctors washed their hands before conducting intimate examinations, the disease all but vanished. "God knows the number of women whom I have consigned prematurely to the grave," he wrote despairingly when he realized it was all a matter of hygiene. Unfortunately, no one at all listened to him. Semmelweis, who was not the most stable of persons at the best of times, lost his job and then his mind and ended up stalking through the streets of Vienna, ranting at thin air. Eventually, he was confined to an asylum where he was beaten to death by his guards. Streets and hospitals should be named for him, poor man.

A commitment to hygiene did gradually catch on, though it was an uphill battle. In Britain, the surgeon Joseph Lister (1827–1912) famously introduced the use of carbolic acid, an extract of coal tar, into operating theaters. He also believed that it was necessary to sterilize the air around patients, so he built a device that put out a mist of carbolic acid all around the operating table, which must have been pretty awful, particularly for anyone wearing spectacles. Carbolic acid was actually a terrible antiseptic. It could be absorbed through the skin of patients and medical practitioners alike and could cause kidney damage. In any case, Lister's practices didn't spread much beyond operating theaters.

In consequence, puerperal fever went on for far longer than it need have. Into the 1930s, it was responsible for four out of every ten maternal hospital deaths in Europe and America. As late as 1932, one mother in every 238 died in (or from) childbirth. (For purposes of comparison, today in Britain it is one in every 12,200; in the United States, it is one in every 6,000.)

Partly for these reasons, women continued to shun hospitals well into the modern era. Into the 1930s, fewer than half of American women gave birth in hospitals. In Britain, it was closer to one in five. Today the proportion in both countries is 99 percent. It was the rise of penicillin, not improved hygiene, that finally conquered puerperal fever.

Even now, however, there is huge variability in maternal mortality rates among countries of the developed world. In Italy, the number of women who die in childbirth is 3.9 per 100,000. Sweden is 4.6, Australia 5.1, Ireland 5.7, Canada 6.6. Britain comes only twenty-third on the list with 8.2 deaths per 100,000 live births, putting it below Hungary, Poland, and Albania. But also doing surprisingly poorly are Denmark (9.4 per 100,000) and France (10.0). Among developed nations, the United States is in a league of its own, with a maternal death rate of 16.7 per 100,000, putting it thirty-ninth among nations.

The good news is that for most women in the world childbirth has become vastly safer. In the first decade of the twenty-first century, only eight countries in the world saw their rates of childbirth deaths increase. The bad news is that the United States was one of those eight.

"Despite its lavish spending, the United States has one of the highest rates of both infant and maternal death among industrialized nations," according to *The New York Times*. The average cost of childbirth in the United States is about $30,000 for a conventional birth and $50,000 for a Cesarean, about three times the cost for either in the Netherlands. Yet American women are 70 percent more likely to die in childbirth than women in Europe and about three times more likely to suffer a pregnancy-related fatality than women in Britain, Germany, Japan, or the Czech Republic. Their infants are no less at risk. One of every 233 newborn babies dies in the United States, compared with just one in 450 in France and one in 909 in Japan. Even countries like Cuba (one in 345) and Lithuania (one in 385) do much better.

The causes in America include higher rates of maternal obesity, greater use of fertility treatments (which produce more failed outcomes), and increased incidence of the rather mysterious disease known as preeclampsia. Formerly known as toxemia, preeclampsia is a condition in pregnancy that leads to high blood pressure in the mother, which can be a danger to both her and her baby. About

3.4 percent of pregnant women get it, so it is not uncommon. It is thought to result from structural deformities in the placenta, but the cause is still largely a mystery. If not headed off, preeclampsia can advance to the more serious condition of eclampsia, when a woman may experience seizures, coma, or death.

If we don't know as much as we would like to about preeclampsia and eclampsia, it is in large part because we don't know as much as we ought to about the placenta. The placenta has been called "the least understood organ in the human body." For years the focus of medical research on childbirth was almost exclusively on the developing baby. The placenta was just a kind of adjunct to the process, useful and necessary but not very interesting. Only belatedly have researchers come to realize that the placenta does much more than just filter wastes and pass on oxygen. It takes an active role in the development of the child: stops toxins from passing from the mother to the fetus, kills parasites and pathogens, distributes hormones, and does everything it can to compensate for maternal deficiencies–if, say, the mother smokes or drinks or stays up too late. It is in a sense a kind of proto-mother for the developing baby. It can't work miracles if the mother is truly deprived or neglectful, but it can make a difference.

At all events, we now know, most miscarriages and other setbacks in pregnancy are because of problems with the placenta, not the fetus. Much of this is not well understood. The placenta acts as a barrier to pathogens, but only to some. The notorious Zika virus, for instance, can cross the placental barrier and cause terrible birth defects, but the very similar dengue virus cannot cross the barrier. No one knows why the placenta stops one but not the other.

The good news is that with intelligent, targeted prenatal care, outcomes for all kinds of conditions can be greatly improved. California addressed preeclampsia and the other leading causes of maternal death in childbirth through a program called the Maternal Quality Care Collaborative, and in just six years reduced the rate of childbirth deaths from 17 per 100,000 to just 7.3 between 2006 and 2013. During the same period, alas, the national rate rose from 13.3 deaths to 22 deaths per 100,000.

The moment of birth, the starting of a new life, really is quite a miracle. In the womb, a fetus's lungs are filled with amniotic fluid, but with exquisite timing at the moment of birth the fluid drains away, the lungs inflate, and blood from the tiny, freshly beating heart is sent on its first circuit around the body. What had until a moment before effectively been a parasite is now on its way to becoming a fully independent, self-maintaining entity.

We don't know what triggers birth. Something must count down the 280 days of human gestation, but no one has worked out where and what that mechanism is or what makes its alarm go off. All that is known is that the body begins to produce hormones called prostaglandins, which normally are involved in dealing with injuries to tissue but now activate the uterus, which begins a series of increasingly painful contractions to move the baby into position for birth. This first stage will go on for about twelve hours on average during a woman's first birth but often becomes faster for subsequent births.

The problem with human childbirth is cephalopelvic disproportion. In simple terms, a baby's head is too big for smooth passage through the birth canal, as any mother will freely attest. The average woman's birth canal is about an inch narrower than the width of the average newborn's head, making it the most painful inch in nature. To squeeze through this constricted space, the baby must execute an almost absurdly challenging ninety-degree turn as it proceeds through the pelvis. If ever there was an event that challenges the concept of intelligent design, it is the act of childbirth. No woman, however devout, has ever in childbirth said, "Thank you, Lord, for thinking this through for me."

The one piece of assistance that nature gives is that the baby's head is a bit compressible because the skull bones have not yet fused into a single plate. The reason for these contortions is that the pelvis had to undergo a number of design adjustments to make upright walking feasible, and that made human birth a much more trying and protracted business. Some species of primates can give birth in literally a couple of minutes. Human females can only dream of such ease.

We have made surprisingly little progress in making the process more bearable. As the journal *Nature* noted in 2016, "Women in labour have pretty much the same pain-relief options as their great grandmothers–namely gas and air, an injection of pethidine (an opioid) or an epidural anaesthetic." According to several studies, women are not terribly good at remembering the severity of the pain of childbirth; almost certainly this is a kind of mental defense mechanism to prepare them for further births.

You leave the womb sterile, or so it is generally thought, but are liberally swabbed with your mother's personal complement of microbes as you move through the birth canal. We are only beginning to understand the importance and nature of a woman's vaginal microbiome. Babies born by Cesarean section are robbed of this initial wash. The consequences for the baby can be profound. Various studies have found that people born by C-section have substantially increased risks for type 1 diabetes, asthma, celiac disease, and even obesity and an eightfold greater risk of developing allergies. Cesarean babies eventually acquire the same mix of microbes as those born vaginally–by a year their microbiota are usually indistinguishable–but there is something about those initial exposures that makes a long-term difference. No one has figured out quite why that should be.

Doctors and their hospitals can charge more for Cesarean births than for vaginal ones, and women understandably often like to know exactly when birth will take place. One-third of women in the United States give birth by Cesarean section now, and more than 60 percent of Cesareans are done for convenience rather than from medical necessity. In Brazil, nearly 60 percent of all births are by C-section; in Britain, it is 23 percent; in the Netherlands, it is 13 percent. If it were done only for medical reasons, the rate would be between 5 and 10 percent.

Other useful microbes are picked up from the mother's skin. Martin Blaser, a doctor and professor at New York University, suggests that the rush to clean up babies as soon as they are born may actually be depriving them of protective microorganisms.

On top of all that, about four women in every ten are given antibiotics during delivery, which means that doctors are declaring war on babies' microbes just as they are acquiring them. We've no idea what consequences this has for their long-term health, but it's unlikely to be good. There are concerns already that certain beneficial bacteria are becoming endangered. *B. infantis,* an important microbe in mother's milk, is found in up to 90 percent of children in developing countries but as little as 30 percent in the developed world.

Whether born by Cesarean or not, by the age of one the average baby has accumulated about a hundred trillion microbes, or so it has been estimated. But by that time, for reasons unknown, it appears to be too late to reverse the predisposition for acquiring certain diseases.

One of the most extraordinary features of early life is that nursing mothers produce over two hundred kinds of complex sugars—the formal name is oligosaccharides—in their milk that their babies cannot digest because humans lack the necessary enzymes. The oligosaccharides are produced purely for the benefit of the baby's gut microbes—as bribes, in effect. As well as nurturing symbiotic bacteria, breast milk is full of antibodies. There is some evidence that a nursing mother absorbs a little of her suckling baby's saliva through her breast ducts and that this is analyzed by her immune system, which adjusts the amount and types of antibodies she supplies to the baby, according to its needs. Isn't life marvelous?

In 1962, only 20 percent of American women breast-fed their babies. By 1977, this had increased to 40 percent, still clearly a minority. Today almost 80 percent of American women breast-feed just after birth, though that number falls to 49 percent after six months and 27 percent after a year. In Britain, the proportion starts at 81 percent but then plunges to 34 percent after six months and just 0.5 percent after a year, the worst rate in the developed world. In the poorer nations, many women were long encouraged by advertising to believe that infant formula was better for their babies than their own milk and so began switching to formula. But formula was expensive, so often they watered it down to make it go further. Sometimes also the only water available to them was less clean than their own breast milk. The result in some places was an increase in childhood mortality.

Although formulas have greatly improved over the years, no formula can fully replicate the immunological benefits of mother's milk. In the summer of 2018, the administration of President Donald Trump provoked dismay among many health authorities by opposing an international resolution to encourage breast-feeding and reportedly threatened Ecuador, the sponsor of the initiative, with trade sanctions if it didn't change its position. Cynics pointed out that the infant formula industry, which is worth $70 billion a year, might have had a hand in determining the U.S. position. A Department of Health and Human Services spokesperson denied that that was the case and said that America was merely "fighting to protect women's abilities to make the best choices for the nutrition of their babies" and to make sure that they were not denied access to formula–something the resolution wouldn't have done anyway.

In 1986, Professor David Barker of the University of Southampton in England proposed what has become known as the Barker hypothesis or, a little less snappily, the theory of fetal origins of adult disease. Barker, an epidemiologist, posited that what happens in the womb can determine health and well-being for the rest of one's life. "For every organ, there is a critical period, often very brief, when it goes through development," he said not long before his death in 2013. "It happens for different organs at different times. After birth only the liver and the brain and the immune system remain plastic. Everything else is done."

Most authorities now extend that period of crucial vulnerability from the moment of your conception to your second birthday–what has become known as the first thousand days. That means that what happens to you in this comparatively brief, formative period of your life can powerfully influence how comfortably alive you are decades later.

A famous example of this tendency was revealed by studies done in the Netherlands of people who lived through a very serious famine in the winter of 1944, when Nazi Germany stopped food from entering the parts of the country that were still in its control. The babies conceived during the famine had miraculously normal birth weights, presumably because their mothers instinctively diverted nutrition to their developing fetuses. And because the famine ended with the fall of Germany the following year, the children grew up eating as healthily and as well as any other children in the world. To the delight of all concerned, they seemed to escape all the effects of the Great Hunger, as it was known, and were indistinguishable from children born elsewhere in less stressful circumstances. But then a disturbing thing happened. As they reached their fifties and sixties, the famine children developed double the rate of heart disease, and increased rates of cancer, diabetes, and other life-compromising maladies, as children born elsewhere at the same time.

These days the legacy newborn babies bring into the world with them isn't a lack of nutrition but the opposite. So they are not only being born into households where people eat more and exercise less, but have an innate and enhanced vulnerability to succumb to the diseases that poor lifestyles bring.

It has been suggested that children growing up today will be the first in modern history to live shorter, less healthy lives than those of their parents. We aren't just eating ourselves into early graves, it seems, but breeding children to jump in alongside us.

19 NERVES AND PAIN

Pain has an element of blank;
It cannot recollect
When it began, or if there were
A day when it was not.

—EMILY DICKINSON

PAIN IS A strange and troublesome thing. Nothing in your life is more necessary and less welcome. It is one of humanity's greatest preoccupations and bewilderments and one of medical science's greatest challenges.

Sometimes it saves us, as we are vividly reminded each time we recoil from a jolt of electricity or try to walk barefoot across hot sand. So sensitive are we to threatening stimuli that our bodies are programmed to react to and withdraw from painful events before our brains have even received the news. All that is unquestionably a good thing. But quite a lot of the time—for up to 40 percent of people, by one calculation—pain just goes on and on and seems to have no purpose at all.

Pain is full of paradoxes. Its most self-evident characteristic is that it hurts—that's what it is there for, after all—but sometimes pain feels slightly wonderful: when your muscles ache after a long run, say, or when you slide into a bath that is at once unbearably hot but also, somehow, deliciously not. Sometimes we cannot explain it at all. One of the most severe and challenging of all pains is said to be phantom limb pain, when the sufferer perceives agonies in a part of the body that has been lost to accident or amputation. It is an obvious irony that one of the greatest pains we feel can be in a part of us that is no longer there. Worse, unlike normal pain, which usually abates as a wound heals, phantom pain may go on for a lifetime. No one can yet explain why. One theory is that in the absence of receiving any signal from the nerve fibers in the missing body part, the brain interprets this as an injury so severe that the cells have died, and so sends out an unending call of distress, like a burglar alarm that won't turn off. If surgeons know they are going to amputate a limb, they now often numb the nerves in the affected limb over a period of days beforehand to prepare the brain for the oncoming loss of feeling. The practice has been found to greatly reduce phantom limb pain.

If phantom pain has a rival, it may be said to be trigeminal neuralgia, named for the principal nerve of the face and historically known as tic douloureux (literally "painful twitch" in French). The condition is associated with a sharp, stabbing pain across the face—"like an electric shock," in the words of one pain specialist. Often there is a clear cause—when, for instance, a tumor presses against the trigeminal nerve—but sometimes no cause can be discerned. Patients may suffer periodic attacks, which can start and stop abruptly, without warning. These can be excruciating, but then they may cease altogether for days or weeks before coming back again. Over time, the pain may wander around the face. Nothing can explain why it wanders or what makes it come and go.

Exactly how pain works is, as you will gather, still largely a mystery. There is no pain center in the brain, no one place where pain signals congregate. A thought must travel through the hippocampus to become a memory, but a pain can surface almost anywhere. Stub your toe and the sensation will register across one set of brain regions; hit it with a hammer and it will light up others. Repeat the experiences, and the patterns may change yet again.

Perhaps the weirdest irony of all is that the brain has no pain receptors itself, yet it is where all pain is felt. "Pain only emerges when the brain gets it," says Irene Tracey, head of the Nuffield Department of Clinical Neurosciences at the University of Oxford and one of the world's leading authorities on pain. "The pain might have started in the big toe, but the brain is the thing that gives you the ouch. Up until then it is not pain."

All pain is private and intensely personal. Meaningful definition is impossible. The International Association for the Study of Pain summarizes pain as "an unpleasant sensory and emotional experience associated with actual or potential tissue damage, or described in terms of such damage," which is to say that it is anything that hurts, or might hurt, or sounds as if it might hurt, or feels as if it might hurt, whether literally or metaphorically. That pretty much covers every bad experience there is, from bullet wounds to the heartache of a failed relationship.

The best-known measure of pain is something called the McGill Pain Questionnaire, devised in 1971 by Ronald Melzack and Warren S. Torgerson at McGill University in Montreal. It is simply a detailed questionnaire that provides subjects with a list of seventy-eight words describing different levels of discomfort—"stabbing," "stinging," "dull," "tender," and so on. Many of the terms are vague or indistinguishable. Who could differentiate between "annoying" and "troublesome" or "miserable" and "horrible"? Largely for that reason, most pain researchers today use a simpler one-to-ten scale.

The whole experience of pain is obviously subjective. "I've had three children and believe me that has changed my experience of where the maximum lies," says Irene Tracey, with a broad and knowing smile, when we meet in her office at the John Radcliffe Hospital in Oxford. Tracey may be the busiest person in Oxford. As well as her extensive departmental and academic duties, at the time of my visit, in late 2018, she had just moved house, just returned from two trips abroad, and was about to take over as warden (or dean) of Merton College.

Tracey's working life is devoted to understanding how we perceive pain and how we might ameliorate it. Understanding pain is the hard part. "We still don't know exactly how the brain constructs the experience of pain," she says. "But we are making a lot of progress, and I think the whole landscape of our understanding of pain is going to change dramatically over the next few years."

One advantage Tracey has over previous generations of pain researchers is the possession of a really powerful magnetic resonance imaging machine. In her lab, Tracey and her assistants gently torment volunteers for the good of science by pricking them with pins or daubing them with capsaicin, the chemical behind the Scoville scale and the heat of chilies, as you may recall from chapter 6. Inflicting pain on innocent people is a delicate business—the pain needs to be genuinely felt but for obvious ethical reasons mustn't inflict serious or lasting damage—but it does allow Tracey and her colleagues to watch in real time how the subjects' brains respond to pain as it is administered.

As you might imagine, lots of people would love, for purely commercial reasons, to be able to peer into other people's brains to know when they are feeling pain, or being untruthful, or even perhaps responding favorably to a marketing ploy. Personal injury lawyers would be overjoyed to have pain profiles that they could present as evidence in court. "We are not at that point yet," says Tracey, with what appears to be a slight air of relief, "but where we are making really rapid progress is in learning how to manage and limit pain, and that is helping lots of people."

The experience of pain begins just beneath the skin in specialized nerve endings known as nociceptors. ("Noci-" is from a Latin word meaning "hurt.") Nociceptors respond to three kinds of painful stimuli: thermal, chemical, and mechanical, or at least so it is universally assumed. Remarkably, scientists have not yet found the nociceptor that responds to mechanical pain. It is extraordinary surely that when you whack your thumb with a hammer or prick yourself with a needle, we don't know what actually happens beneath your outer surface. All that can be said is that signals from all types of pain are conveyed on to the spinal cord and brain by two different types of fibers–fast-conducting A delta fibers (they're coated in myelin, so slicker, as it were) and slower-acting C fibers. The swift A delta fibers give you the sharp ouch of a hammer blow; the slower C fibers give you the throbbing pain that follows. Nociceptors only respond to disagreeable (or potentially disagreeable) sensations. Normal touch signals–the feel of your feet against the ground, your hand on a doorknob, your cheek on a satin pillow–are conveyed by different receptors on a separate set of A-beta nerves.

Nerve signals are not particularly swift. Light travels at 300 million meters per second, while nerve signals move at a decidedly more stately 120 meters a second–about 2.5 million times slower. Still, 120 meters a second is nearly 270 miles an hour, quite fast enough over the space of a human frame to be effectively instantaneous in most circumstances. Even so, as an aid to responding quickly, we have reflexes, which means that the central nervous system can intercept a signal and act on it before passing it on to the brain. That's why if you touch something very undesirable, your hand recoils before your brain knows what's going on. The spinal cord, in short, is not just a length of impassive cabling carrying messages between the body and the brain but an active and literally decisive part of your sensory apparatus.

Several of your nociceptors are polymodal, which means they are triggered by different stimuli. That's why spicy foods taste hot, for instance. They chemically activate the same nociceptors in your mouth that respond thermally to real heat. Your tongue can't tell the difference. Even your brain is a little confused. It realizes, at a rational level, that your tongue isn't literally on fire, but it sure feels that way. What is oddest of all is that the nociceptors somehow allow you to perceive a stimulus as pleasurable if it's a vindaloo and yelp inducing if it's a hot match head, even though both activate the same nerves.

The person who first identified nociceptors–who can indeed fairly be called patriarch of the central nervous system altogether–was Charles Scott Sherrington (1857–1952), one of the greatest and most inexplicably forgotten British scientists of the modern era. Sherrington's life seems to have been lifted straight out of a nineteenth-century boys' adventure story. A gifted athlete, he played soccer for Ipswich Town while still in school and had a distinguished rowing career at Cambridge. He was above all a brilliant student, winning many honors while impressing all who met him with his modest manner and keen intellect.

After graduating in 1885, he studied bacteriology under the great German Robert Koch, then embarked on a dazzlingly varied and productive career in which he did seminal work on tetanus, industrial fatigue, diphtheria, cholera, bacteriology, and hematology. He proposed the law of reciprocal innervation for muscles, which states that when one muscle contracts, a companion muscle must relax—essentially explaining how muscles work.

While studying the brain, he developed the concept of the synapse, coining the term "synapse" in the process. This in turn led to the idea of proprioception—another Sherrington coinage—which is the body's ability to know its own orientation in space. (Even with your eyes closed, you know whether you are lying down or whether your arms are outstretched and so on.) And this, in further turn, led to the discovery in 1906 of nociceptors, the nerve endings that alert you to pain. Sherrington's landmark book on the subject, *The Integrative Action of the Nervous System,* has been compared to Newton's *Principia* and Harvey's *De motu cordis* (*On the Motion of the Heart*) in terms of its revolutionary importance to its field.

But Sherrington's admirable qualities don't stop there. He was, by all accounts, a pretty wonderful person: devoted husband, gracious host, delightful company, beloved teacher. Among his students were Wilder Penfield, the authority on memory whom we met in chapter 4; Howard Florey, who won a Nobel Prize for his role in developing penicillin; and Harvey Cushing, who went on to become one of America's leading neurosurgeons.

In 1925, Sherrington astonished even his closest friends by

producing a volume of poetry, which was widely praised. Seven years later, he won a Nobel Prize for his work on reflexes. He was a distinguished president of the Royal Society, a benefactor of museums and libraries, and a devoted bibliophile with a world-class collection of books. At the age of eighty-three in 1940 he wrote a bestselling work, *Man on His Nature,* which went through several editions and was voted one of the hundred best books of modern Britain at the Festival of Britain in 1951. In it, he invented the expression "the enchanted loom" as a metaphor for the mind. And now, unaccountably, he is almost completely forgotten outside his field and not hugely remembered even there.

The nervous system is divided in various ways depending on whether you are looking at its structure or its function. Anatomically, it has two divisions. The central nervous system is the brain and spinal cord. The nerves radiating out from this central hub–the ones that reach out to the other parts of your body–are the peripheral nervous system. The nervous system is additionally divided by function into the somatic nervous system, which is the part that controls voluntary actions (like scratching your head), and the autonomic nervous system, which controls all those things like heartbeats that you don't have to think about because they are automatic. The autonomic nervous system is further divided into sympathetic and parasympathetic systems. The sympathetic is the part that responds when the body needs sudden actions–what is generally referred to as the fight-or-flight response. The parasympathetic is sometimes referred to as the "rest and digest" or "feed and breed" system and looks after a miscellany of other, generally less urgent matters like digestion and waste disposal, the production of saliva and tears, and sexual arousal (which may be intense but not urgent in the fight-or-flight sense).

An oddity of human nerves is that those in the peripheral nervous system can heal and regrow when damaged, whereas the more vital ones in the brain and spinal cord cannot. If you cut your finger, the nerves can grow back, but damage your spinal cord and you are out of luck. Spinal cord injuries are dismayingly common. More than one million people in the United States are paralyzed from them. More than half of spinal cord injuries in America result from car accidents or gunshot wounds, so, as you might expect, men are four times more likely to get a spinal cord injury than women. They are especially susceptible between the ages of sixteen and thirty–just when they are old enough to have guns and cars and foolish enough to misuse them.

Pain, like the nervous system itself, is classified in a multiplicity of ways, and these vary in type and number from authority to authority. The most common category is nociceptive pain, which simply means stimulated pain. It's the pain you get when you stub a toe or break your shoulder in a fall. It is sometimes referred to as "good" pain, in the sense that it is the kind of pain that tells you to rest the affected part and give it a chance to heal. A second type is inflammatory pain, for when tissue becomes swollen and red. A third category is dysfunctional pain, which is pain without external stimulus and that causes no nerve damage or inflammation. It is pain without evident purpose. A fourth kind of pain is neuropathic pain, in which nerves are damaged or grow sensitive, sometimes as a result of trauma, sometimes for no apparent reason.

When pains don't go away, pain goes from being acute to chronic. Some twenty years ago, Patrick Wall, a leading British neuroscientist, in an influential book called *Pain: The Science of Suffering,* maintained that pain beyond a certain level and duration is almost entirely pointless. He noted that nearly every textbook he had ever seen contained an illustration showing a hand recoiling from a flame or hot surface to demonstrate the usefulness of pain as a protective reflex.

"I despise that diagram for its triviality," he wrote with somewhat startling passion. "I would estimate that we spend a few seconds in an entire lifetime successfully withdrawing from a threatening stimulus. Unfortunately, we spend days and months in pain during our lifetime, none of which is explained by that silly diagram."

Wall singled out cancer pain as "the apogee of pointlessness."

Most cancers don't cause pain in their early stages when it might usefully alert us to take remedial action. Instead, all too often cancer pain becomes evident only when it is too late to be useful. Wall's observations came from the heart. He was dying of prostate cancer at the time. The book was published in 1999, and Wall died two years later. From the perspective of pain research, the two events together marked the end of an era.

Irene Tracey has been studying pain for twenty years—coincidentally almost exactly the period since Wall died—and has seen a complete transformation in that time in how pain is clinically regarded.

"Patrick Wall was in an era when people kept trying to hypothesize a *purpose* for chronic pain," she says. "Acute pain has an obvious point: it tells you that something is wrong and needs attention. They wanted chronic pain to have that kind of point, too—to exist for a purpose. But chronic pain has no purpose. It's just a system gone wrong, in the same way that cancer is a system gone wrong. We now believe that many types of chronic pain are diseases in their own right, something quite separate from acute pain."

There is a paradox at the heart of pain that makes its treatment particularly intractable. "When most parts of the body are damaged, they stop working—they switch off," Tracey says. "But when nerves are damaged, they do exactly the opposite—they switch on. Sometimes they just won't switch off, and that is when you get chronic pain." In the worst cases, as Tracey puts it, it is as if the volume knob on their pain has been turned all the way up. Figuring out how to turn that volume down has proved to be one of the greatest frustrations in medical science.

Generally, we don't feel pain in most of our internal organs. Any pain that arises from them is known as referred pain because it is "referred" to another part of the body. So the pain of coronary heart disease, for instance, may be felt in the arms or neck, sometimes in the jaw. The brain is also without feelings, which raises the natural question of where do headaches come from? The answer is that the scalp, the face, and the other outer parts of the head all have plenty of nerve endings—more than enough to account for most headaches. Even if it feels as if it were coming from deep within your head, a routine headache is almost certain to be a surface feature. Inside your skull, the meninges, the protective covering of the brain, also have nociceptors, and pressure on the meninges is what causes pain from brain tumors, but luckily that is something most of us will never have to experience.

You would think that if any condition is universal, it is the headache, but 4 percent of people say they have never had one. The International Classification of Headache Disorders recognizes fourteen categories of headaches—migraine, trauma-induced headache, infection-induced headache, disorder of homeostasis, and so on. However, most authorities divide headaches into two broader categories: primary headaches, such as migraine and tension headaches, which have no direct, identifiable cause, and secondary headaches, which arise from some other precipitating event, like an infection or tumor.

Among the most puzzling of headaches are migraines. Migraine (the word is a corruption of the French *demi-craine,* meaning "half the head") affects 15 percent of people but is three times more common in women than in men. Migraines are almost wholly a mystery. They are highly individual. Oliver Sacks in a book on migraines described nearly one hundred varieties of migraine. Some people feel surprisingly wonderful before migraines. The novelist George Eliot said she always felt "dangerously well" just before a migraine started. Others are indisposed for days and left feeling starkly suicidal.

Pain is curiously mutable. It can be increased, attenuated, or even ignored by the brain depending on the situation. In extreme circumstances, it may not register at all. A famous instance was at the Battle of Aspern-Essling during the Napoleonic Wars, when an Austrian colonel, directing operations from horseback, was informed by his aide-de-camp that his right leg had been shot away.

"*Donnerwetter,* so it has," replied the colonel phlegmatically, and kept on fighting.

Being depressed or worried will almost always increase perceived levels of pain. But equally pain is decreased by pleasant aromas, soothing images, pleasurable music, good food, and sex. Just having a sympathetic and loving partner cuts the reported pain of angina by half, according to one study. Expectation is hugely important, too. In one experiment done by Tracey and her team, when subjects in pain received morphine without being told, its analgesic effects were greatly lessened. In many ways, we feel the pain we expect to feel.

For millions of people, pain is a nightmare that cannot be escaped. According to the U.S. Institute of Medicine, part of the National Academy of Sciences, about 40 percent of adult Americans—100 million people—are experiencing chronic pain at any given moment. One-fifth of them will suffer it for more than twenty years. Altogether chronic pain affects more people than cancer, heart disease, and diabetes combined. It can be hugely debilitating. As the French novelist Alphonse Daudet noted in his classic *In the Land of Pain* (*La doulou* in French) almost a century ago, the pain that racked him as he was slowly ravaged by the effects of syphilis left him "deaf and blind to other people, to life, to everything except my wretched body."

Medical science offered very little in the way of safe, lasting relief back then. We are not much further along now. As Andrew Rice, a pain researcher at Imperial College London, told *Nature* in 2016, "The drugs we have relieve 50 percent of pain in somewhere between one in four and one in seven of the patients we treat. That's for the best drugs." In other words, some 75 percent to 85 percent of people get no benefit at all from even the best pain drugs, and those who do get benefit don't usually get much. Pain relief, as Tracey puts it, has been "a pharmacological graveyard." Pharmaceutical companies have poured billions and billions into drug development but have not come up with a drug that controls pain effectively and doesn't cause addictions.

One unhappy result has been the infamous opioid crisis. Opioids, as surely everyone knows by now, are painkillers that act in much the same way as heroin and come from the same addictive source: opiates. For a long time, they were mostly used sparingly, primarily for short-term relief after surgery or in the treatment of cancer. But in the late 1990s, pharmaceutical companies began pushing them as a long-term solution to pain. A promotional video made by Purdue Pharma, the maker of the opioid OxyContin, featured a physician who specializes in pain treatment looking straight into the camera and claiming with evident sincerity that opioids were perfectly safe and hardly ever addictive. "We doctors were wrong in thinking that opioids can't be used long term. They can be and they should be," he added.

The reality was rather different. People across America were becoming rapidly addicted and often dying. Between 1999 and 2014, by one estimate, a quarter of a million Americans died from opioid overdoses. Opioid abuse remains for the most part a peculiarly American problem. The United States has 4 percent of the world's population but consumes 80 percent of its opioids. About two million Americans are thought to be opioid addicts. Another ten million or so are users. The cost to the economy has been put at over $500 billion a year in lost earnings, medical treatments, and criminal proceedings. Opioid use has become such big business that we have now reached the surreal situation that pharmaceutical companies are producing drugs to alleviate the side effects of opioid overuse. Having helped to create millions of addicts, the industry is now profiting from medications designed to make their addiction a little more comfortable. So far the crisis doesn't seem to be going away. Every year opioids (both legal and illegal) claim forty-five thousand or so American lives, far more than are killed in car crashes.

The one positive to come out of the experience is that opioid fatalities have led to a rise in organ donations. In 2000, according to *The Washington Post,* fewer than 150 organ donors were opioid addicts; today the number is over 3,500.

* * *

In the absence of pharmaceutical perfection, Irene Tracey focuses on what she calls "free analgesia"—understanding how people can manage their pain through cognitive-behavioral therapies and exercise.

"It's been really interesting to me," she says, "how useful neuroimaging has been to persuade people to engage with the brain to recognize that it does seem to have a big role in making pain bearable. You can achieve a lot with that alone."

One of the great advantages with pain management is that we are marvelously suggestible, which is of course why the well-known placebo effect works. The concept of the placebo effect has been around for a very long time. In the modern medical sense of something given for psychological benefit, it is first recorded in a British medical text as far back as 1811, but the word "placebo" itself has existed in English since the Middle Ages. For most of its history, it meant a flatterer or sycophant. (Chaucer used it in that sense in *The Canterbury Tales.*) It comes from a Latin term meaning "to please."

Nobody knows quite why placebos work, but they do. In one experiment, people who had just had a wisdom tooth extracted had their faces massaged with an ultrasound device and overwhelmingly reported feeling better. What was interesting was that the treatment worked as well with the machine turned off as on. Other studies have shown that people given a colored tablet with corners will report feeling better than when given a plain white tablet. Red pills are deemed more fast acting than white pills. Green and blue pills have a more soothing effect. Patrick Wall, in his book on pain, reported how one doctor got good results from handing his patients pills held in a forceps, explaining that they were too potent to be held by bare fingers. Extraordinarily, placebos are even effective when people know they are placebos. Ted Kaptchuk of the Harvard Medical School gave people suffering from irritable bowel syndrome sugar pills and told them that that's all they were. Even so, 59 percent of those tested reported relief of symptoms.

The one problem with placebos is that while they are often

effective for matters over which our mind has some control, they can't help with problems that lie below the conscious level. Placebos don't shrink tumors or banish plaque from narrowed arteries. But then, come to that, neither do more aggressive painkillers, and placebos at least have never sent anyone to an early grave.

20 WHEN THINGS GO WRONG: DISEASES

I came to typhoid fever—read the symptoms—discovered that I had typhoid fever, must have had it for months without knowing it—wondered what else I had got; turned up St. Vitus's Dance—found, as I expected, that I had that too,—began to get interested in my case, and determined to sift it to the bottom, and so started alphabetically—read up ague, and learnt that I was sickening for it, and that the acute stage would commence in about another fortnight. Bright's disease, I was relieved to find, I had only in a modified form, and, so far as that was concerned, I might live for years.

—JEROME K. JEROME ON READING A MEDICAL TEXTBOOK

I

IN THE AUTUMN of 1948, people in the small city of Akureyri, on the north coast of Iceland, began to come down with an illness that was at first taken to be poliomyelitis, but then proved not to be. Between October 1948 and April 1949, almost five hundred people, out of a population of ninety-six hundred, grew ill. The symptoms were wondrously diverse—muscle aches, headaches, nervousness, restlessness, depression, constipation, disturbed sleep, loss of memory, and generally being out of sorts but in a pretty serious way. The illness didn't kill anyone, but it did make nearly every victim feel wretched, sometimes for months. The cause of the outbreak was a mystery. All tests for pathogens came back negative. The disease was so peculiarly specific to the vicinity that it became known as the Akureyri disease.

For about a year, nothing more happened. Then outbreaks began to occur in other, curiously distant places—in Louisville, Kentucky; in Seward, Alaska; in Pittsfield and Williamstown, Massachusetts; in a little farming community in the far north of England called Dalston. Altogether through the 1950s ten outbreaks were recorded in the United States and three in Europe. The symptoms everywhere were broadly similar but often with local peculiarities. People in some places said they felt unusually depressed or sleepy or had very specific muscle tenderness. As the disease proliferated, it attracted other names: post-viral syndrome, atypical poliomyelitis, and epidemic neuromyasthenia, by which it is most commonly known now. Why outbreaks didn't radiate outward to neighboring communities but rather leaped across great geographical expanses was just one of many puzzling aspects of the disease.*

All the outbreaks attracted little more than local attention, but in 1970, after several years of quiescence, the epidemic reappeared at Lackland Air Force Base in Texas, and now at last medical investigators began to look at it closely—though not, it must be said, much more productively. The Lackland outbreak made 221 people sick, most for about a week but some for up to a year. Sometimes just one person in a department came down with it; sometimes nearly everyone did. Most victims recovered completely, but a few experienced relapses weeks or months later. As usual nothing about the outbreak fit into a logical pattern, and all tests for bacterial or viral agents came back negative. Many of the victims were children too small to be suggestible, ruling out hysteria—the most common explanation for otherwise

* Because of the similarity of symptoms and difficulty of diagnosis, it is sometimes lumped in with chronic fatigue syndrome (CFS) but is really quite different. CFS (formally myalgic encephalomyelitis) tends to affect individuals, while epidemic neuromyasthenia hits populations.

unexplained mass outbreaks. The epidemic lasted for a little over two months, then ceased (apart from the relapses) and has never returned. A report in *The Journal of the American Medical Association* concluded that the victims had been suffering from a "subtle but nevertheless primarily organic illness whose effects may include exacerbation of underlying psychogenic illness." Which is another way of saying, "We have no idea."

Infectious diseases, as you will gather, are curious things. Some flit about like Akureyri disease, popping up seemingly at random, then going quiet for a time before popping up somewhere else. Others advance across landscapes like a conquering army. West Nile virus surfaced in New York in 1999 and within four years had covered the whole of America. Some diseases wreak havoc and then quietly withdraw, sometimes for years, occasionally forever. Between 1485 and 1551, Britain was repeatedly ravaged by a terrifying malady called the sweating sickness, which killed untold thousands. Then it abruptly stopped and was never seen there again. Two hundred years later, a very similar illness appeared in France, where it was called the Picardy sweats. Then it too vanished. We have no idea where and how it incubated, why it disappeared when it did, or where it might be now.

Baffling outbreaks, particularly small ones, are more common than you might think. Every year in the United States about six people, preponderantly in northern Minnesota, grow ill with Powassan virus. Some victims suffer only mild flu-like symptoms, but others are left with permanent neurological damage. About 10 percent die. There is no cure or treatment. In Wisconsin in the winter of 2015–16, fifty-four people, from twelve different counties, fell ill from a little-known bacterial infection called Elizabethkingia. Fifteen of the victims died. Elizabethkingia is a common soil microbe, but it only rarely infects people. Why it suddenly became rampant across a wide area of the state, and then stopped, is anyone's guess. Tularemia, an infectious disease spread by ticks, kills 150 or so people a year in America, but with unaccountable variability. In the eleven years from 2006 to 2016, it killed 232 people in Arkansas, but only one person in neighboring Alabama despite abundant similarities in climate, ground cover, and tick populations. The list goes on and on.

Perhaps no case has been harder to explain than Bourbon virus, named for the county in Kansas where it first appeared in 2014. In the spring of that year, John Seested, a healthy, middle-aged man from Fort Scott, about ninety miles south of Kansas City, was working on his property when he noticed he had been bitten by a tick. After a while he began to grow achy and feverish. When his symptoms didn't improve, he was admitted to a local hospital and given doxycycline, a drug for tick-bite infections, but it had no effect. Over the next day or two, Seested's condition steadily worsened. Then his organs began to fail. On the eleventh day he died.

Bourbon virus, as it became known, represented a whole new class of virus. It came from a group called thogotoviruses, which are endemic to regions of Africa, Asia, and eastern Europe, but this particular strain was entirely novel. Why it appeared suddenly in the very middle of the United States is a mystery. No one else got the disease in Fort Scott or anywhere else in Kansas, but a year later a man 250 miles away in Oklahoma came down with it. At least five other cases have since been reported. The Centers for Disease Control is curiously reticent about numbers. It says only that "as of June 2018, a limited number of Bourbon virus disease cases have been identified in the Midwest and southern United States," a somewhat odd way of putting it because there is clearly no limit on the number of infections any disease can cause. The most recent confirmed case, at the time of writing, was a fifty-eight-year-old woman who was bitten by a tick while working in Meramec State Park in eastern Missouri and died soon afterward.

It may be that all of these elusive diseases infect lots more people, but not seriously enough to be noticed. "Unless doctors are doing laboratory tests specifically for this infection, they'll miss it," a CDC scientist told a reporter for National Public Radio in 2015, in reference to Heartland virus, yet another mysterious pathogen. (There really are a lot of these.) As of late 2018, the Heartland virus had infected some twenty people and killed an unknown number since it first appeared near St. Joseph, Missouri, in 2009. But so far all that can be said for sure is that these diseases only infect a very unlucky few people far removed from each other with no known connections between them.

Sometimes it turns out that what seems to be a new disease is not new at all. Such proved to be the case in 1976 when delegates to an American Legion convention at the Bellevue-Stratford Hotel in Philadelphia began to fall ill from a disease no authority could identify. Soon many of them were dying. Within a few days, 34 were dead and another 190 or so were ill, some gravely. An additional puzzle was that about one-fifth of the victims had not set foot in the hotel, but had only walked past it. Epidemiologists from the Centers for Disease Control took two years to identify the culprit, a novel bacterium from a genus they called *Legionella*. It had spread through the hotel's air-conditioning ducts. The unlucky passersby had been infected by walking through exhaust fumes.

Only much later was it realized that *Legionella* was almost certainly responsible for similarly unexplained outbreaks in Washington, D.C., in 1965 and in Pontiac, Michigan, three years later. Indeed, it turned out that the Bellevue-Stratford Hotel had suffered a smaller, less lethal cluster of pneumonia cases two years earlier during a convention of the Independent Order of Odd Fellows, but that had attracted little attention because no one died. We now know that *Legionella* is widely distributed in soil and freshwater, and Legionnaires' disease has become more common than most people suppose. A dozen or so outbreaks are reported each year in America, and about eighteen thousand people become sick enough to need hospitalization, but the CDC thinks that that number is probably underreported.

Much the same thing happened with Akureyri disease where further investigations showed that there had been similar outbreaks in Switzerland in 1937 and 1939 and probably in Los Angeles in 1934 (where it was taken to be a mild form of poliomyelitis). Where, if anywhere, it was before that is unknown.

Whether or not a disease becomes epidemic is dependent on four factors: how lethal it is, how good it is at finding new victims, how easy or difficult it is to contain, and how susceptible it is to vaccines. Most really scary diseases are not actually very good at all four; in fact, the qualities that make them scary often render them ineffective at spreading. Ebola, for instance, is so terrifying that people in the area of infection flee before it, doing everything in their powers to escape exposure. In addition, it incapacitates its victims swiftly, so most are removed from circulation before they can spread the disease widely anyway. Ebola is almost ludicrously infectious—a single droplet of blood no bigger than this *o* may contain a hundred million Ebola particles, every one of them as lethal as a hand grenade—but it is held back by its clumsiness at spreading.

A successful virus is one that doesn't kill too well and can circulate widely. That's what makes flu such a perennial threat. A typical flu renders its victims infectious for about a day before they get symptoms and for about a week after they recover, which turns every victim into a vector. The great Spanish flu of 1918 racked up a global death toll of tens of millions—some estimates put it as high as a hundred million—not by being especially lethal but by being persistent and highly transmissible. It killed only about 2.5 percent of victims, it is thought. Ebola would be more effective—and in the long run more dangerous—if it mutated a milder version that didn't strike such panic into communities and made it easier for victims to mingle with unsuspecting others.

That is, of course, no grounds for complacency. Ebola was only formally identified in the 1970s, and until recently all its outbreaks were isolated and short-lived, but in 2013 it spread to three countries—Guinea, Liberia, and Sierra Leone—where it infected twenty-eight thousand people and killed eleven thousand. That's a big outbreak.

On several occasions, thanks to air travel, it escaped to other countries, though fortunately in each instance it was contained. We may not always be so lucky. Hypervigilance makes it less likely diseases will spread, but it's no guarantee that they won't.

It's remarkable that bad things don't happen more often. According to one estimate reported by Ed Yong in *The Atlantic,* the number of viruses in birds and mammals that have the potential to leap the species barrier and infect us may be as high as 800,000. That is a lot of potential danger.*

II

IT IS SOMETIMES said, only partly in jest, that the worst health initiative in history was the invention of agriculture. Jared Diamond has called it "a catastrophe from which we have never recovered."

Perversely, farming didn't bring improved diets but almost everywhere poorer ones. Focusing on a narrower range of staple foods meant most people suffered at least some dietary deficiencies, without necessarily being aware of it. Moreover, living in proximity to domesticated animals meant that their diseases became our diseases. Leprosy, plague, tuberculosis, typhus, diphtheria, measles, influenzas—all vaulted from goats and pigs and cows and the like straight into us. By one estimate, about 60 percent of all infectious diseases are zoonotic (that is, from animals). Farming led to the rise of commerce and literacy and the fruits of civilization but also gave us millennia of rotten teeth, stunted growth, and diminished health.

We forget how devastating many diseases were until quite recent times. Take diphtheria. Into the 1920s, before the introduction of a vaccine, it struck down more than 200,000 people a year in America, killing 15,000 of them. Children were especially susceptible. It usually

* When talking of diseases, people often use "infectious" and "contagious" interchangeably, but there is a difference. An infectious disease is one caused by a microbe; a contagious disease is one transmitted by contact.

started with a mild temperature and a sore throat, so at first was easily mistaken for a cold, but it soon became much more serious as dead cells accumulated in the throat, forming a leathery coating (the term "diphtheria" comes from the Greek for "leather"; the disease, incidentally, is correctly pronounced "diff-theria," not "dip-theria") that made breathing increasingly difficult, and the disease spread through the body, shutting down organs one by one. Death tended to follow swiftly. There were many cases of parents losing all their children in a single outbreak. Today diphtheria has become so rare—just five cases in the United States in the most recent decade measured—that many doctors would struggle to recognize it.

Typhoid fever was no less frightening and caused at least as much distress. The great French microbiologist Louis Pasteur understood pathogens better than anyone of his day but still lost three of his five children to typhoid fever.

Typhoid and typhus have similar names and symptoms but are different diseases. Both are bacterial in origin and marked by sharp abdominal pain, listlessness, and a tendency to grow confused. Typhus is caused by a rickettsia bacillus; typhoid is caused by a type of salmonella bacillus and is the more serious of the two. A small proportion of people infected with typhoid—between 2 and 5 percent—are infectious but have no symptoms of illness, making them highly effective, if nearly always unwitting, vectors. The most famous such carrier was a shadowy cook and housekeeper named Mary Mallon who became notorious in the early years of the twentieth century as Typhoid Mary.

Almost nothing is known of her beginnings. She was variously reported in her own day as being from Ireland, England, or the United States. All that can be said for certain is that from young adulthood Mary worked in a number of well-to-do households, mostly in the New York City area, and wherever she went, two things always happened: people came down with typhoid, and Mary abruptly disappeared. In 1907, after a particularly bad outbreak, she was tracked down and tested and in the process became the first person to be confirmed as an asymptomatic carrier—that is, was infectious but had no symptoms herself. So fearsome did this make her that she was held in protective custody, very much against her will, for three years.

She was released when she promised never again to take a job handling food. Mary, alas, was not the most trustworthy of souls. Almost immediately she began working in kitchens again, spreading typhoid to a number of new locations. She managed to elude capture until 1915, when twenty-five people developed typhoid at the Sloane Hospital for Women in Manhattan, where Mary had been working under an assumed name as a cook. Two of the victims died. Mary fled but was recaptured and spent the remaining twenty-three years of her life under house arrest on North Brother Island in the East River until her death in 1938. She was personally responsible for at least fifty-three cases of typhoid and three confirmed deaths, but possibly many more. The particular tragedy of it is that she could have spared her unfortunate victims if she had just washed her hands before handling food.

Typhoid may not worry people as it once did, but it still affects more than 20 million people a year around the world and kills between 200,000 and 600,000, depending on whose figures you rely on. The United States has an estimated 5,750 cases each year, about two-thirds brought in from abroad but nearly 2,000 acquired domestically.

If you want to imagine what a disease might do if it became bad in every possible way, you could do no better than consider the case of smallpox. Smallpox is almost certainly the most devastating disease in the history of humankind. It infected nearly everyone who was exposed to it and killed about 30 percent of victims. The death toll in the twentieth century alone is thought to have been around 500 million. Smallpox's astounding infectiousness was vividly demonstrated in Germany in 1970 after a youthful tourist developed it upon returning home from a trip to Pakistan. He was placed in hospital quarantine but opened his window one day to sneak a cigarette. This, it has been reported, was enough to infect seventeen others, some two floors away.

Smallpox only infects humans, and that proved to be its fatal weakness. Other infectious diseases—flus notably—can disappear from human populations but rest up, as it were, among birds or pigs or other animals. Smallpox had no such reservoir to retreat to as humans gradually persecuted it into smaller and smaller patches of the planet. At some point in the distant past, it had lost the ability to infect other animals in order to focus exclusively on humans. As it turned out, it chose the wrong enemy.

Now the only way any human can get smallpox is if we inflict it upon ourselves. Unfortunately, that has happened. In 1978, at the University of Birmingham in England, a medical photographer named Janet Parker went home from work early one afternoon in late summer complaining of a blinding headache. Soon she was very ill indeed—fevered, delirious, and covered in pustules. She had contracted smallpox via an air duct from a lab one floor below her office. There, a virologist named Henry Bedson had been studying one of the last smallpox samples on Earth still allowed for research. He was frantically working against a deadline before his own stocks were to be destroyed and evidently grew careless in keeping them safe. Poor Janet Parker died about two weeks after being exposed and in so doing became the last person on Earth to be killed by smallpox. She had actually been vaccinated against the disease twelve years earlier, but smallpox vaccine doesn't last. When Bedson learned that smallpox had escaped from his lab and killed an innocent person, he went out to his garden shed and committed suicide, so in a sense he was smallpox's last victim. The hospital ward on which Parker was treated was sealed off for five years.

Two years after Parker's terrible death, on May 8, 1980, the World Health Organization announced that smallpox had been eradicated from Earth, the first and so far only human disease to be made extinct. Officially just two stocks of smallpox remain in the world now—in government freezers at the Centers for Disease Control in Atlanta, Georgia, and at a Russian virology institute near Novosibirsk in Siberia. Both countries have several times promised to destroy the remaining stocks but never have. In 2002, the CIA claimed there were probably also stocks in France, Iraq, and North Korea. No one can say whether, or how many, samples may survive accidentally as well. In 2014, someone looking through a storage area at a Food and Drug Administration facility in Bethesda, Maryland, found vials of smallpox dating from the 1950s but still viable. The vials were destroyed, but it was an unnerving reminder of how easily such samples can be overlooked.

With smallpox gone, tuberculosis is today the deadliest infectious disease on the planet. Between 1.5 and 2 million people die of it every year. It is another disease that we have mostly forgotten, but only a couple of generations ago it was devastating. Lewis Thomas, writing in *The New York Review of Books* in 1978, recalled how hopeless all treatments for TB were in the 1930s when he was a medical student. Anyone could catch it, he noted, and there was really nothing you could do to make yourself safe from infection. If you got it, that was it. "The hardest part of the disease, for both the patient and the family, was that it took so long to die," Thomas wrote. "The only relief was a curious phenomenon near the end, known as spes phthisica, when the patient suddenly became optimistic and hopeful, even mildly elated. This was the worst of signs; spes phthisica meant that death was coming soon."

As a scourge, TB actually got worse as time passed. Until late in the nineteenth century, it was known as consumption and was believed to be inherited. But when Robert Koch discovered the tubercle bacillus in 1882, the medical community realized beyond doubt that it was infectious—a far more unnerving proposition to loved ones and carers alike—and it became more widely known as tuberculosis. Victims were previously sent to sanatoriums entirely for their own sake; now there was a more urgent sense of exile.

Almost everywhere patients were subjected to harsh regimens. At some institutions, doctors reduced patients' lung capacity by cutting nerves to their diaphragm (a process known as a phrenic crush) or by injecting gas into their chest cavity so that the lungs couldn't fully inflate. At Frimley Sanatorium in England, authorities tried the opposite tack. Inmates were given pickaxes and made to do hard, pointless labor in the belief that that would strengthen their wearied lungs. None of these did, or possibly could do, the slightest bit of good. In most places, however, the approach was simply to keep patients very quiet to try to stop the disease from spreading from their lungs to other parts of their bodies. Patients were forbidden to talk, write letters, or even read books or newspapers for fear that the content would unnecessarily excite them. Betty MacDonald, in her popular and still very readable 1948 book, *The Plague and I,* about her own experiences in a TB sanatorium in Washington State, recorded that she and other inmates were allowed visits by their children just once a month for ten minutes and by spouses and other adults for two hours on Thursdays and Sundays. Patients were not allowed to talk or laugh unnecessarily or to sing ever. They were ordered to lie perfectly still for most of their waking day and not permitted to bend over or reach for things.

If TB is off the radar for most of us, that's because 95 percent of its more than a million and a half annual deaths are in low- or middle-income countries. About one in every three people on the planet carries the TB bacterium, but only a small proportion of those contract the disease. But it is still around. About seven hundred people a year die from tuberculosis in America. Some boroughs of London now have rates of infection that nearly match those of Nigeria or Brazil. No less alarmingly, drug-resistant strains of TB now account for 10 percent of new cases. It is entirely possible that we could one day in the not too distant future be facing an epidemic of TB that medicine cannot treat.

Lots of historically formidable diseases are still out there, not quite entirely vanquished. Even bubonic plague is still around, believe it or not. The United States averages seven cases a year. Most years there are one or two deaths. And there are lots of diseases in the wider world from which most of us in the developed world are spared—diseases like leishmaniasis, trachoma, and yaws, which few of us have even heard of. Those three and fifteen others, known collectively as neglected tropical diseases, affect more than a billion people worldwide. More than 120 million people, to take just one example, suffer from lymphatic filariasis, a disfiguring parasitic infection. What is particularly unfortunate is that a simple compound added to table salt could eliminate the filariasis wherever it appears. Many of the other neglected tropical diseases are beyond horrible. Guinea worms grow up to a meter long inside the bodies of their victims, then escape by burrowing out of their skin. The only treatment, even now, is to speed the process of exit by winding the worms onto a stick as they emerge.

To say that much of our progress against these diseases has been hard won is to put it mildly. Consider the contribution of the great German parasitologist Theodor Bilharz (1825–62), who is often called the father of tropical medicine. His entire career was devoted, at constant risk to himself, to trying to understand and conquer some of the world's worst infectious diseases. Wishing to better understand the truly horrid disease schistosomiasis—also now sometimes called bilharzia in his honor—Bilharz bandaged the pupae of cercariae worms to his stomach and took careful notes over the following days as they burrowed through his skin en route to invading his liver. He survived that experience but died soon afterward, aged just thirty-seven, while trying to help stop a typhus epidemic in Cairo. Similarly, Howard Taylor Ricketts (1871–1910), the American discoverer of the bacterial group rickettsia, went to Mexico to study typhus but contracted the disease himself and died. His fellow American Jesse Lazear (1866–1900), from the Johns Hopkins Medical School, went to Cuba in 1900 to try to prove that yellow fever was spread by mosquitoes, caught the disease—probably by intentionally infecting himself—and died. Stanislaus von Prowazek (1875–1915), of Bohemia, traveled the world studying infectious diseases, and found the agent behind trachoma, before succumbing to typhus himself in 1915 while working on an outbreak at a German prison. I could go on and on. Medical science has never produced a more noble and selfless group of investigators than the pathologists and parasitologists who risked and all too often lost their lives in trying to conquer the most pernicious of the world's diseases in the late nineteenth and early twentieth centuries. There ought to be a monument to them somewhere.

III

IF WE DON'T die so much from communicable diseases anymore, plenty of other maladies have stepped in to fill the gap. Two types of diseases in particular are more visible now than they were in times past, in part at least because we aren't being killed off by other things first.

One is genetic diseases. Twenty years ago, about five thousand genetic diseases were known. Today it is seven thousand. The number of genetic diseases is constant. What has changed is our ability to identify them. Sometimes one rogue gene can cause a breakdown, as with Huntington's disease, which used to be known as Huntington's chorea, from the Greek for "dance," a strange and decidedly insensitive reference to the jerky movements of Huntington's sufferers. It is a thoroughly wretched disease, affecting about one person in every ten thousand. Symptoms usually first appear when the victim is in his or her thirties or forties, and progress ineluctably to senility and premature death. It is all because of one mutation in the HTT gene, which produces a protein called huntingtin, one of the largest and most complex proteins in the human body, and we have no idea what huntingtin is for.

Far more often, multiple genes are at play, usually in ways too complex to fully understand. The number of genes that have been implicated in inflammatory bowel disease, for instance, is comfortably over a hundred. At least forty have been linked to type 2 diabetes, and that is before you start to factor in other determinants like health and lifestyle. Most diseases have a complex array of triggers.

That means that it is often impossible to pinpoint a cause. Take multiple sclerosis, a disease of the central nervous system in which sufferers experience a gradual onset of paralysis and loss of motor control, nearly always beginning before the age of forty. It is indubitably genetic, but it also has a geographical element that no one can quite explain. People from northern Europe get it much more often than people from warmer climes. As David Bainbridge has observed, "Why a temperate climate should make you attack your own spinal cord is not so obvious. Yet the effect is clear, and it has even been shown that if you are a northerner you can reduce your risk by relocating southward before puberty." It also affects women disproportionately, again for no reason that anyone has yet determined.

Mercifully, most genetic diseases are quite rare, often vanishingly so. One of the more famous sufferers of a rare genetic disorder was the artist Henri de Toulouse-Lautrec, who is thought to have suffered from pycnodysostosis. Toulouse-Lautrec was normally proportioned until puberty, but then his legs stopped growing while his trunk continued growing to normal adult size. In consequence, when standing, he looked as if he were on his knees. Only about two hundred cases of the disorder have ever been recorded.

Rare diseases are defined as diseases that afflict no more than one person in two thousand, and there is a paradox at their heart, which is that although each disease doesn't affect many people, collectively they affect a lot. Altogether there are about seven thousand rare diseases—so many that about one person in seventeen in the developed world has one, which isn't very rare at all. But, sadly, so long as a disease affects only a small number of people, it is unlikely to get much research attention. For 90 percent of rare diseases, there are no treatments at all.

A second category of disorders that have become more common in modern times, and represent a much greater risk for most of us, is what Professor Daniel Lieberman of Harvard calls mismatch diseases—that is, diseases brought on by our indolent and overindulgent modern lifestyles. The idea, roughly, is that we are born with the bodies of hunter-gatherers but pass our lives as couch potatoes. If we want to be healthy, we need to eat and move about a little more like our ancient ancestors did. That doesn't mean we have to eat tubers and hunt wildebeest. It means we should consume a lot less processed and sugary foods and get more exercise. Failure to do that, however, is what is giving us the disorders like type 2 diabetes and cardiovascular disease that are killing us in great numbers. Indeed, as Lieberman notes, medical care is actually making things worse by treating the symptoms of mismatch diseases so effectively that we "unwittingly perpetuate their causes." As Lieberman puts it with chilling bluntness, "You are most likely going to die from a mismatch disease." Even more chillingly, he believes that 70 percent of the diseases that kill us could easily be preventable if we would just live more sensibly.

When I met Washington University's Michael Kinch in St. Louis, I asked him what he believed was the greatest disease risk to us now.

"Flu," he said without hesitation. "Flu is way more dangerous than people think. For a start, it kills a lot of people already–about thirty to forty thousand every year in the United States–and that's in a so-called good year. But it also evolves very rapidly, and that's what makes it especially dangerous."

Every February, the World Health Organization and the Centers for Disease Control get together and decide what to make the next flu vaccine from, usually based on what's going on in eastern Asia. The problem is that flu strains are extremely variable and really hard to predict. You are probably aware that all flus have names like H5N1 or H3N2. That is because every flu virus has two types of proteins on its surface–hemagglutinin and neuraminidase–and these account for the *H* and *N* in their names. H5N1 means that the virus combines the fifth known iteration of hemagglutinin with the first known iteration of neuraminidase, and for some reason that is a particularly nasty combination. "H5N1 is the version commonly known as bird flu, and it kills between 50 to 90 percent of victims," says Kinch. "Luckily, it isn't readily transmissible between humans. So far this century, it has killed about four hundred people–roughly 60 percent of those it has infected. But look out if it mutates."

Based on all the available information, the WHO and CDC

announce their decision on February 28, and all the flu vaccine manufacturers in the world begin working on the same strain. Says Kinch, "From February to October they make the new flu vaccine, in the hope that we will be ready for the next big flu season. But when a really devastating new flu emerges, there's no guarantee that we will actually have targeted the right virus."

In the 2017–18 flu season, to take one recent example, people who had been vaccinated were only 36 percent less likely to get flu than those who hadn't been vaccinated. In consequence, it was a bad year for flu in America, with a death toll estimated at eighty thousand. In the event of a really catastrophic epidemic—one that killed children or young adults in large numbers, say—Kinch believes we wouldn't be able to produce vaccine fast enough to treat everyone, even if the vaccine was effective.

"The fact is," he says, "we are really no better prepared for a bad outbreak today than we were when Spanish flu killed tens of millions of people a hundred years ago. The reason we haven't had another experience like that isn't because we have been especially vigilant. It's because we have been lucky."

21 WHEN THINGS GO VERY WRONG: CANCER

We are bodies. They go wrong.

—TOM LUBBOCK, *UNTIL FURTHER NOTICE, I AM ALIVE*

I

CANCER IS THE malady most of us fear more than any other, yet much of that dread is fairly recent.* In 1896, when the newly founded *American Journal of Psychology* asked people to name the health crises they most feared, hardly any mentioned cancer. Diphtheria, smallpox, and tuberculosis were the most worrying afflictions, but even lockjaw, drowning, being bitten by a rabid animal or caught in an earthquake were more fearsome to the average person than cancer.

Partly this was because people in the past often didn't live long enough to get cancer in great numbers. As a colleague told Siddhartha Mukherjee, author of *The Emperor of All Maladies,* a history of cancer, "The early history of cancer is that there is very little early history of cancer." It isn't that cancer didn't exist at all, but more that it

* Originally "cancer" described any non-healing sore, from which it is related to "canker." In its more specific modern sense, it dates from the sixteenth century. The word comes from the Latin for "crab" (which is why the celestial constellation and its associated zodiac sign are called Cancer). It is said that Hippocrates, the Greek physician, used the term for tumors because their shape reminded him of crabs.

didn't register with people as something probable and fearful. In that sense it was rather like pneumonia now. Pneumonia is still the ninth most common cause of death, yet few of us greatly fear dying from it because we tend to associate it with frail elderly people who are about to shuffle off anyway. So it was for a very long time with cancer.

All that changed in the twentieth century. Between 1900 and 1940, cancer jumped from eighth place to second place (just behind heart disease) as a cause of death, and it has cast a long shadow over our perceptions of mortality ever since. Today some 40 percent of us will discover we have cancer at some point in our lives. Many, many more will have it without knowing it and will die of something else first. Half of men over sixty and three-quarters over seventy, for instance, have prostate cancer at death without being aware of it. It has been suggested, in fact, that if all men lived long enough, they would all get prostate cancer.

Cancer in the twentieth century became not only the great dread but the great stigma. A survey of physicians in America in 1961 found that nine out of ten did not inform patients when they had cancer because the shame and horror of it were so great. Surveys in Britain at about the same time found that roughly 85 percent of cancer patients wished to know if they were dying but that between 70 and 90 percent of doctors declined to tell them anyway.

We tend to think of cancer as something we catch, like a bacterial infection. In fact, cancer is entirely internal, a case of the body turning on itself. In 2000, a landmark paper in the journal *Cell* listed six attributes in particular that all cancer cells have, namely:

They divide without limit.
They grow without direction or influence from outside agents like hormones.
They engage in angiogenesis, which is to say they trick the body into giving them a blood supply.

They disregard any signals to stop growing.
They fail to succumb to apoptosis, or programmed cell death.
They metastasize, or spread to other parts of the body.

What it comes down to really is cancer is, appallingly, your own body doing its best to kill you. It is suicide without permission.

"That's why cancers aren't contagious," says Dr. Josef Vormoor, founding clinical director of pediatric hemato-oncology at the new Princess Máxima Center for childhood cancers in Utrecht, the Netherlands. "They are you attacking you."

Vormoor is an old friend, whom I first met when he was in a previous post as director of the Northern Institute for Cancer Research at Newcastle University in England. He joined the Princess Máxima Center shortly before its opening in the summer of 2018.

Cancer cells are just like normal cells except that they are proliferating wildly. Because they are so seemingly normal, the body sometimes fails to detect them and doesn't invoke an inflammatory response as it would with a foreign agent. That means that most cancers in their early stages are painless and invisible. It is only when tumors grow big enough to press on nerves or form a lump that we become aware that something is wrong. Some cancers can quietly accrete for decades before they become evident. Others never become evident at all.

Cancer is quite unlike other maladies. It is often relentless in its attacks. Victory against it is nearly always hard won and often at great cost to the victim's overall health. It will retreat under an onslaught, regroup, and return in a more potent form. Even when seemingly defeated, it may leave behind "sleeper" cells that can lie dormant for years before springing to life again. Above all, cancer cells are selfish. Normally, human cells do their job, then die on command when instructed to by other cells for the good of the body. Cancer cells don't. They proliferate entirely in their own interests.

"They have evolved to avoid detection," says Vormoor. "They can hide from drugs. They can develop resistance. They can recruit other cells to help them. They can go into hibernation and wait for better conditions. They can do any number of things that make it hard for us to kill them."

Something we have only recently realized is that before cancers metastasize, they are able to prepare the ground for an invasion in distant target organs, probably through some form of chemical signaling. "What this means," Vormoor says, "is that when cancer cells spread to other organs, they don't just turn up and hope for the best. They already have a base camp in the destination organ. Why certain cancers go to certain organs, often in distant parts of the body, has always been a mystery."

We need to remind ourselves from time to time that these are brainless cells we are considering here. They are not willfully malevolent. They are not plotting to kill us. All they are doing is what all cells try to do—survive. "The world is a challenging place," says Vormoor. "All cells have evolved a repertoire of programs that they use to help protect themselves from DNA damage. They are just doing what they are programmed to do."

Or as one of Vormoor's colleagues, Olaf Heidenreich, explained it to me, "Cancer is the price we pay for evolution. If our cells couldn't mutate, we would never get cancer, but we also couldn't evolve. We would be fixed forever. What this means in practice is that although evolution is sometimes tough on the individual, it's beneficial for the species."

Cancer is actually not one disease, but a suite of more than two hundred with lots of different causes and prognoses. Eighty percent of cancers, known as carcinomas, arise in epithelial cells—that is, the cells that make up the skin and the linings of organs. Breast cancers, for instance, don't just grow randomly within the breast, but normally begin in the milk ducts. Epithelial cells are assumed to be particularly susceptible to cancers because they divide rapidly and often. Only about 1 percent of cancers are found in connective tissue; these are known as sarcomas.

Cancer is above all an age thing. Between birth and the age of forty, men have just a one in seventy-one chance of getting cancer and women one in fifty-one, but over sixty the odds drop to one in three for men and one in four for women. An eighty-year-old person is a thousand times more likely than a teenager to develop cancer.

Lifestyle is a huge factor in determining which of us get cancer. More than half of cases, by some calculations, are caused by things we can do something about—smoking, drinking to excess, and overeating primarily. The American Cancer Society found a "significant association" between being overweight and incidence of cancer of the liver, breast, esophagus, prostate, colon, pancreas, kidney, cervix, thyroid, and stomach—just about everywhere, in short. How exactly weight tips the balance is not at all understood, but it certainly seems to.

Environmental exposures are also a significant source of cancers—more perhaps than most of us realize. The first person to notice a connection between environment and cancers was a British surgeon, Percivall Pott, who in 1775 noted that scrotal cancer was disproportionately prevalent among chimney sweeps—indeed, was so particular to the profession that the disorder was called chimney sweep's cancer. Pott's investigation into their plight, in a work called *Chirurgical Observations: Relative to the Cataract, the Polypus of the Nose, the Cancer of the Scrotum, the Different Kinds of Ruptures, and the Mortification of the Toes and Feet,* was notable not only for identifying an environmental source for a cancer but in showing some compassion for the poor chimney sweeps, for even in that hard and neglectful age they were a forlorn group. From earliest childhood, Pott recorded, sweeps were "frequently treated with great brutality, and almost starved with cold and hunger; they are thrust up narrow and sometimes hot chimnies, where they are bruised, burned, and almost suffocated; and when they get to puberty, become peculiarly liable to a most noisome, painful, and fatal disease." The cause of the cancer, Pott discovered, was an accumulation of soot in the sweeps' scrotal folds. A good wash once a week stopped the cancer from arising, but most sweeps didn't get a weekly wash, and scrotal cancer remained a problem until late in the nineteenth century.

No one knows, because it is essentially impossible to determine, to what extent environmental factors contribute to cancers now. More than eighty thousand chemicals are produced commercially in the world today, and by one calculation 86 percent of them have never been tested for their effects on humans. We don't even know much about the good or neutral chemicals around us. As Pieter Dorrestein of the University of California at San Diego told a journalist from the journal *Chemistry World* in 2016, "If one asks the question what are the ten most abundant molecules in the human habitat, no one can answer." Of the things that might harm us, only radon, carbon dioxide, tobacco smoke, and asbestos have been studied really extensively. The rest is mostly speculation. We inhale a lot of formaldehyde, which is used in flame retardants and the glues that hold together our furniture. We also produce and breathe in a lot of nitrogen dioxide, polycyclic hydrocarbons, semi-organic compounds, and miscellaneous particulates. Even the cooking of food and the burning of candles can throw off particulates that may do us no good at all. Although no one can say to what extent pollutants in air and water contribute to cancers, it has been estimated that it may be as much as 20 percent.

Viruses and bacteria cause cancers, too. The World Health Organization in 2011 estimated that 6 percent of cancers in the developed world but 22 percent in low- and middle-income countries are attributable to viruses alone. This was once a very radical idea. In 1911, when Peyton Rous, a recently qualified researcher at the Rockefeller Institute in New York, found that a virus caused cancer in chickens, the discovery was universally dismissed. In the face of opposition and even some ridicule, Rous dropped the idea and turned to other research. It was not until 1966, more than half a century after his discovery, that he was formally vindicated with the award of a Nobel Prize. We now know that pathogens are responsible for cervical cancer (caused by the human papillomavirus), Burkitt's lymphoma, hepatitis B and C, and several others. Altogether, it has been estimated, pathogens may account for a quarter of all cancers globally.*

And sometimes cancer just seems to be cruelly random. About 10 percent of men and 15 percent of women who get lung cancer are not smokers, have not been exposed to known environmental hazards, or have not faced any other increased risks, as far as can be told. They are just, it seems, very, very unlucky, but whether they are unlucky in a fateful sense or a genetic one is usually impossible to say.

One thing is common to all cancers, however. Treatment is rough.

II

IN 1810, THE English novelist Fanny Burney, while living in France, developed breast cancer at the age of fifty-eight. It is almost impossible to imagine how horrifying this must have been. Two hundred years ago, every form of cancer was horrible, but breast cancer especially so. Most victims suffered years of torment and often unspeakable embarrassment as a tumor slowly devoured their breast and replaced it with an open hole from which seeped foul fluids that made it impossible for the poor victim to mix with others, sometimes even with her own family. Surgery was the only possible treatment, but in the days before anesthetics it was at least as painful and distressing as the cancer itself and was nearly always lethal.

Burney was told that her only hope was to undergo a mastectomy. She recounted the ordeal—"a terror that surpasses all description"—in a letter to her sister Esther. Even now it makes painful reading. On

* The alert reader will note that all these percentages taken together add up to more than 100 percent. That's partly because they are estimates—in some cases little more than guesses—and come from different sources, and partly because of double or triple counting. A retired coal miner's fatal lung cancer could, for instance, be attributed to his working environment or the fact that he had smoked for forty years, or both. More often than not, the cause of a cancer is anyone's guess.

a September afternoon, Burney's surgeon, Antoine Dubois, came to her house with six assistants—four other doctors and two students. A bed had been moved to the middle of the room and space around it cleared for the team to work.

"M. Dubois placed me upon the mattress, and spread a cambric handkerchief upon my face," Burney reported to her sister. "It was transparent, however, and I saw through it that the bedstead was instantly surrounded by seven men and my nurse. I refused to be held; but when, bright through the cambric, I saw the glitter of polished steel—I closed my eyes. When the dreadful steel was plunged into the breast—cutting through veins—arteries—flesh—nerves—I needed no injunctions to restrain my cries. I began a scream that lasted intermittingly during the whole time of the incision—and I almost marvel that it rings not in my ears still, so excruciating was the agony. . . . I felt the instrument—describing a curve—cutting against the grain, if I may say so, while the flesh resisted in a manner so forcible as to oppose and tire the hand of the operator, who was forced to change from the right to the left—then, indeed, I thought I must have expired. I attempted no more to open my eyes."

She thought the surgery was over, but Dubois found that the breast was still attached by the tumor, so cutting recommenced. "Oh heaven! I then felt the knife rackling against the breast bone—scraping it!" For some minutes, the surgeon cut away at muscle and diseased tissue until he was confident that he had got as much as he could. Burney endured this final part in silence—"in utterly speechless torture."

The whole procedure took seventeen and a half minutes, though it must have seemed a lifetime to poor Fanny Burney. Remarkably, it worked. Burney lived another twenty-nine years.

Although the development of anesthetics in the mid-nineteenth century did much to remove the immediate pain and horror of surgery, treatment for breast cancer became, if anything, even more brutal as we moved into the modern age. And the person almost single-handedly responsible for that was one of the most extraordinary figures in the history of modern surgery, William Stewart Halsted (1852–1922).

The son of a wealthy businessman in New York, Halsted studied medicine at Columbia University and upon graduating quickly distinguished himself as a deft and innovative surgeon. You will recall him from chapter 8, where we noted that he was one of the first people daring enough to perform gallbladder surgery, on his mother on a kitchen table in the family home in upstate New York. He also attempted the first appendectomy in New York (the patient died) and, more happily, one of the first successful transfusions in America—on his sister Minnie after she suffered a severe hemorrhage in childbirth. As she lay near death, Halsted transferred two pints of blood from his arm into hers and saved her life. This was before anyone understood the need for blood type compatibility, but luckily they were a match.

Halsted became the first professor of surgery at the new Johns Hopkins Medical School in Baltimore after its founding in 1893. There he trained a generation of leading surgeons and made many worthwhile advances in surgical techniques. Among much else, he invented the surgical glove. He became famous for instilling in his students the need for the most exacting standards of surgical care and hygiene—an approach so influential that it soon became universally known as "Halstedian technique." People commonly referred to him as the father of American surgery.

What makes Halsted's achievements all the more remarkable is that for much of his career he was a drug addict. While investigating methods for providing pain relief, he experimented with cocaine and soon found himself helplessly attached to it. As his addiction took over his life, he became conspicuously more reserved in manner—most of his colleagues thought he was simply being more thoughtful and reflective—but in print he became positively manic. Here is the opening of a paper he wrote in 1885, just four years after he operated on his mother: "Neither indifferent as to which of how many possibilities may best explain, nor yet at a loss to comprehend, why surgeons have, and that so many, quite without discredit, could have exhibited scarcely any interest in what, as a local anaesthetic, had been supposed, if not declared, by most so very sure to prove, especially to them, attractive, still I do not think that this circumstance, or some sense of obligation . . ."–and so it goes on for several lines more without straying at any point to within sight of coherence.

In an effort to remove him from temptation and break the habit, Halsted was sent on a Caribbean cruise but was there caught searching for drugs in the ship medicine locker. Then he was committed to an institution in Rhode Island where unfortunately doctors tried to wean him off cocaine by giving him morphine. He ended up addicted to both. He lived out his life with almost everyone except one or two immediate superiors unaware that he was completely dependent on drugs to get through the day. There is some evidence that his wife became an addict, too.

In 1894 at a conference in Maryland, and at the height of his addiction, Halsted introduced his most revolutionary innovation–the concept of the radical mastectomy. Halsted believed, wrongly, that breast cancer spread by radiating outward, like wine spilled on a tablecloth, and that the only effective treatment was to cut out not just the tumor but as much surrounding tissue as one dared. The radical mastectomy wasn't so much surgery as excavation. It involved removing the whole breast and surrounding chest muscles, lymph nodes, and sometimes ribs–whatever could be taken away without causing immediate death. The excision was so extensive that the only way to close the wound was to take a large skin graft from the thigh, giving yet more pain and an additional site of disfigurement to the poor, battered patient.

But it got good results. About a third of Halsted's patients survived for at least three years, a proportion that astounded other cancer specialists. Many more patients gained at least a few months of reasonably comfortable life without the embarrassing stench and seepage that made so many previous sufferers into recluses.

Not everyone was convinced that Halsted's approach was the right one. In Britain, a surgeon named Stephen Paget (1855–1926) looked at 735 cases of breast cancer and found that cancers didn't spread like a stain at all, but rather cropped up in distant locations. More often than not, breast cancers migrated to the liver–and, moreover, to specific sites within the liver. Though Paget's findings were correct and incontestable, no one paid any attention to them for about a hundred years, during which time tens of thousands of women were disfigured to a far greater degree than was necessary.

Meanwhile, elsewhere in the world of medicine researchers were developing other cancer treatments, which generally proved just as taxing to the patients and sometimes to those who treated them. One of the great excitements of the early twentieth century was radium, discovered by Marie and Pierre Curie in France in 1898. Quite early on, it was realized that radium accumulated in the bones of people exposed to it, but this was thought to be a good thing because it was believed that radiation was wholly beneficial. Radioactive products were liberally added to many medications, with sometimes devastating consequences. A popular over-the-counter painkiller called Radithor was made with diluted radium. An industrialist in Pittsburgh named Eben M. Byers treated it as a tonic and drank a bottle every day for three years until he discovered that the bones in his head were slowly softening and dissolving, like a stick of blackboard chalk left in the rain. He lost most of his jaw and parts of his skull en route to dying a slow and hideous death.

For many others, radium was an occupational hazard. In 1920, four million radium watches were sold in America, and the watchmaking industry employed two thousand women to paint the dials. It was delicate work, and the simplest way to keep a fine point on the brush was to roll it gently between one's lips. As Timothy J. Jorgensen notes in his superb history, *Strange Glow: The Story of Radiation,* it was subsequently calculated that the average dial painter swallowed about a teaspoon of radioactive material a week in this way. There was so much radium dust in the air that some of the factory girls noticed that they glowed in the dark themselves. Not surprisingly, some of the women soon began to sicken and die. Others developed strange fragilities; one young woman's leg broke spontaneously while she was on the dance floor.

One of the very first people to take an interest in radiation therapy was a medical student at the Hahnemann Medical College in Chicago named Emil H. Grubbe (1875–1960). In 1896, just a month after Wilhelm Röntgen announced his discovery of X-rays, Grubbe decided to try X-rays out on cancer patients, even though he was not actually qualified to do so. All Grubbe's early patients died quickly—all were near death anyway so probably beyond saving even with today's treatments, and Grubbe was only guessing dosages—but the young medical student persevered and had more success as he gained experience. Unfortunately, he did not understand the need to limit his own exposures. By the 1920s, he had begun to develop tumors all over, most notably on his face. Surgery to remove these growths left him disfigured. His medical practice failed as his patients abandoned him. "By 1951," writes Jorgensen, "he was so badly disfigured by his multiple surgeries that his landlord asked him to vacate his apartment because his grotesque appearance was scaring away tenants."

Sometimes, happily, better outcomes were achieved. In 1937, Gunda Lawrence, a teacher and homemaker from South Dakota, lay close to death from abdominal cancer. Doctors at the Mayo Clinic in Minnesota had given her three months to live. Luckily, Mrs. Lawrence had two exceptional and devoted sons—John, a gifted physician, and Ernest, one of the most brilliant physicists of the twentieth century. Ernest was head of the new Radiation Laboratory at the University of California at Berkeley and had just invented the cyclotron, a particle accelerator that generated massive amounts of radioactivity as a side effect of energizing protons. They had in effect the most powerful X-ray machine in the country at their disposal, capable of generating a million volts of energy. Without any certainty what the consequences would be—no one had ever tried anything remotely like this on humans before—the brothers aimed a deuteron beam directly into their mother's belly. It was an agonizing experience, so painful and distressing to poor Mrs. Lawrence that she begged her sons to let her die. "At times I felt very cruel in not giving in," John recorded later. Happily, after a few treatments, Mrs. Lawrence's cancer went into remission and she lived another twenty-two years. More important, a new field of cancer treatment had been born.

It was also at the Radiation Lab at Berkeley that researchers finally and belatedly began to grow concerned about the dangers of radiation after the body of a mouse was found beside the machine after one set of experiments. It occurred to Ernest Lawrence that the huge amounts of radioactivity generated by the machine might be dangerous to human tissue. So protective barriers were installed and operators retreated to another room when the machine was running. It was subsequently discovered that the mouse had died of asphyxiation, not irradiation, but it was decided to proceed with safety measures anyway, and thank goodness.

Chemotherapy, the third main prong in cancer treatment after surgery and radiation, came about by similarly unlikely means. Although chemical weapons had been outlawed by international treaty after World War I, several nations still produced them, if only as a precaution in the event that others did likewise. The United States was among the transgressors. For obvious reasons, this was kept secret, but in 1943 a U.S. Navy supply ship, the SS *John Harvey*, carrying mustard gas bombs as part of its cargo, was caught in a German bombing raid on the Italian port of Bari. The *Harvey* was blown up, releasing a cloud of mustard gas over a wide area, killing an unknown number of people. Realizing that this was an excellent, if accidental, test of the mustard gas's efficacy as a killing agent, the navy dispatched a chemical expert, Lieutenant Colonel Stewart Francis Alexander, to study the effects of the mustard gas on the ship's crew and others nearby. Luckily for posterity, Alexander was an astute and diligent investigator, for he noticed something that might have been overlooked: mustard gas dramatically slowed the creation of white blood cells in those exposed to it. From this, it was realized that some derivative of mustard gas might be useful in treating some cancers. Thus was born chemotherapy.

"What is quite remarkable," one cancer specialist told me, "is that we are basically still using mustard gases. They are refined, of course, but they are really not that much different from what armies were using on each other in the First World War."

III

IF YOU WISHED to see how far cancer therapies have come in recent years, you could do much worse than visit the new Princess Máxima Center in Utrecht. The largest children's cancer center in Europe, it was created through the merging of the children's oncology units of seven university hospitals in the Netherlands, to bring all treatments and research in the country under one roof. It is a bright, generously resourced, and surprisingly lively place. As Josef Vormoor showed me around, we had to step aside from time to time as small children on pedal go-karts—each child bald and with a breathing tube in his or her nostrils—shot around or through us at breakneck speeds. "We sort of let them have the run of the place," Josef apologized happily.

Cancer is actually rare among children. Of the fourteen million cases of cancer diagnosed in the world each year, only about 2 percent are in people aged nineteen or younger. The principal cause of childhood cancers, accounting for about 80 percent of cases, is acute lymphoblastic leukemia. Fifty years ago it was a death sentence. Drugs could put it into remission for a while, but it soon came back. The five-year survival rate was less than 0.1 percent. Today the survival rate is about 90 percent.

The breakthrough moment was in 1968 when Donald Pinkel of St. Jude Children's Research Hospital in Memphis, Tennessee, tried a new approach. Pinkel was convinced that giving drugs in moderate dosages, which was then the standard practice, allowed some leukemic cells to escape and to bounce back after treatment stopped. That's why remissions were always temporary. Pinkel blasted the leukemic cells with the full range of available drugs, frequently in combinations, always at the highest possible dosages, accompanied by bouts of radiation. It was a punishing regime, lasting up to two years, but it worked. Survival rates improved dramatically.

"We're still essentially following the approach of the early pioneers of leukemia therapy," Josef says. "All we have done in the years since is fine-tune things. We have better ways of dealing with the side effects of chemotherapy and of fighting infections, but basically we are still doing what Pinkel did."

And that is hard on any human body, not least young ones that are still forming. A significant fraction of childhood cancer deaths come not from the cancer itself but from the treatments for it. "There's a lot of collateral damage," Josef tells me. "Treatments don't affect just cancer cells, but many healthy cells as well." The most visible manifestation of this is damage to hair cells, which causes patients' hair to fall out. More critically, there is also often long-term damage to the heart and other organs. Girls who have had chemotherapy have a greater chance of experiencing menopause earlier and run an enhanced risk of suffering ovarian failures later in life. For both sexes, fertility may be compromised. Much depends on the type of cancer and form of treatment.

Still, the story is mostly positive, and not just for childhood cancers but for cancers at all ages. In the developed world, death rates from lung, colon, prostate, Hodgkin's disease, testicular cancer, and breast cancer have all fallen sharply—by between 25 and 90 percent—in twenty-five years or so. In the United States alone, 2.4 million fewer people have died of cancer in the last thirty years than would have if the rates had stayed unchanged.

The dream of many researchers is to find some way of detecting tiny changes in the chemistry of blood or urine or perhaps saliva that would betray the early onset of a cancer when it could be more easily treated.

"The problem," Josef says, "is that even when we can detect cancer early now, we cannot tell whether it is aggressive or benign.

Overwhelmingly, we focus on trying to cure cancers when they happen rather than prevent them from happening in the first place." Globally, by one reckoning, no more than 2 to 3 percent of cancer research money is spent on prevention.

"You can't imagine how much things have improved in a generation," Josef reflected as we came to the end of our tour. "It's the most satisfying thing in the world to know that most of these children will be cured and can go home and resume their lives. But wouldn't it be even more wonderful if they didn't have to come here in the first place? That's the dream."

22 MEDICINE GOOD AND BAD

Doctor: What did you operate on Jones for?
Surgeon: A hundred pounds.
Doctor: No, I mean what had he got?
Surgeon: A hundred pounds.

—*PUNCH* CARTOON, 1925

I SHOULD LIKE to say a word about Albert Schatz, for if ever there was a man who deserved a moment's grateful attention, it is he. Schatz, who lived from 1920 to 2005, was from a poor farming family in Connecticut. He studied soil biology at Rutgers University in New Jersey not because he had a passion for soil but because, as a Jew, he was subject to university admission quotas and he couldn't get into a better institution. He reasoned that whatever he learned about soil fertility would at least be useful back on the family farm.

It was an unfairness that ended up saving lives, for in 1943, still a student, Schatz followed a hunch that soil microbes might provide an additional antibiotic to put alongside the new drug penicillin, which, for all its value, didn't work against bacteria of a type known as Gram-negative. This included the microbe responsible for tuberculosis. Schatz patiently tested hundreds of samples and in just under a year came up with streptomycin, the first drug to vanquish Gram-negative bacteria. It was one of the most important microbiological breakthroughs of the twentieth century.*

Schatz's supervisor, Selman Waksman, immediately saw the potential of Schatz's discovery. He took charge of the clinical trials of the drug and, in the process, had Schatz sign an agreement ceding patent rights to Rutgers. Soon afterward, Schatz discovered that Waksman was taking full credit for the discovery and keeping Schatz from being invited to meetings and conferences where he would have received praise and attention. With the passage of time, Schatz also discovered that Waksman had not relinquished patent rights himself, but was pocketing a generous share of profits, which were soon running into millions of dollars a year.

Unable to get any satisfaction, Schatz eventually sued Waksman and Rutgers, and won. In settlement, he was given a portion of the royalties and credit as co-discoverer, but the lawsuit ruined him: it was considered very bad form to sue a superior in academia in those days. For many years, the only work Schatz could find was at a small agricultural college in Pennsylvania. His papers were repeatedly rejected by leading journals. When he wrote an account of the discovery of streptomycin as it had really happened, the only publication he could find that would accept it was the *Pakistan Dental Review*.

In 1952, in one of the supreme injustices of modern science, Selman Waksman was awarded the Nobel Prize in Physiology or Medicine. Albert Schatz received nothing. Waksman continued to take the credit for the discovery for the rest of his life. He didn't mention Schatz in his Nobel acceptance speech or in his 1958 autobiography, in which he merely noted in passing that he had been assisted in his discovery by a graduate student. When Waksman died in 1973, he was

* The "Gram" in Gram-negative and Gram-positive bacteria has nothing to do with weights and measures. It is named for a Danish bacteriologist, Hans Christian Gram (1853–1938), who in 1884 developed a technique for distinguishing the two major types of bacteria by what color they turned when stained on a microscopic slide. The difference between the two types has to do with the thickness of their cell walls and how easily or not they are penetrated by antibodies.

described in many obituaries as "the father of antibiotics," which he most assuredly was not.

Twenty years after Waksman's death, the American Society for Microbiology made a somewhat belated attempt at amends by inviting Schatz to address the society on the occasion of the fiftieth anniversary of streptomycin's discovery. In recognition of his achievements, and presumably without giving the matter a lot of thought, it bestowed on him its highest award: the Selman A. Waksman medal. Life sometimes really is very unfair.

If there is a hopeful moral to the story, it is that medical science progresses anyway. Thanks to thousands and thousands of mostly unsung heroes like Albert Schatz, our armory against assaults of nature has grown stronger and stronger with every passing generation—a fact happily reflected in dramatically improved life spans across the planet.

By one reckoning, life expectancy on Earth improved by as much in the twentieth century as in the whole of the preceding eight thousand years. The average life span for an American male went from 46 years in 1900 to 74 by century's end. For American women, the improvement was better still—from 48 to 80. Elsewhere, the improvements have been little short of breathtaking. A woman born in Singapore today can expect to live for 87.6 years, more than double what her great-grandmother could have counted on. Across the planet as a whole, life expectancy grew from 48.1 years for men in 1950 (which is as far back as global records reliably go) to 70.5 today; for women the rise was from 52.9 to 75.6 years. In more than two dozen countries, life expectancy today is over 80 years. At the top is Hong Kong at 84.3 years, closely followed by Japan at 83.8 and Italy at 83.5. The United Kingdom does quite well at 81.6 years, while the United States, for reasons that will be discussed below, comes in at a decidedly mediocre life expectancy of 78.6 years. Globally, however, the story is one of success, with most countries, even in the developing world, recording improvements of 40 to 60 percent in life spans in just a generation or two.

Nor do we die as we used to. Consider the lists below of principal causes of death in 1900 and now. (The accompanying numbers indicate deaths per 100,000 of population in each category.)

1900	TODAY
Pneumonia and flu, 202.2	Heart disease, 192.9
Tuberculosis, 194.4	Cancer, 185.9
Diarrhea, 142.7	Respiratory disease, 44.6
Heart disease, 137.4	Stroke, 41.8
Stroke, 106.9	Accidents, 38.2
Kidney disease, 88.6	Alzheimer's disease, 27.0
Accidents, 72.3	Diabetes, 22.3
Cancer, 64.0	Kidney disease, 16.3
Senility, 50.2	Pneumonia and flu, 16.2
Diphtheria, 40.3	Suicide, 12.2

The most striking difference between the two eras is that nearly half of deaths in 1900 were from infectious diseases compared with just 3 percent now. Tuberculosis and diphtheria have disappeared from the modern top ten but been replaced by Alzheimer's and diabetes. Accidents as a cause of death have jumped from seventh place to fifth, not because we have grown clumsier, but because other causes have been eliminated from the top tier. In the same way, heart disease in 1900 killed 137.4 people per 100,000 per year, while today it kills 192.9 per 100,000, a 40 percent increase, but that's almost entirely because other things used to kill people first. The same goes for cancer.

There are, it must be said, problems with life expectancy figures. All death lists are in some measure arbitrary, particularly with respect to the elderly, who may have lots of debilitating conditions, any one of which may finish them off and all of which are bound to contribute. In 1993, two American epidemiologists, William Foege and Michael McGinnis, wrote a famous paper for *The Journal of the American Medical Association* arguing that the leading causes of death recorded on mortality tables—heart attacks, diabetes, cancer, and so on—were very often outcomes of other conditions and that the real causes were factors like smoking, poor diet, illicit use of drugs, and other behaviors overlooked on death certificates.

A separate problem is that deaths in the past were often recorded in strikingly vague and imaginative terms. When the writer and traveler George Borrow died in England in 1881, to cite one example, the cause of death was listed as "decay of nature." Who can say what that might have been? Others were recorded as being carried off by "nervous fevers," "stagnation of the fluids," "sore teeth," and "fright," among many other causes of a wholly uncertain nature. Such ambiguous terms make it nearly impossible to produce reliable comparisons between causes of death now and in the past. Even for the two lists above, there is no telling how much correspondence may exist between senility in 1900 and Alzheimer's disease today.

It is also important to bear in mind that historic life expectancy figures were always skewed by childhood deaths. When we read that life expectancy was forty-six years for American men in 1900, that doesn't mean that most men got to forty-six and then keeled over. Life expectancies were short because so many children died in infancy, and that dragged the average down for everyone. If you got past childhood, the chances of living to a reasonably advanced age weren't bad. Lots of people died early, but it was by no means a cause of wonder when people lived into old age. As the American academic Marlene Zuk has put it, "Old age is not a recent invention, but its commonness is." The most heartening advance of recent times, however, is the striking improvement in mortality rates for the very young. In 1950, 216 children in every thousand—nearly a quarter—died before the age of five. Today the figure is just 38.9 early childhood deaths in a thousand—one-fifth what it was seventy years ago.

Even allowing for all the uncertainties, there is no question that early in the twentieth century people in the developed world began to enjoy much better prospects for living longer lives in better health. As the Harvard physiologist Lawrence Henderson famously remarked, "At some point between 1900 and 1912, a random patient with a random disease, consulting a doctor chosen at random, had for the first time in history a better than fifty-fifty chance of profiting from the encounter." The more or less universal consensus among historians and academics was that medical science somehow turned a corner when it entered the twentieth century and just kept getting better and better as the century progressed.

Any number of reasons have been proposed for the improvement. The rise of penicillin and other antibiotics like Albert Schatz's streptomycin had an obvious and significant impact on infectious diseases, but other medicines flooded the market, too, as the century proceeded. By 1950, half of the medicines available for prescription had been invented or discovered in just the previous ten years. Another huge boost can be attributed to vaccines. In 1921, America had about 200,000 cases of diphtheria; by the early 1980s, with vaccination, that had fallen to just 3. In roughly the same period, whooping cough and measles infections fell from about 1.1 million cases a year to just 1,500. Before vaccines, 20,000 Americans a year got polio. By the 1980s, that had dropped to 7 a year. According to the British Nobel laureate Max Perutz, vaccinations might have saved more lives in the twentieth century even than antibiotics.

The one thing no one doubted was that practically all the credit for the great advances lay securely with medical science. But then, in the early 1960s, a British epidemiologist named Thomas McKeown (1912–88) looked at the records again and noted some curious anomalies. Deaths from a large number of maladies—tuberculosis, whooping cough, measles, and scarlet fever notably—had begun to decline well before effective treatments had become available. Tuberculosis deaths in Britain dropped from four thousand per million in 1828 to twelve hundred in 1900, and to just eight hundred per million in 1925—a fall of 80 percent in a century. Medicine could account for none of that. Childhood scarlet fever deaths went from twenty-three per ten thousand in 1865 to just one per ten thousand in 1935, again without vaccines or other effective medical interventions. All told, McKeown suggested, medicine could account for no more than perhaps 20 percent of the improvements. All the rest were the result of improved sanitation and diet, healthier lifestyles, and even things like the rise of the railways, which improved food distribution, bringing fresher meat and vegetables to city dwellers.

McKeown's thesis attracted a good deal of criticism. Opponents maintained that McKeown was carefully selective in the diseases he used to illustrate his thesis and that he underplayed the role of improved medical care. Max Perutz, one of his critics, argued persuasively that hygiene standards in the nineteenth century hadn't advanced at all, but were continually eroded by the hordes of people crowding into newly industrializing cities and living in squalid conditions. The quality of drinking water in New York City, for one, declined steadily, and dangerously, in the nineteenth century—so much so that by 1900 residents of Manhattan were being instructed to boil all water before using it. The city didn't get its first filtration plant until just before World War I. It was the same in almost every other major urban area in America as growth outpaced municipalities' abilities or willingness to provide safe water and efficient sewerage.

However we decide to apportion the credit for our improved life spans, the bottom line is that nearly all of us are better able today to resist the contagions and afflictions that commonly sickened our great-grandparents, while having massively better medical care to call on when we need it. In short, we have never had it so good.

Or at least we have never had it so good if we are reasonably well-off. If there is one thing that should alarm and concern us today, it is how unequally the benefits of the last century have been shared. British life expectancies might have soared overall, but as John Lanchester noted in an essay in the *London Review of Books* in 2017, males in the East End of Glasgow today have a life expectancy of just fifty-four years—nine years less than a man in India. In exactly the same way, a thirty-year-old black male in Harlem, New York, is at much greater risk of dying than a thirty-year-old male Bangladeshi—and not, as you might think, from drugs or street violence but from stroke, heart disease, cancer, or diabetes.

Climb aboard a bus or subway train in almost any large city in the

Western world and you can experience similar vast disparities with a short journey. In Paris, travel five stops on the Metro's B line from Port-Royal to La Plaine–Stade de France and you will find yourself among people who have an 82 percent greater chance of dying in a given year than those just down the line. In London, life expectancy drops reliably by one year for every two stops traveled eastward from Westminster on the District Line of the Underground. In St. Louis, Missouri, make a twenty-minute drive from prosperous Clayton to the inner-city Jeff-Vander-Lou neighborhood and life expectancy drops by one year for every minute of the journey, a little over two years for every mile.

Two things can be said with confidence about life expectancy in the world today. One is that it is really helpful to be rich. If you are middle-aged, exceptionally well-off, and from almost any high-income nation, the chances are excellent that you will live into your late eighties. Someone who is otherwise identical to you but poor–exercises as devotedly, sleeps as many hours, eats a similarly healthy diet, but just has less money in the bank–can expect to die between ten and fifteen years sooner. That's a lot of difference for an equivalent lifestyle, and no one is sure how to account for it.

The second thing that can be said with regard to life expectancy is that it is not a good idea to be an American. Compared with your peers in the rest of the industrialized world, even being well-off doesn't help you here. A randomly selected American aged forty-five to fifty-four is more than twice as likely to die, from any cause, as someone from the same age-group in Sweden. Just consider that. If you are a middle-aged American, your risk of dying before your time is more than double that of a person picked at random off the streets of Uppsala or Stockholm or Linköping. It is much the same when other nationalities are brought in for comparison. For every 400 middle-aged Americans who die each year, just 220 die in Australia, 230 in Britain, 290 in Germany, and 300 in France.

These health deficits begin at birth and go right on through life. Children in the United States are 70 percent more likely to die in childhood than children in the rest of the wealthy world. Among rich countries, America is at or near the bottom for virtually every measure of medical well-being—for chronic disease, depression, drug abuse, homicide, teenage pregnancies, HIV prevalence. Even sufferers of cystic fibrosis live ten years longer on average in Canada than in the United States. What is perhaps most surprising is that all these poorer outcomes apply not just to underprivileged citizens but to prosperous white college-educated Americans when compared with their socioeconomic equivalents abroad.

This is all a touch counterintuitive when you consider that America spends more on health care than any other nation—two and a half times more per person than the average for all the other developed nations of the world. One-fifth of all the money Americans earn—$10,209 a year for every citizen, $3.2 trillion altogether—is spent on health care. It is the nation's sixth-largest industry and provides one-sixth of its employment. You can't get health care any higher on a national agenda without putting everyone in a white coat or uniform.

Yet despite the generous spending, and the undoubted high quality of American hospitals and health care generally, the United States comes just thirty-first in global rankings of life expectancy, behind Cyprus, Costa Rica, and Chile, and just ahead of Cuba and Albania.

How to explain such a paradox? Well, to begin with, and most inescapably, Americans lead more unhealthy lifestyles than most other people, and that is true at all levels of society. As Allan S. Detsky observed in *The New Yorker*, "Even wealthy Americans are not isolated from a lifestyle filled with oversized food portions, physical inactivity, and stress." The average Dutch or Swedish citizen consumes about 20 percent fewer calories than the average American, for instance. That doesn't sound massively excessive, but it adds up to 250,000 calories over the course of a year. You would get a similar boost if you sat down about twice a week and ate an entire cheesecake.

Life in America is also much riskier, especially for young people. A U.S. teenager is twice as likely to be killed in a car accident as a young person in a comparable country abroad and is eighty-two times more likely to be killed by a gun. Americans drink and drive more often than almost anybody else and wear seat belts less devotedly than everyone in the rich world but the Italians. Nearly all advanced nations require helmets for all motorcyclists and passengers. In America, 60 percent of states don't. Three states have no helmet requirements at any age, and sixteen others require them only for riders aged twenty or under. Once citizens of those states reach their maturity, they can let the wind, and all too often the pavement, run through their hair. A helmeted rider is 70 percent less likely to suffer a brain injury and about 40 percent less likely to die in a crash. In consequence of all these factors, the United States records a really quite spectacular 11 traffic deaths per 100,000 people every year, compared with 3.1 in the United Kingdom, 3.4 in Sweden, and 4.3 in Japan.

Where America really differs from other countries is in the colossal costs of its health care. An angiogram, a survey by *The New York Times* found, costs an average of $914 in the United States, $35 in Canada. Insulin costs about six times as much in America as it does in Europe. The average hip replacement costs $40,364 in America, almost six times the cost in Spain, while an MRI scan in the United States is, at $1,121, four times more than in the Netherlands. The entire system is notoriously unwieldy and cost-heavy. America has about 800,000 practicing physicians but needs twice that number of people to administer its payments system. The inescapable conclusion is that higher spending in America doesn't necessarily result in better medicine, just higher costs.

One commonly accepted yardstick for quality of health care is five-year cancer survival rates, and here there are great disparities. For colon cancer, five-year survival rates are 71.8 percent in South Korea and 70.6 percent in Australia, but just 64.9 percent in the United States. For cervical cancer, Japan comes out on top at 71.4 percent, closely followed by Denmark at 69.1 percent, with the United States at a middling 67 percent. For breast cancer, the United States tops the world rankings with 90.2 percent of victims still alive after five years, just ahead of Australia at 89.1 percent and considerably ahead of Britain at

85.6 percent. It is worth noting that overall survival figures can mask a lot of troubling ethnic disparities. For cervical cancer, for instance, white women in the United States have a 69 percent five-year survival rate, which puts them near the top of world rankings, while black women have just a 55 percent survival rate, leaving them close to the bottom. (That is all black women, rich and poor alike.)

The upshot is that Australia, New Zealand, the Nordic countries, and the wealthier nations of the Far East all do really well, and other European nations do pretty well. For the United States, the result is, at best, decidedly mixed. For Britain, cancer survival rates are grim and ought to be a matter of national concern.

Nothing in medicine is simple, however, and there is an additional consideration that profoundly complicates results almost everywhere: overtreatment.

It hardly needs pointing out that for most of history the focus of medicine has been to make sick people better, but now increasingly doctors devote their energies to trying to head off problems before they even arise, through programs of screening and the like, and that changes the dynamics of care entirely. There is an old joke in medicine that seems especially apt here:

Q. What is the definition of a well person?

A. Someone who hasn't been examined yet.

The thinking behind a great deal of modern health care is that you cannot be too careful and you cannot have too many tests. Surely it is better, the logic runs, to check out and deal with or eliminate any potential problems, however remote, before they have a chance to turn into something bad. The drawback with this approach is what are known as false positives. Consider screening for breast cancer. Studies show that between 20 and 30 percent of women given the all clear after a breast cancer screening actually had tumors. But equally, and contrarily, screenings often catch tumors that needn't cause concern, and result in interventions that aren't actually necessary. Oncologists use a concept called sojourn time, which is the interval between when a cancer is caught by screening and when it would become evident anyway. Many cancers have long sojourn times and progress so slowly that the victims almost always die of something else before the cancer gets them. A study in Britain found that as many as one in three women with breast cancer receive treatments that may leave them mutilated and even possibly shorten their lives quite unnecessarily.

Mammograms are in fact fuzzy things. Reading them accurately is a challenging task—much more challenging than even many medical professionals realize. As Timothy J. Jorgensen has noted, when 160 gynecologists were asked to assess the likelihood of a fifty-year-old woman having breast cancer if her mammogram was positive, 60 percent of them thought the chances were 8 or 9 out of 10. "The truth is that the odds the woman actually has cancer are only 1 in 10," writes Jorgensen. Remarkably, radiologists do little better.

The unfortunate bottom line is that breast cancer screening doesn't save a lot of lives. For every thousand women screened, four will die of breast cancer anyway (either because the cancer was missed or because it was too aggressive to be treated successfully). For every thousand women who are not screened, five will die of breast cancer. So screening saves one life in every thousand.

Men face similarly unhappy prospects with prostate screening. The prostate is a small gland, about the size of a walnut and weighing just one ounce, which is chiefly involved in producing and distributing seminal fluid. It is tucked neatly—not to say inaccessibly—up against the bladder and wrapped around the urethra like a neckerchief ring. Prostate cancer is the second leading cause of cancer death among men (after lung cancer) and grows more common as men get to their fifties and beyond. The problem is that the test for prostate cancer, called a PSA test, is not trustworthy. It measures levels in the blood of a chemical called prostate-specific antigen (PSA). A high PSA reading indicates a possibility of cancer, but only a possibility. The only way of confirming if cancer exists is with a biopsy, which involves sticking a long needle into the prostate via the rectum and withdrawing multiple tissue samples—not a procedure any man is likely to undertake eagerly. Because the needle can only be randomly inserted into the prostate, it is a matter of luck whether it strikes a tumor or not. If it does find a tumor, there is no telling with current technology if the cancer is aggressive or benign. On the basis of this uncertain information, a decision must be made on whether to surgically remove the prostate—a tricky operation with frequently dispiriting consequences—or treat it with radiation. Between 20 and 70 percent of men suffer impotence or incontinence after treatments. One in five experience complications from the biopsy alone.

The PSA test is "hardly more effective than a coin toss," Professor Richard J. Ablin of the University of Arizona has written, and he should know. He was the man who discovered the prostate-specific antigen in 1970. Noting that American men spend at least $3 billion a year on prostate tests, he added, "I never dreamed that my discovery four decades ago would lead to such a profit-driven disaster."

A meta-analysis of six randomized control trials involving 382,000 men found that for every 1,000 men screened for prostate cancer, about one life was saved—great news for that individual, but not so good for the large numbers of others who may spend the rest of their lives incontinent or impotent, the majority of them having undergone serious but possibly ineffectual treatments.

All this isn't to say that men should absolutely avoid PSA tests or women breast cancer screening. For all their flaws, they are the best tools available, and they do indubitably save lives. But those undergoing screenings should perhaps be made more aware of the shortcomings. As with any serious medical issue, if you have concerns you should consult a trusted physician.

Accidental discoveries made during routine investigations happen so often that doctors have a word for them: "incidentalomas." The National Academy of Medicine in the United States has estimated that $765 billion a year—a quarter of all health-care spending—is wasted on pointless precautionary maneuvers. A similar study in Washington State put the amount of waste even higher, at nearly 50 percent, and concluded that as much as 85 percent of preoperative lab tests are completely unnecessary.

The problem of overtreatment is exacerbated in many places by fear of litigation and, it must be said, by a desire of some doctors to inflate their earnings. According to the author and physician Jerome Groopman, most doctors are "less concerned about healing and more worried about being sued or maximizing their income." Or as another commentator put it more drolly, "One person's overtreatment is another's income stream."

The pharmaceutical industry has a lot to answer for in this respect. Drug companies commonly offer generous rewards to doctors to promote their drugs. Marcia Angell of the Harvard Medical School, writing in *The New York Review of Books*, has said that "most doctors take money or gifts from drug companies in one way or another." Some companies pay for doctors to attend conferences at luxury resorts where they do little more than play golf and enjoy themselves. Others pay doctors to put their names to papers that they haven't in fact written or reward them for "research" that they didn't really do. Altogether, Angell has estimated, drug companies in America spend "tens of billions" of dollars on direct and indirect payments to doctors every year.

We have reached the decidedly bizarre point in health care in which pharmaceutical companies are producing drugs that do exactly what they are designed to do but without necessarily doing any good. A case in point is the drug atenolol, a beta-blocker designed to lower blood pressure, which has been widely prescribed since 1976. A study in 2004, involving a total of twenty-four thousand patients, found that atenolol did indeed reduce blood pressure but did not reduce heart attacks or fatalities compared with giving no treatment at all. People on atenolol expired at the same rate as everyone else, but, as one observer put it, "they just had better blood-pressure numbers when they died."

Drug companies have not always behaved in the most ethical of ways. Purdue Pharma paid $600 million in fines and penalties in 2007 for marketing the opioid OxyContin with fraudulent claims. Merck paid $950 million in fines for failing to disclose problems with its anti-inflammatory drug Vioxx, which was withdrawn from sale, but not before it had caused perhaps as many as 140,000 avoidable heart attacks. GlaxoSmithKline currently owns the record for a penalty—$3 billion for a raft of transgressions. But to quote Marcia Angell again, "These kinds of fines are just the cost of doing business." For the most part, they come nowhere near offsetting the huge profits made by the errant companies before they are hauled into court.

Even in the best and most diligent circumstances, drug development is an inherently hit-or-miss undertaking. Laws almost everywhere require researchers to test drugs on animals before they try them out on humans, but animals don't necessarily make good surrogates. They have different metabolisms, respond differently to stimuli, contract different diseases. As a tuberculosis researcher observed years ago, "Mice don't cough." The point was frustratingly well illustrated on tests of drugs to fight Alzheimer's. Because mice don't get Alzheimer's naturally, they must be genetically engineered to accumulate in their brains a specific protein, beta-amyloid, associated with Alzheimer's in humans. When such doctored mice were treated with a class of drugs called BACE inhibitors, their beta-amyloid accumulations melted away, much to the excitement of researchers. But when the same drugs were tried on humans, they actually worsened the dementia in test subjects. In late 2018, three companies announced they were abandoning clinical trials of BACE inhibitors.

Another problem of clinical trials is that test subjects are nearly always excluded if they have any other medical conditions or are on other medications because those considerations could complicate results. The idea is to get rid of what are known as confounding variables. The problem is that real life is full of confounding variables even if drug tests are not. That means that lots of possible consequences are not tested for. We rarely know, for instance, what happens when various medications are taken in combination. One study found that 6.5 percent of hospital admissions in the U.K. were because of side effects from drugs, often taken in combination with other drugs.

All drugs come with a mixture of benefits and risks, and these are often not well studied. Everyone has heard that taking a low-dose aspirin daily may help prevent a heart attack. That is true, but only up to a point. According to one study of people who had taken low-dose aspirin daily for five years, 1 in 1,667 had been spared a cardiovascular problem, 1 in 2,002 had been spared a nonfatal heart attack, and 1 in 3,000 spared a nonfatal stroke, while 1 in 3,333 suffered major gastrointestinal bleeding that they would not otherwise have experienced. So for most people there is about as much chance of suffering dangerous internal bleeding from taking a daily aspirin as there is of avoiding a heart attack or stroke, but in all cases actually there is very little risk of either.

In the summer of 2018, matters became even more confused when Peter Rothwell, professor of clinical neurology at Oxford University, and colleagues found that low-dose aspirin actually is not effective at all in reducing cardiac or cancer risk in anyone weighing 154 pounds or more—but does still pose a risk of serious internal bleeding. Because about 80 percent of men and 50 percent of women exceed that threshold, it appears that a lot of people are getting no possible benefit from a daily aspirin while preserving all the risk. Rothwell suggested that people over 154 pounds should double the dose, perhaps by taking the pills twice a day rather than once, but that was really only an educated guess.

I don't wish to minimize the enormous, and undoubted, benefits of modern medicine, but it is an inescapable fact that it is far from perfect and in ways that aren't always widely appreciated. In 2013, an international team of researchers investigated common medical practices and found 146 in which "a current standard practice either had no benefit at all or was inferior to the practice it replaced." A similar study in Australia found 156 common medical practices "that are probably unsafe or ineffective."

The simple fact is that medical science alone cannot do it all—but then it doesn't need to. Other factors can significantly affect outcomes, sometimes in surprising ways. Just being kind, for instance. A study in New Zealand of diabetic patients in 2016 found that the proportion suffering severe complications was 40 percent lower among patients treated by doctors rated high for compassion. As one observer put it, that is "comparable to the benefits seen with the most intensive medical therapy for diabetes."

In short, everyday attributes like empathy and common sense can be just as important as the most technologically sophisticated equipment. In that sense at least, perhaps Thomas McKeown was on to something.

23 THE END

> Exercise regularly. Eat sensibly. Die anyway.
>
> —ANONYMOUS

I

IN 2011, AN interesting milestone in human history was passed. For the first time, more people globally died from non-communicable diseases like heart failure, stroke, and diabetes than from all infectious diseases combined. We live in an age in which we are killed, more often than not, by lifestyle. We are in effect choosing how we shall die, albeit without much reflection or insight.

About one-fifth of all deaths are sudden, as with a heart attack or car crash, and another fifth come quickly, following a short illness. But the great majority, about 60 percent, are the result of a protracted decline. We live long lives; we also die long deaths. "Nearly a third of Americans who die after 65 will have spent time in an intensive care unit in their final three months of life," *The Economist* noted grimly in 2017.

There's no question that people are living longer than ever. If you are a seventy-year-old man in America today, you have only a 2 percent chance of dying in the next year. In 1940, that probability was reached at age fifty-six. In the developed world at large, 90 percent of people reach their sixty-fifth birthday, the great majority of them in a healthy condition.

But now it seems we have reached a point of diminishing returns. By one calculation, if we found a cure for all cancers tomorrow, it would add just 3.2 years to overall life expectancy. Eliminating every last form of heart disease would add only 5.5 years. That's because people who die of these things tend to be old already, and if cancer or heart disease doesn't get them, something else will. Of nothing is that more true than Alzheimer's disease. Eradicating it altogether, according to the biologist Leonard Hayflick, would add just nineteen days to life expectancy.

Our extraordinary improvements in life span have come at a price. As Daniel Lieberman has noted, "For every year of added life that has been achieved since 1990, only 10 months is healthy." Already nearly half of people aged fifty or more suffer from some chronic pain or disability. We have become much better at extending life, but not necessarily better at extending quality of life. Older people cost the economy a lot. In the United States, the elderly constitute just over a tenth of the population but fill half the hospital beds and consume a third of all the medicines. Falls among the elderly alone cost the U.S. economy $31 billion a year, according to the Centers for Disease Control.

The time we spend in retirement has grown substantially, but the amount of work we do to fund it has not. The average person born before 1945 could expect to enjoy only about eight years of retirement before being permanently eliminated from the living, but someone born in 1971 can expect more like twenty years of retirement, and someone born in 1998 can, on current trends, expect perhaps thirty-five years—but all funded in each case by roughly forty years of labor. Most nations haven't begun to face up to the long-term costs of all these unwell, unproductive people who just go on and on. We have, in short, a lot of problems ahead of us all, both personally and societally.

Slowing down, losing vigor and resiliency, experiencing a steady, ineluctable diminution in the ability to self-repair—in a word, aging—is universal across all species, and it is intrinsic: that is, it is initiated from within the organism. At some point, your body will decide to grow senescent and then to die. You can slow the process a little by following a carefully virtuous lifestyle, but you can't escape it indefinitely. Put another way, we are all dying. Some of us are just doing it more quickly than others.

We don't have any idea why we age, or actually we have lots of ideas; we just don't know if any of them are correct. Almost thirty years ago, Zhores Medvedev, a Russian biogerontologist, counted some three hundred serious scientific theories to explain why we age, and the number has not shrunk in the decades since. As Professor José Viña and colleagues from the University of Valencia put it in a summary of current thinking, the theories fall into three broad categories: the genetic mutation theories (your genes malfunction and kill you), the wear-and-tear theories (the body just wears out), and the cellular waste accumulation theories (your cells clog up with toxic by-products). It may be that all three factors work together, or it may be that any two of the above are side effects of the third. Or it may be something else altogether. No one knows.

In 1961, Leonard Hayflick, then a young researcher at the Wistar Institute in Philadelphia, made a discovery that nearly everyone in his field found impossible to accept. He discovered that cultured human stem cells—that is, cells grown in a lab, as opposed to in a living body—can divide only about fifty times before they mysteriously lose their power to go on. In essence, they appear to be programmed to die of old age. The phenomenon became known as the Hayflick limit. It was a milestone moment for biology because it was the first time anyone had shown that aging was a process happening within cells. Hayflick also found that the cells he cultured could be frozen and kept in storage for any length of time and when thawed would resume their decline from precisely where they had left off. Clearly something within them was serving as a kind of tallying device to keep track of how many times they had divided. The idea that cells possess some form of memory and can count down toward their own extermination was so wildly radical that it was almost universally rejected.

For about a decade, Hayflick's findings languished. But then a team of researchers at the University of California at San Francisco discovered that stretches of specialized DNA at the end of each chromosome called telomeres fulfill the role of tallying device. With each cell division, telomeres shorten until eventually they reach a predetermined length (which varies markedly from one cell type to another) and the cell dies or becomes inactive. With this finding, the Hayflick limit suddenly became credible. It was hailed as the secret of aging. Arrest the shortening of telomeres and you could stop cell aging in its tracks. Gerontologists everywhere became very excited.

Alas, years of subsequent research have shown that telomere shortening can account for only a small part of the process. After the age of sixty, the risk of death doubles every eight years. A study by geneticists at the University of Utah found that telomere length may account for as little as 4 percent of that additional risk. As the gerontologist Judith Campisi told *Stat* in 2017, "If all aging was due to telomeres, we would have solved the aging problem a long time ago."

Aging, it turns out, not only involves much more than telomeres, but telomeres are involved in much more than aging. Telomere chemistry is regulated by an enzyme called telomerase, which switches off the cell when it has reached its preset quota of divisions. In cancerous cells, however, telomerase doesn't instruct the cells to stop dividing, but rather lets them go on proliferating endlessly. This has raised the possibility that a way to fight cancer would be to target telomerase in the cells. In sum, it's clear that telomeres are important not just for understanding aging but also for understanding cancer, but unfortunately we are still a long way from fully understanding either.

Two other terms encountered commonly, if no more productively, in discussions of aging are "free radicals" and "antioxidants." Free radicals are wisps of cellular waste that build up in the body in the process of metabolism. They are a by-product of our breathing oxygen. As one toxicologist has put it, "The biochemical price of breathing is aging."

Antioxidants are molecules that neutralize free radicals, so the thinking is that if you take a lot of them in the form of supplements, you can counter the effects of aging. Unfortunately, there is no scientific evidence to support that.

Most of us would almost certainly never have heard of either free radicals or antioxidants if a research chemist in California named Denham Harman had not, in 1945, read an article about aging in his wife's *Ladies' Home Journal* and developed a theory that free radicals and antioxidants are at the heart of human aging. Harman's idea was never anything more than a hunch, and subsequent research proved it to be wrong, but nonetheless the idea has taken hold and will not go away. The sale of antioxidant supplements alone is now worth well over $2 billion a year.

"It is a massive racket," David Gems of University College London told *Nature* in 2015. "The reason the notion of oxidation and ageing hangs around is because it is perpetuated by people making money out of it."

"Some studies have even suggested that antioxidant supplements can be harmful," *The New York Times* has noted. The principal learned journal of the field, *Antioxidants and Redox Signaling*, noted in 2013 that "antioxidant supplementation did not lower the incidence of many age-associated diseases but, in some cases, increased the risk of death."

In the United States, there is the additional, rather extraordinary consideration that the Food and Drug Administration exercises practically no oversight on supplements. As long as supplements don't contain any prescription medications and don't obviously kill or seriously harm anybody, manufacturers can sell pretty much whatever they want, with "no guarantees of purity or potency, no established guidelines on dosage, and often no warnings about side effects that may result when the products are taken along with approved medications," as an article in *Scientific American* noted. The products might be beneficial; it's just that no one has to prove it.

Although Dr. Harman didn't have anything to do with the supplements industry and was not a spokesman for antioxidant theories, he did follow a lifelong regime of taking high doses of the antioxidant vitamins C and E, and eating large quantities of antioxidant-rich fruits and vegetables, and it must be said it didn't do him any harm at all. He lived to be ninety-eight.

Even if you enjoy robust health, aging has inescapable consequences for us all. As we age, the bladder becomes less elastic and cannot hold as much, which is why one of the curses of aging is being forever on the lookout for a restroom. Skin loses elasticity, too, and becomes drier and more leathery. The blood vessels break more readily and create bruises. The immune system fails to detect intruders as reliably as it once did. The number of pigment cells usually decreases, but those that remain sometimes enlarge, producing age spots, or liver spots, which of course have nothing to do with the liver. The layer of fat directly associated with skin also thins, making it harder for elderly people to stay warm.

More seriously, the amount of blood pushed out with each heartbeat falls gradually as we age. If nothing else gets you first, your heart will eventually give out. That is a certainty. And because the amount of blood being moved around by the heart falls, your organs get less blood, too. After the age of forty, the volume of blood going to the kidneys decreases by an average of 1 percent a year.

Women are vividly reminded of the aging process when they reach menopause. Most animals die soon after they cease to be reproductive, but not (and thank goodness, of course) human females, who spend roughly a third of their lives in a postmenopausal state. We are the only primates that undergo menopause, and one of only a very few animals. The Florey Institute in Melbourne, for instance, studies menopause using sheep for the simple reason that sheep are almost the only land-based creatures known to experience menopause, too. At least two species of whales also go through it. Why any animals get it is a question yet to be answered.

The bad news is that menopause can be a terrible ordeal. Hot flashes are experienced by about three-quarters of women during menopause. (It is a feeling of sudden warmth, generally in the chest or above, induced by hormonal changes for unknown reasons.) Menopause is related to a fall in production of estrogen, but even now there isn't any test that can definitively confirm the condition. The best indicators for a woman that she is entering menopause (a stage known as perimenopause) are that her periods become irregular and she is likely to find herself experiencing a "sense that things aren't quite right," as Rose George wrote for the Wellcome Trust publication *Mosaic*.

Menopause is as much a mystery as aging itself. Two principal theories have been advanced, known rather neatly as the mother hypothesis and the grandmother hypothesis. The mother hypothesis is that childbearing is dangerous and exhausting, and it becomes more of both as women age. So menopause may simply be a kind of protection strategy. By no longer having the wear and distraction of further childbirth, a woman can better focus on maintaining her own health while completing the rearing of her children just as they are entering their most productive years. This leads naturally to the grandmother hypothesis, which is that women stop breeding in middle age so that they can help their offspring raise their children.

It is a myth, incidentally, that menopause is triggered by women exhausting their supply of eggs. They still have eggs. Not many, to be sure, but more than enough to remain fertile. So it isn't the literal running out of eggs that triggers the process (as even many doctors appear to believe). No one knows exactly what is the trigger.

II

A STUDY BY the Albert Einstein College of Medicine in New York in 2016 concluded that however much medical care may advance, it is unlikely that many people will ever live past about 115 years. On the other hand, Matt Kaeberlein, a University of Washington biogerontologist, thinks that young people alive today may routinely live up to

50 percent longer than people do now, and Dr. Aubrey de Grey, chief science officer of the SENS Research Foundation of Mountain View, California, believes that some people alive right now will live to be one thousand. Richard Cawthon, a geneticist at the University of Utah, has suggested that such a span is at least theoretically possible.

We'll have to wait and see. What can be said is that at present only about one person in ten thousand lives to be even a hundred. We don't know much at all about people who live beyond that, partly because there aren't many of them. The Gerontology Research Group in Los Angeles keeps track, as well as it can, of all the world's supercentenarians–that is, people who have reached their 110th birthday. But because records in much of the world are poor and because a lot of people for various reasons would like the world to think they are older than they really are, the GRG researchers tend to be cautious in admitting candidates to this most exclusive of clubs. Usually about seventy confirmed supercentenarians are on the group's books, but that is probably only about half the actual number in the world.

The chances of reaching your 110th birthday are about one in seven million. It helps a lot to be a woman; they are ten times more likely to reach 110 than a man. It is an interesting fact that women have always outlived men. This is a little counterintuitive when you consider that no man has ever died in childbirth. Nor, through much of history, have men been as closely exposed to contagions through nursing the sick. Yet in every period in history, in every society examined, women have always lived several years longer on average than men. And they still do now, even though men and women are subjected to more or less identical health care.

The longest-lived person that we know of was Jeanne Louise Calment of Arles, in Provence, who died at the decidedly ripe age of 122 years and 164 days in 1997. She was the first person to reach not only 122 but also 116, 117, 118, 119, 120, and 121. Calment had a leisurely life: her father was a rich shipbuilder and her husband a prosperous businessman. She never worked. She outlived her husband by more than half a century and her only child, a daughter, by sixty-three years. Calment smoked all her life—at the age of 117, when she finally gave up, she was still smoking two cigarettes a day—and ate two pounds of chocolate every week but was active up to the very end and enjoyed robust health. Her proud and charming boast in old age was, "I've never had but one wrinkle, and I'm sitting on it."

Calment was also a beneficiary of one of the most delightfully misjudged deals ever made. In 1965, when she ran into financial difficulties, she agreed to leave her apartment to a lawyer in return for a payment of 2,500 francs a months until she died. Because Calment was then ninety, it seemed a pretty good deal for the lawyer. In fact, it was the lawyer who died first, thirty years after signing the deal, having paid Calment more than 900,000 francs for an apartment he was never able to occupy.

The oldest man, meanwhile, was Jiroemon Kimura of Japan, who died aged 116 years and 54 days in 2013, after a quiet life as a government communications worker followed by a very long retirement in a village near Kyoto. Kimura lived a healthy lifestyle, but then millions of Japanese do. What enabled him to live so much longer than the rest of us is a question to which there is no answer, but family genes seem to play a significant role. As Daniel Lieberman told me, reaching 80 is largely a consequence of following a healthy lifestyle, but after that it is almost entirely a matter of genes. Or as Bernard Starr, a professor emeritus at City University of New York, put it, "The best way to assure longevity is to pick your parents."

At the time of writing, there were three people on Earth with a confirmed age of 115 (two in Japan, one in Italy) and three aged 114 (two in France, one in Japan).

Some people live longer than they ought to by any known measures. As Jo Marchant notes in her book *Cure*, Costa Ricans have only about one-fifth the personal wealth of Americans, and have poorer health care, but live longer. Moreover, people in one of the poorest regions of Costa Rica, the Nicoya Peninsula, live longest of all, even though they have much higher rates of obesity and hypertension. They also have longer telomeres. The theory is that they benefit from closer social bonds and family relationships. Curiously, it was found that if they live alone or don't see a child at least once a week, the telomere length advantage vanishes. It is an extraordinary fact that having good and loving relationships physically alters your DNA. Conversely, a 2010 U.S. study found, not having such relationships doubles your risk of dying from any cause.

III

IN NOVEMBER 1901, at a psychiatric hospital in Frankfurt am Main, Germany, a woman named Auguste Deter presented herself to the pathologist and psychiatrist Alois Alzheimer (1864–1915) complaining of persistent and worsening forgetfulness. She could feel her personality draining away, like sand from an hourglass. "I have lost myself," she explained sadly.

Alzheimer, a gruff but kindly Bavarian with pince-nez spectacles and a cigar perpetually plugged into the side of his mouth, was fascinated and frustrated by his inability to do anything to slow the unfortunate woman's deterioration. This was a sad time for Alzheimer himself. His wife of just seven years, Cäcilia, had died earlier in the year, leaving him with three children to raise, so when Frau Deter came into his life, he had to deal with his profoundest grief and greatest clinical impotence at the same time. Over the following weeks, the woman became increasingly confused and agitated, and nothing Alzheimer tried provided even slight relief.

Alzheimer moved to Munich the following year to take up a new post but continued to follow Frau Deter's decline from a distance, and when at last she died in 1906, he had her brain sent to him for autopsy. Alzheimer found that the poor woman's brain was riddled with clumps of destroyed cells. He reported these findings in a lecture and a paper, and in so doing became permanently associated with the disease, though in fact it was a colleague who first called it Alzheimer's disease in 1910. Remarkably, the tissue samples Alzheimer took from Frau Deter survived and have been restudied using modern techniques, and it turns out that she was suffering from a genetic mutation unlike any ever seen in another Alzheimer's patient. It appears that she might have been suffering not from Alzheimer's at all but rather from another genetic condition known as metachromatic leukodystrophy. Alzheimer didn't live long enough to fully understand the importance of his findings. He died from complications of a severe cold in 1915 aged just fifty-one.

We now know that Alzheimer's begins with an accumulation of a protein fragment called beta-amyloid in the sufferer's brain. Nobody is quite sure what amyloids do for us when they are working properly, but it is thought they may have a role in forming memories. In any case, they are normally cleared away after they have been used and are no longer needed. In Alzheimer's victims, however, they aren't flushed away but accumulate in clusters known as plaques and stop the brain from functioning as it should.

Later in the disease, victims also accumulate tangled fibrils of tau proteins, which are invariably referred to as tau tangles. How tau proteins relate to amyloids and how both relate to Alzheimer's are also uncertain, but the bottom line is that sufferers experience steady, irreversible memory loss. In its normal progression, Alzheimer's first demolishes short-term memories, then moves on to all or most memories, leading to confusion, shortness of temper, loss of inhibition, and eventually loss of all bodily functions, including how to breathe and swallow. As one observer has put it, in the end "one forgets, on a muscular level, how to exhale." People with Alzheimer's, it could be said, die twice—first in the mind, then in the body.

This much has been known for a century, but beyond that nearly all is confusion. The bewildering fact is that it is possible to have dementia without having buildups of amyloid and tau, and it is equally possible to have amyloid and tau buildups without having dementia. One study found that about 30 percent of elderly people have substantial beta-amyloid accumulations but no hint of cognitive decline.

It may be that plaques and tangles aren't the cause of the disease but simply its "signature"—the detritus left behind by the disease itself. In short, nobody knows if amyloid and tau are there because the victim is making too much of them or is simply failing to clear them adequately. The absence of consensus means that researchers fall into two camps: those who principally blame beta-amyloid proteins (and who are wryly known as baptists) and those who blame tau (known as tauists).

One thing that is known is that plaques and tangles accumulate slowly and begin their buildup long before signs of dementia become evident, so clearly the key to treating Alzheimer's will be to get to accumulations early, before they start doing real damage. So far we lack the technology to do so. We can't even definitively diagnose Alzheimer's. The only certain way to identify the condition is postmortem—after the patient dies.

The greatest mystery of all is why some people get Alzheimer's and others don't. Several genes have been found to be associated with Alzheimer's, but none has been directly implicated as a root cause. Just getting old vastly increases your susceptibility to Alzheimer's, but then the same could be said of almost all bad things. The more education you have had, the less likely you are to get Alzheimer's, though having an active and questing mind, as opposed to just racking up a lot of classroom hours in one's youth, is almost certainly what keeps Alzheimer's at bay. Dementias of all types are considerably rarer in people who eat a healthy diet, exercise at least moderately, maintain a sound weight, and don't smoke at all or drink to excess. Virtuous living doesn't eliminate the risk of Alzheimer's, but it does reduce it by about 60 percent.

Alzheimer's accounts for between 60 and 70 percent of all dementia cases and is thought to affect some fifty million people around the world, but Alzheimer's is only one of about a hundred types of dementias, and it is often difficult to distinguish among them. Lewy body dementia, for instance, is highly similar to Alzheimer's in that it involves a disturbance of neural proteins. (It's named for Dr. Friedrich H. Lewy, who worked alongside Alois Alzheimer in Germany.) Frontotemporal dementia arises from damage to the frontal and temporal lobes of the brain, often because of a stroke. It is often highly distressing to loved ones because victims frequently lose inhibitions and the ability to control impulses, so they tend to do embarrassing things—shed clothes in public, eat food abandoned by strangers, steal from supermarkets, and so on. Korsakoff's syndrome, named for a nineteenth-century Russian investigator, Sergei Korsakoff, is a dementia that arises most often from chronic alcoholism.

Altogether, one-third of all people over the age of sixty-five will die with some form of dementia. The cost to society is huge, yet almost everywhere research is curiously underfunded. In Britain, dementias cost the National Health Service £26 billion a year, but receive only £90 million annually in research funding, compared with £160 million for heart disease and £500 million for cancer.

Few diseases have been more resistant to treatment than Alzheimer's. It is the third most common cause of death among older people, exceeded only by heart disease and cancer, and we have no effective treatment for it at all. In clinical trials, Alzheimer's drugs have a 99.6 percent failure rate, one of the highest in the whole field of pharmacology. In the late 1990s, many researchers were suggesting that a cure was imminent, but that proved premature. One promising treatment was withdrawn after four people taking part in trials developed encephalitis, an inflammation of the brain. Part of the problem, as mentioned in chapter 22, is that Alzheimer's trials must be done on laboratory mice, and mice don't get Alzheimer's. They must be bred to grow plaques inside their brains, and that means they respond to drugs in different ways than humans would. Many pharmaceutical companies have now given up altogether. In 2018, Pfizer announced that it was withdrawing from research into Alzheimer's and Parkinson's disease and cutting three hundred jobs from two research facilities in New England. It is a sobering thought that poor Auguste Deter, if she presented herself to a doctor today, would be no better off now than she was with Alois Alzheimer almost 120 years ago.

IV

IT HAPPENS TO us all. Every day, around the world 160,000 people die. That's about 60 million fresh bodies a year, roughly equivalent to killing off the populations of Sweden, Norway, Belgium, Austria, and Australia year after year. On the other hand, it's only about 0.7 deaths per 100 people, which means that considerably less than one person in a hundred dies in any given year. Compared with other types of animals, we are awfully good at surviving.

Getting old is the surest route to dying. In the Western world, 75 percent of deaths from cancer, 90 percent from pneumonia, 90 percent from flu, and 80 percent from all causes occur in people sixty-five years of age or older. Interestingly, in the United States no one has died of old age since 1951, at least not officially, for in that year old age was banished as a cause from death certificates. In Britain, it is still allowed, though not much used.

Death is, for most of us, the most terrifying event imaginable. Jenny Diski, facing impending death (in 2016) from cancer, wrote movingly in a series of essays for the *London Review of Books* about the "excruciating terror" of knowing one is soon to die—"the razor-sharp claws digging into that interior organ where all dreaded things come to scrape and gnaw and live in me." But we do seem to have some measure of defense mechanism built into us. According to a 2014 study in the *Journal of Palliative Medicine*, between 50 and 60 percent of terminally ill patients report having intense but highly comforting dreams about their impending passing. A separate study found evidence of a surge of chemicals in the brain at death, which may account for the intense experiences often reported by survivors of near-death incidents.

Most dying people lose any desire to eat or drink in the last day or two of life. Some lose the power of speech. When the ability to cough or swallow goes, they often make a rasping sound commonly known as a death rattle. It can sound distressing but seems not to be to those experiencing it. However, another kind of labored breathing at death, called agonal breathing, may very well be. Agonal breathing, in which the sufferer can't get enough breath because of a failing heart, may last only for a few seconds, but it can go on for forty minutes or more and be extremely distressing to both victim and loved ones at the bedside. It can be stopped with a neuromuscular blocking agent, but many doctors won't administer it, because it inevitably hastens death and is therefore thought unethical or even possibly illegal, even though death is just around the corner anyway.

We are extraordinarily sensitive about dying, it seems, and often take the most desperate steps to put off the inevitable. Almost everywhere, overtreatment of dying people is routine. Among those dying of cancer in America, one in eight receives chemotherapy right up to the last two weeks of their lives, long past the point where it is effective. Three separate studies have shown that cancer sufferers receiving palliative care in their final weeks rather than chemotherapy actually live longer and suffer much less.

Predicting deaths, even among the dying, is not easy. As Dr. Steven Hatch of the University of Massachusetts Medical School has written, "One review found that, even among terminally ill patients whose median survival is only four weeks, doctors were correct to within a week of survival in only 25 percent of cases, and in another 25 percent their predictions were wrong by more than four weeks!"

Death becomes apparent very quickly. Almost at once the blood begins to drain from the capillaries near the surface, leading to the ghostly pallor associated with death. "A man's corpse looks as though his essence has left him, and it has. He is flat and toneless, no longer inflated by the vital spirit the Greeks called *pneuma*," wrote Sherwin Nuland in *How We Die*. Even to someone unused to dead bodies, death is usually instantly recognizable.

Tissue deterioration starts almost at once, which is why "harvesting" (surely the ugliest term in medicine) organs for transplant is such an urgent business. Blood pools in the lowest parts of the body, as gravity demands, turning the skin there purple in a process known as *livor mortis*. Internal cells rupture and enzymes spill out and begin a self-digesting process known as autolysis. Some organs function longer than others. The liver will continue to break down alcohol after death, even though it has absolutely no need to do so. Cells, too, die at different rates. Brain cells go quickly, in no more than about three or four minutes, but muscle and skin cells may last for hours—perhaps a whole day. The famous muscle stiffening known as rigor mortis (literally "stiffness of death") sets in between thirty minutes and four hours after death, starting in the facial muscles and moving downward through the body and outward to the extremities. Rigor mortis lasts for a day or so.

A corpse is still very much alive. It's just not *your* life any longer. It's the bacteria you leave behind, plus any others that flock in. As they devour the body, gut bacteria produce a range of gases, among them methane, ammonia, hydrogen sulfide, and sulfur dioxide, as well as the self-explanatorily named compounds cadaverine and putrescine. The smell of a rotting corpse usually becomes horrible within two to three days, less if the weather is hot. Then, gradually, the smells begin to ease until there's no remaining flesh and thus nothing left to cause odor. Of course, the process can be disrupted if the body falls into a glacier or peaty bog, where bacteria can't survive and proliferate, or is kept very dry so that the body mummifies. It is a myth, and physiological impossibility, incidentally, that hair and nails continue to grow after death. Nothing grows after death.

For those who choose to be buried, decomposition in a sealed coffin takes a long time—between five and forty years, according to one estimate, and that's only for those who are not embalmed. The average grave is visited for only about fifteen years, so most of us take a lot longer to vanish from the earth than from others' memories. A century ago, only about one person in a hundred was cremated, but today three-quarters of Britons and 40 percent of Americans are. If you are cremated, your ashes will weigh about five pounds.

And that's you gone. But it was good while it lasted, wasn't it?

ACKNOWLEDGMENTS

I don't believe I have ever been indebted to more people for expert help and guidance, more generously given, than I have with this book. In particular I wish to thank two people for their especially close help: my son Dr. David Bryson, pediatric orthopedic fellow at Alder Hey Children's Hospital in Liverpool, and my good friend Ben Ollivere, clinical associate professor of trauma surgery at the University of Nottingham and a consultant trauma surgeon at Queen's Medical Centre in Nottingham.

I am also much indebted to the following:

In England: Dr. Katie Rollins, Dr. Margy Pratten, and Dr. Siobhan Loughna of the University of Nottingham and Queen's Medical Centre, Nottingham; Professor John Wass, Professor Irene Tracey, and Professor Russell Foster of Oxford University; Professor Neil Pearce of the London School of Hygiene and Tropical Medicine; Dr. Magnus Bordewich of the Department of Computer Science at Durham University; Dr. Karen Ogilvie of the Royal Society of Chemistry; Daniel M. Davis, professor of immunology and director of research at the Manchester Collaborative Centre for Inflammation Research at the University of

Manchester, and his colleagues Jonathan Worboys, Poppy Simmonds, Pippa Kennedy, and Karoliina Tuomela; Professor Rod Skinner of Newcastle University; Dr. Charles Tomson, consultant nephrologist at Newcastle upon Tyne Hospitals; and Dr. Mark Gompels of North Bristol NHS Trust. Special thanks also to my good friend Joshua Ollivere.

In the United States: Professor Daniel Lieberman of Harvard University; Professor Nina Jablonski of Penn State University; Dr. Leslie J. Stein and Dr. Gary Beauchamp of the Monell Chemical Senses Center in Philadelphia; Dr. Allan Doctor and Dr. Michael Kinch of Washington University in St. Louis; Dr. Matthew Porteus and Professor Christopher Gardner of Stanford University; and Patrick Losinski and his helpful staff at the Columbus Metropolitan Library in Columbus, Ohio.

In the Netherlands: Drs. Josef and Britta Vormoor, Professor Hans Clevers, Dr. Olaf Heidenreich, and Dr. Anne Rios of the Princess Máxima Center for Pediatric Oncology in Utrecht. Special thanks also to Johanna and Benedikt Vormoor.

I am also much indebted to Gerry Howard, Dame Gail Rebuck, Susanna Wadeson, Larry Finlay, Amy Black, and Kristin Cochrane at Penguin Random House, to the brilliant artist Neil Gower, to Camilla Ferrier and her colleagues at the Marsh Agency in London, and to my children Felicity, Catherine, and Sam for much willing assistance. Above all, and as always, my greatest thanks go to my dear and saintly wife, Cynthia.

NOTES ON SOURCES

The following is intended as a quick guide for those who wish to check a fact or do further reading. Where a fact is commonly known or widely reported–the functions of the liver, for instance–I have not cited the source. On the whole, sources are listed only where assertions are specific, arguable, or otherwise distinctively notable.

CHAPTER 1: HOW TO BUILD A HUMAN

2 Altogether, according to RSC calculations: The information on the cost of building a replica Benedict Cumberbatch was supplied by Karen Ogilvie of the Royal Society of Chemistry, London.

2 We need, for instance, just 20 atoms: Emsley, *Nature's Building Blocks,* 4.

3 We now know that selenium makes two vital enzymes: Ibid., 379–80.

3 you can irremediably poison your liver: *Scientific American,* July 2015, 31.

4 in 2012 *Nova,* the long-running science program on PBS: "Hunting the Elements," *Nova,* April 4, 2012.

4 Well, you blink fourteen thousand times a day: McNeill, *Face,* 27.

5 The length of all your blood vessels: West, *Scale,* 152.

5 if you formed all the DNA in your body: Pollack, *Signs of Life,* 19.

6 You would need twenty billion strands of DNA: Ibid.

7 Its chemical name is 189,819 letters long: Ball, *Stories of the Invisible,* 48.

7 Nobody knows how many types of proteins: Challoner, *Cell,* 38.

7 All humans share 99.9 percent of their DNA: *Nature,* June 26, 2014, 463.

7 My DNA and your DNA will differ: Arney, *Herding Hemingway's Cats*, 184.
7 about a hundred personal mutations: *New Scientist*, Sept. 15, 2012, 30–33.
8 One particular short sequence, called an Alu element: Mukherjee, *Gene*, 322; Ben-Barak, *Invisible Kingdom*, 174.
8 Five out of every six smokers: *Nature*, March 24, 2011, S2.
8 between one and five of your cells turns cancerous: Samuel Cheshier, neurosurgeon and Stanford professor, quoted on *Naked Scientist*, podcast, March 21, 2017.
9 Our bodies are a universe of 37.2 trillion cells: "An Estimation of the Number of Cells in the Human Body," *Annals of Human Biology*, Nov.–Dec. 2013.
9 There are thousands of things that can kill us: *New Yorker*, April 7, 2014, 38–39.
9 We undertake every part of the process: Hafer, *Not-So-Intelligent Designer*, 132.

CHAPTER 2: THE OUTSIDE: SKIN AND HAIR

11 "Our seams don't burst": Jablonski interview, State College, Pa., Feb. 29, 2016.
12 We shed skin copiously, almost carelessly: Andrews, *Life That Lives on Man*, 31.
12 We each trail behind us: Ibid., 166.
12 acne, a word of very uncertain derivation: *Oxford English Dictionary*.
13 They detect light touch: Ackerman, *Natural History of the Senses*, 83.
13 if you sink a spade into gravel or sand: Linden, *Touch*, 46.
13 Curiously, we don't have any receptors for wetness: "The Magic of Touch," *The Uncommon Senses*, BBC Radio 4, March 27, 2017.
13 Women are much better than men at tactile sensitivity: Linden, *Touch*, 73.
15 skin gets its color from a variety of pigments: Jablonski interview.
15 Its production slows dramatically as we age: Challoner, *Cell*, 170.
15 "Melanin is a superb natural sunscreen": Jablonski interview.
15 Melanin often responds to sunlight: Jablonski, *Living Color*, 14.
16 The red of sunburn: Jablonski, *Skin*, 17.
16 The formal name for sunburn is erythema: Smith, *Body*, 410.
16 The process is known as melasma: Jablonski, *Skin*, 90.
17 some 50 percent of people globally: *Journal of Pharmacology and Pharmacotherapeutics*, April/June 2012; *New Scientist*, Aug. 9, 2014, 34–37.
17 As people evolved lighter skin: University College London press release, "Natural Selection Has Altered the Appearance of Europeans over the Past 5000 Years," March 11, 2014.
17 Skin color has been changing: Jablonski, *Living Color*, 24.
17 Indigenous populations in South America: Jablonski, *Skin*, 91.
17 Rather harder to explain have been the KhoeSan people: "Rapid

Evolution of a Skin-Lightening Allele in Southern African KhoeSan," *Proceedings of the National Academy of Sciences,* Dec. 26, 2018.

18 Using DNA analysis: "First Modern Britons Had 'Dark to Black' Skin," *Guardian,* Feb. 7, 2018.

18 suggested that the DNA used in the analysis: *New Scientist,* March 3, 2018, 12.

18 We are actually as hairy as our cousins the apes: Jablonski, *Skin,* 19.

18 Altogether we are estimated to have five million hairs: Linden, *Touch,* 216.

19 it provides warmth, cushioning, and camouflage: "The Naked Truth," *Scientific American,* Feb. 2010.

19 In furry mammals, it adds a useful layer: Ashcroft, *Life at the Extremes,* 157.

19 Horripilation also makes mammalian hair stand up: *Baylor University Medical Center Proceedings,* July 2012, 305.

19 genetic studies that dark pigmentation: "Why Are Humans So Hairy?," *New Scientist,* Oct. 17, 2017.

19 "because it increases the thickness of the space": Jablonski interview.

20 humans don't seem to have pheromones: "Do Human Pheromones Actually Exist?," *Science News,* March 7, 2017.

20 secondary hair is for display: Bainbridge, *Teenagers,* 44–45.

20 We each grow about twenty-five feet of hair: *The Curious Cases of Rutherford and Fry,* podcast, BBC Radio 4, Aug. 22, 2016.

21 The system introduced the concept of the mug shot: Cole, *Suspect Identities,* 49.

21 The uniqueness of fingerprints was first established: Smith, *Body,* 409.

22 They are assumed to aid in gripping: Linden, *Touch,* 37.

22 why our fingers wrinkle when we have long baths: "Why Do We Get Prune Fingers?," Smithsonian.com, Aug. 6, 2015.

22 a condition known as adermatoglyphia: "Adermatoglyphia: The Genetic Disorder of People Born Without Fingerprints," *Smithsonian,* Jan. 14, 2014.

22 Most quadrupeds cool by panting: Daniel E. Lieberman, "Human Locomotion and Heat Loss: An Evolutionary Perspective," *Comprehensive Physiology* 5, no. 1 (Jan. 2015).

22 "The loss of most of our body hair": Jablonski, *Living Color,* 26.

23 a man who weighs 155 pounds: Stark, *Last Breath,* 283–85.

23 Although salt is only a tiny part: Ashcroft, *Life at the Extremes,* 139.

23 Sweating is activated by the release of adrenaline: Ibid., 122.

24 Emotional sweating is what is measured: Tallis, *Kingdom of Infinite Space,* 23.

24 The two chemicals that account for the odor: Bainbridge, *Teenagers,* 48.

24 the number of bacteria on you: Andrews, *Life That Lives on Man,* 11.

24 To make one's hands safely clean: Gawande, *Better,* 14–15; "What Is the Right Way to Wash Your Hands?," *Atlantic,* Jan. 23, 2017.

25 One volunteer harbored a microbe: National Geographic News, Nov. 14, 2012.

25 The problem with antibacterial soaps: Blaser, *Missing Microbes*, 200.
25 They have lived with us for so long: David Shultz, "What the Mites on Your Face Say About Where You Came From," *Science*, Dec. 14, 2015, www.sciencemag.org.
26 Studies of scratching showed: Linden, *Touch*, 185.
26 the most extraordinary case of unappeasable suffering: Ibid., 187–89.
26 We have about 100,000: Andrews, *Life That Lives on Man*, 38–39.
27 a hormone called dihydrotestosterone: *Baylor University Medical Center Proceedings*, July 2012, 305.
27 considering how easily some of us lose it: Andrews, *Life That Lives on Man*, 42.

CHAPTER 3: MICROBIAL YOU

28 For nitrogen to be useful to us: Ben-Barak, *Invisible Kingdom*, 58.
28 Humans produce twenty digestive enzymes: Interview with Professor Christopher Gardner of Stanford University, Palo Alto, Jan. 29, 2018.
29 the average bacterium weighs about one-trillionth: *Baylor University Medical Center Proceedings*, July 2014; West, *Scale*, 1.
29 But bacteria can swap genes: Crawford, *Invisible Enemy*, 14.
29 A single parent bacterium could in theory: Lane, *Power, Sex, Suicide*, 114.
29 In three days, its progeny: Maddox, *What Remains to Be Discovered*, 170.
29 If you put all Earth's microbes in one heap: Crawford, *Invisible Enemy*, 13.
29 you are likely to have something like 40,000 species: "Learning About Who We Are," *Nature*, June 14, 2012; "Molecular-Phylogenetic Characterization of Microbial Community Imbalances in Human Inflammatory Bowel Diseases," *Proceedings of the National Academy of Sciences*, Aug. 15, 2007.
30 Altogether your private load of microbes: Blaser, *Missing Microbes*, 25; Ben-Barak, *Invisible Kingdom*, 13.
30 In 2016, researchers from Israel and Canada: *Nature*, June 8, 2016.
31 Microbial communities can be surprisingly specific: "The Inside Story," *Nature*, May 28, 2008.
31 just 1,415 are known to cause disease in humans: Crawford, *Invisible Enemy*, 15–16; Pasternak, *Molecules Within Us*, 143.
32 all these microbes have almost nothing in common: "The Microbes Within," *Nature*, Feb. 25, 2015.
32 The herpes virus has endured: "They Reproduce, but They Don't Eat, Breathe, or Excrete," *London Review of Books*, March 9, 2001.
32 If you blew one up to the size of a tennis ball: Ben-Barak, *Invisible Kingdom*, 4.
33 he called the mysterious agent *contagium vivum fluidum:* Roossinck, *Virus*, 13.
33 Of the hundreds of thousands of viruses: *Economist*, June 24, 2017, 76.

33 Proctor found that the average quart of seawater: Zimmer, *Planet of Viruses*, 42–44.

33 ocean viruses alone if laid end to end: Crawford, *Deadly Companions*, 13.

34 Colds unquestionably are more frequent in winter: "Cold Comfort," *New Yorker*, March 11, 2002, 42.

34 The common cold is not a single illness: "Unraveling the Key to a Cold Virus's Effectiveness," *New York Times*, Jan. 8, 2015.

34 In one, a volunteer was fitted with a device: "Cold Comfort," 45.

35 In a similar study at the University of Arizona: *Baylor University Medical Center Proceedings*, Jan. 2017, 127.

35 In the real world, such infestations: "Germs Thrive at Work, Too," *Wall Street Journal*, Sept. 30, 2014.

35 Where microbes thrive is in the fabrics: *Nature*, June 25, 2015, 400.

36 *Cryptococcus gattii* was for decades: *Scientific American*, Dec. 2013, 47.

37 A most arresting illustration of that: "Giant Viruses," *American Scientist*, July–Aug. 2011; Zimmer, *Planet of Viruses*, 89–91; "The Discovery and Characterization of Mimivirus, the Largest Known Virus and Putative Pneumonia Agent," *Emerging Infections*, May 21, 2007; "Ironmonger Who Found a Unique Colony," *Daily Telegraph*, Oct. 15, 2004; *Bradford Telegraph and Argus*, Oct. 15, 2014; "Out on a Limb," *Nature*, Aug. 4, 2011.

39 Max von Pettenkofer was so vehemently offended: Le Fanu, *Rise and Fall of Modern Medicine*, 179.

39 Salvarsan was effective against only a few things: *Journal of Antimicrobial Chemotherapy* 71 (2016).

41 The principal investigator at Oxford: Lax, *Mould in Dr. Florey's Coat*, 77–79.

41 He was an unlikely candidate: *Oxford Dictionary of National Biography*, s.v. "Chain, Sir Ernst Boris."

41 By early 1941, they had just enough to trial the drug: Le Fanu, *Rise and Fall of Modern Medicine*, 3–12; *Economist*, May 21, 2016, 19.

42 a lab assistant in Peoria named Mary Hunt: "Penicillin Comes to Peoria," *Historynet*, June 2, 2014.

42 Every bit of penicillin made since that day: Blaser, *Missing Microbes*, 60; "The Real Story Behind Penicillin," *PBS NewsHour* website, Sept. 27, 2013.

42 The British discoverers found to their chagrin: *Oxford Dictionary of National Biography*, s.v. "Florey, Howard Walter."

43 Chain, despite sharing the Nobel Prize: *Oxford Dictionary of National Biography*, s.v. "Chain, Sir Ernst Boris."

43 By attacking a broad spectrum of bacteria: *New Yorker*, Oct. 22, 2012, 36.

44 Grant ended up in Yale New Haven Hospital: Interview with Michael Kinch, Washington University of St. Louis, April 18, 2018.

45 antibiotics are prescribed for 70 percent of acute bronchitis cases: "Superbug: An Epidemic Begins," *Harvard Magazine*, May–June 2014.

45 most Americans consume secondhand antibiotics: Blaser, *Missing*

Microbes, 85; *Baylor University Medical Center Proceedings*, July 2012, 306.

45 Sweden banned the agricultural use of antibiotics: Blaser, *Missing Microbes*, 84.

45 In 1977, the Food and Drug Administration: *Baylor University Medical Center Proceedings*, July 2012, 306.

45 In consequence, the death rate: Bakalar, *Where the Germs Are*, 5–6.

45 They not only have grown steadily more resistant: "Don't Pick Your Nose," *London Review of Books*, July 2004.

46 Methicillin-resistant Staphylococcus aureus: "World Super Germ Born in Guildford," *Daily Telegraph*, Aug. 26, 2001; "Squashing Superbugs," *Scientific American*, July 2009.

46 Today, MRSA and its cousins kill: "A Dearth in Innovation for Key Drugs," *New York Times*, July 22, 2014.

46 CRE kills about half of all those it sickens: *Nature*, July 25, 2013, 394.

46 "It's just too expensive for them": Kinch interview; "Resistance Is Futile," *Atlantic*, Oct. 15, 2011.

46 all but two of the eighteen largest: "Antibiotic Resistance Is Worrisome, but Not Hopeless," *New York Times*, March 8, 2016.

46 At the current rate of spread: *BBC Inside Science*, BBC Radio 4, June 9, 2016; *Chemistry World*, March 2018, 51.

47 produce quorum-sensing drugs: *New Scientist*, Dec. 14, 2013, 36.

47 the most abundant bioparticles on Earth: "Reengineering Life," *Discovery*, BBC Radio 4, May 8, 2017.

CHAPTER 4: THE BRAIN

48 The consistency of the brain: "Thanks for the Memory," *New York Review of Books*, Oct. 5, 2006; Lieberman, *Evolution of the Human Head*, 211.

49 Altogether, the human brain is estimated to hold: "Solving the Brain," *Nature Neuroscience*, July 17, 2013.

49 It makes up just 2 percent of our body weight: Allen, *Lives of the Brain*, 188.

49 the brain is by far the most expensive of our organs: Bribiescas, *Men*, 42.

50 The most efficient brains: Winston, *Human Mind*, 210.

50 the number is more like 86 billion: "Myths That Will Not Die," *Nature*, Dec. 17, 2015.

50 "in a single cubic centimeter of brain tissue": Eagleman, *Incognito*, 2.

51 It is divided into two hemispheres: Ashcroft, *Spark of Life*, 227; Allen, *Lives of the Brain*, 19.

51 six patches on the temporal lobe: "How Your Brain Recognizes All Those Faces," Smithsonian.com, June 6, 2017.

52 Although the cerebellum occupies just 10 percent: Allen, *Lives of the Brain*, 14; Zeman, *Consciousness*, 57; Ashcroft, *Spark of Life*, 228–29.

53 how slowly or rapidly we age: "A Tiny Part of the Brain Appears to Orchestrate the Whole Body's Aging," *Stat*, July 26, 2017.

53 People whose amygdalae are destroyed: O'Sullivan, *Brainstorm,* 91.

53 Your nightmares may simply be: "What Are Dreams?," *Nova,* PBS, Nov. 24, 2009.

54 The eyes send a hundred billion signals: "Attention," *New Yorker,* Oct. 1, 2014.

54 only about 10 percent of the information: *Nature,* April 20, 2017, 296.

55 "While we have the overwhelming impression": Le Fanu, *Why Us?,* 199.

57 implant entirely false memories in people's heads: *Guardian,* Dec. 4, 2003, 8.

57 One year later, the psychologists asked: *New Scientist,* May 14, 2011, 39.

58 The mind breaks each memory: Bainbridge, *Beyond the Zonules of Zinn,* 287.

58 A single fleeting thought: Lieberman, *Evolution of the Human Head,* 183.

58 these fragments of memory: Le Fanu, *Why Us?,* 213; Winston, *Human Mind,* 82.

58 "It's a little more like a *Wikipedia* page": *The Why Factor,* BBC World Service, Sept. 6, 2013.

59 the United States has a national memory championship: *Nature,* April 7, 2011, 33.

60 The idea arose principally from a series: Draaisma, *Forgetting,* 163–70; "Memory," *National Geographic,* Nov. 2007.

60 The person from whom we learned: "The Man Who Couldn't Remember," *Nova,* PBS, June 1, 2009; "How Memory Speaks," *New York Review of Books,* May 22, 2014; *New Scientist,* Nov. 28, 2015, 36.

61 "Rarely in the history of neuroscience": *Nature Neuroscience,* Feb. 2010, 139.

61 Brodmann was repeatedly overlooked: *Neurosurgery,* Jan. 2011, 6–11.

62 Both white matter and gray matter: Ashcroft, *Spark of Life,* 229.

62 the idea that we use only 10 percent: *Scientific American,* Aug. 2011, 35.

62 A teenager's brain is only: "Get Knitting," *London Review of Books,* Aug. 18, 2005.

63 The leading cause of deaths among teenagers: *New Yorker,* Aug. 31, 2015, 85.

63 The difficulty is that there is no certain way: "Human Brain Make New Nerve Cells," *Science News,* April 5, 2018; *All Things Considered* transcript, National Public Radio, March 17, 2018.

64 The remaining third of his brain: Le Fanu, *Why Us?,* 192.

64 "If you were designing an organic machine": "The Mystery of Consciousness," *New York Review of Books,* Nov. 2, 1995.

65 In the 1880s, in a series of operations: Dittrich, *Patient H.M.,* 79.

65 Moniz provided an almost perfect demonstration: "Unkind Cuts," *New York Review of Books,* April 24, 1986.

66 The procedure was so crude: "The Lobotomy Files: One Doctor's Legacy," *Wall Street Journal,* Dec. 12, 2013.

67 Freeman was a psychiatrist with no surgical certification: El-Hai, *Lobotomist,* 209.

67 About two-thirds of Freeman's subjects: Ibid., 171.

67 His most notorious failure was Rosemary Kennedy: Ibid., 173–74.

68 the very fact that the brain is so snugly encased: Sanghavi, *Map of the Child*, 107; Bainbridge, *Beyond the Zonules of Zinn*, 233–35.

68 known as contrecoup injuries: Lieberman, *Evolution of the Human Head*, 217.

68 In Britain, epilepsy remained on the statute books: *Literary Review*, Aug. 2016, 36.

69 "The history of epilepsy can be summarised": *British Medical Journal* 315 (1997).

69 Capgras syndrome is a condition: "Can the Brain Explain Your Mind?," *New York Review of Books*, March 24, 2011.

69 In Klüver-Bucy syndrome, the victims: "Urge," *New York Review of Books*, Sept. 24, 2015.

69 Perhaps the most bizarre of all: Sternberg, *NeuroLogic*, 133.

70 Locked-in syndrome is different again: Owen, *Into the Grey Zone*, 4.

70 No one knows how many: "The Mind Reader," *Nature Neuroscience*, June 13, 2014.

71 It may be simply that a less robust: Lieberman, *Evolution of the Human Head*, 556; "If Modern Humans Are So Smart, Why Are Our Brains Shrinking?," *Discover*, Jan. 20, 2011.

CHAPTER 5: THE HEAD

72 Mary, Queen of Scots, needed three hearty whacks: Larson, *Severed*, 13.

72 Charlotte Corday, guillotined in 1793: Ibid., 246.

74 Davis became so celebrated: *Australian Indigenous Law Review*, no. 92 (2007); *New Literatures Review*, University of Melbourne, Oct. 2004.

74 He was convinced that a person's intellect: *Anthropological Review*, Oct. 1868, 386–94.

74 he referred to it as "Mongolism": Blakelaw and Jennett, *Oxford Companion to the Body*, 249; *Oxford Dictionary of National Biography*.

75 In one case, cited by Stephen Jay Gould: Gould, *Mismeasure of Man*, 138.

75 In 1861, during an autopsy on a stroke victim: Le Fanu, *Why Us?*, 180; "The Inferiority Complex," *New York Review of Books*, Oct. 22, 1981.

76 No two authorities seem to agree: See McNeill, *Face*, 180; Perrett, *In Your Face*, 21; "A Conversation with Paul Ekman," *New York Times*, Aug. 5, 2003.

77 Babies fresh from the womb: McNeill, *Face*, 4.

77 Although the change was too slight: Ibid., 26.

77 the French anatomist G.-B. Duchenne de Boulogne: *New Yorker*, Jan. 12, 2015, 35.

77 we all indulge in "microexpressions": "Conversation with Paul Ekman."

78 in favor of our small, active eyebrows: "Scientists Have an Intriguing New Theory About Our Eyebrows and Foreheads," *Vox*, April 9, 2018.

78 One of the reasons the *Mona Lisa*: Perrett, *In Your Face*, 18.

79 external nose and intricate sinuses: Lieberman, *Evolution of the Human Head,* 312.

79 we have as many as thirty-three systems: *The Uncommon Senses,* BBC Radio 4, March 20, 2017.

80 your own white blood cells: "Blue Sky Sprites," *Naked Scientists,* podcast, May 17, 2016; "Evolution of the Human Eye," *Scientific American,* July 2011, 53.

80 muscae volitantes, or "hovering flies": "Meet the Culprits Behind Bright Lights and Strange Floaters in Your Vision," Smithsonian.com, Dec. 24, 2014.

80 If you held a human eyeball: McNeill, *Face,* 24.

81 The lens, which gets all the credit: Davies, *Life Unfolding,* 231.

81 Tears not only keep our eyelids: Lutz, *Crying,* 67–68.

82 you produce about five to ten ounces of tears: Ibid., 69.

82 Our scleras are unique: Lieberman, *Evolution of the Human Head,* 388.

83 Their main problem isn't that their world is pallid: "Outcasts of the Islands," *New York Review of Books,* March 6, 1997.

83 Much later, primates re-evolved the ability: *National Geographic,* Feb. 2016, 56.

83 The movements of the eye are called saccades: *New Scientist,* May 14, 2011, 356; Eagleman, *Brain,* 60.

84 Victorian naturalists sometimes cited this: Blakelaw and Jennett, *Oxford Companion to the Body,* 82; Roberts, *Incredible Unlikeliness of Being,* 114; Eagleman, *Incognito,* 32.

85 They were jawbones in our ancient ancestors: Shubin, *Your Inner Fish,* 160–62.

85 A pressure wave that moves the eardrum: Goldsmith, *Discord,* 6–7.

86 From the quietest detectable sound to the loudest: Ibid., 161.

86 This means that all sound waves: Bathurst, *Sound,* 28–29.

86 The term was coined by Colonel Sir Thomas Fortune Purves: Ibid., 124.

87 The reason we feel dizzy: Bainbridge, *Beyond the Zonules of Zinn,* 110.

87 When loss of balance is prolonged: Francis, *Adventures in Human Being,* 63.

88 half of people under the age of thirty: "World Without Scent," *Atlantic,* Sept. 12, 2015.

88 "Smell is something of an orphan science": Interview with Gary Beauchamp, Monell Chemical Senses Center, Philadelphia, 2016.

89 the receptors are activated: Al-Khalili and McFadden, *Life on the Edge,* 158–59.

89 A banana, for example, contains three hundred volatiles: Shepherd, *Neurogastronomy,* 34–37.

89 Tomatoes have four hundred: Gilbert, *What the Nose Knows,* 45.

89 The smell of burned almonds: Brook, *At the Edge of Uncertainty,* 149.

89 The smell of licorice: "Secret of Liquorice Smell Unravelled," *Chemistry World,* Jan. 2017.

89 it was first suggested way back in 1927: Holmes, *Flavor,* 49.

90 In 2014, researchers at the Université: *Science*, March 21, 2014.
90 why certain odors are so powerfully evocative of memories: Monell website, "Olfaction Primer: How Smell Works."
90 researchers at the University of California: "Mechanisms of Scent-Tracking in Humans," *Nature*, Jan. 4, 2007.
91 For five of fifteen smells tested: Holmes, *Flavor*, 63.
91 Babies and mothers are similarly skillful: Gilbert, *What the Nose Knows*, 63.
91 One of the early symptoms of Alzheimer's: Platoni, *We Have the Technology*, 39.
91 Ninety percent of people who lose smell: Blodgett, *Remembering Smell*, 19.

CHAPTER 6: DOWN THE HATCH: THE MOUTH AND THROAT

92 Midway through the entertainment: "Profiles," *New Yorker*, Sept. 9, 1953; Vaughan, *Isambard Kingdom Brunel*, 196–97.
94 he was the person who first postulated: Birkhead, *Most Perfect Thing*, 150.
94 The anatomist's word for swallowing: Collis, *Living with a Stranger*, 20.
96 choking is the fourth most common: Lieberman, *Evolution of the Human Head*, 297.
96 Henry Heimlich was something of a showman: "The Choke Artist," *New Republic*, April 23, 2007; *New York Times* obituary, April 23, 2007.
97 2,374 imprudently ingested objects: Cappello, *Swallow*, 4–6; *New York Times*, Jan. 11, 2011.
98 Jackson was a cold and friendless man: *Annals of Thoracic Surgery* 57 (1994): 502–5.
98 A typical adult secretes: "Gut Health May Begin in the Mouth," *Harvard Magazine*, Oct. 20, 2017.
98 we secrete about 31,700 quarts: Tallis, *Kingdom of Infinite Space*, 25.
98 a powerful painkiller called opiorphin: "Natural Painkiller Found in Human Spit," *Nature*, Nov. 13, 2006.
99 We produce very little saliva while we sleep: Enders, *Gut*, 22.
99 150 different chemical compounds: *Scientific American*, May 2013, 20.
100 Altogether, about a thousand species of bacteria: Ibid.
100 Dawson's team found that candle blowing: Clemson University press release, "A True Food Myth Buster," Dec. 13, 2011.
100 teeth have been called "ready-made fossils": Ungar, *Evolution's Bite*, 5.
101 if you are a typical adult male: Lieberman, *Evolution of the Human Head*, 226.
101 the most regenerative of all cells in the body: *New Scientist*, March 16, 2013, 45.
101 In fact, that is a myth, traced to a textbook: *Nature*, June 21, 2012, S2.
101 the body has taste receptors in the gut and throat: Roach, *Gulp*, 46.
102 Taste receptors have also been found: *New Scientist*, Aug. 8, 2015, 40–41.

102 These contain a poison called tetrodotoxin: Ashcroft, *Life at the Extremes*, 54; "Last Supper?," *Guardian*, Aug. 5, 2016.

103 the British author Nicholas Evans: "I Wanted to Die. It Was So Grim," *Daily Telegraph*, Aug. 2, 2011.

103 We have about ten thousand taste receptors: "A Matter of Taste?," *Chemistry World*, Feb. 2017; Holmes, *Flavor*, 83; "Fire-Eaters," *New Yorker*, Nov. 4, 2013.

104 A purified version of a Moroccan spurge plant: Holmes, *Flavor*, 85.

104 Chinese adults who ate a lot of capsaicin: *Baylor University Medical Center Proceedings*, Jan. 2016, 47.

105 Some authorities believe we also have: *New Scientist*, Aug. 8, 2015, 40–41.

105 Today Ajinomoto is a behemoth: Mouritsen and Styrbaek, *Umami*, 28.

106 Smell is said to account for: Holmes, *Flavor*, 21.

107 The students without exception listed: *BMC Neuroscience*, Sept. 18, 2007.

107 if an orange-flavored drink is colored red: *Scientific American*, Jan. 2013, 69.

108 "are perhaps more extensively debated": Lieberman, *Evolution of the Human Head*, 315.

108 Within or around it are nine cartilages: Ibid., 284.

110 Johann Dieffenbach, one of Germany's most eminent surgeons: "The Paralysis of Stuttering," *New York Review of Books*, April 26, 2012.

CHAPTER 7: THE HEART AND BLOOD

112 "Stopped": Quoted in "In the Hands of Any Fool," *London Review of Books*, July 3, 1997.

112 That symbol first appeared: Peto, *Heart*, 30.

113 Every hour your heart dispenses: Nuland, *How We Die*, 22.

113 It has been calculated: Morris, *Body Watching*, 11.

113 Of all the blood pumped out: Blakelaw and Jennett, *Oxford Companion to the Body*, 88–89.

114 Every time you stand up: *The Curious Cases of Rutherford and Fry*, podcast, BBC Radio 4, Sept. 13, 2016.

114 Much of the early research on blood pressure: Amidon and Amidon, *Sublime Engine*, 116; *Oxford Dictionary of National Biography*, s.v. "Hales, Stephen."

115 Well into the twentieth century: "Why So Many of Us Die of Heart Disease," *Atlantic*, March 6, 2018.

115 in 2017 the American Heart Association: "New Blood Pressure Guidelines Put Half of US Adults in Unhealthy Range," *Science News*, Nov. 13, 2017.

115 At least 50 million Americans: Amidon and Amidon, *Sublime Engine*, 227.

115 In the United States alone: Health, United States, 2016, DHSS Publication No. 2017-1232, May 2017.

116 A heart attack and a cardiac arrest: Wolpert, *You're Looking Very Well,* 18; "Don't Try This at Home," *London Review of Books,* Aug. 29, 2013.

116 For about a quarter of victims: *Baylor University Medical Center Proceedings,* April 2017, 240.

117 A woman is more likely to experience: Brooks, *At the Edge of Uncertainty,* 104–5.

117 the Hmong people of Southeast Asia: Amidon and Amidon, *Sublime Engine,* 191–92.

117 Hypertrophic cardiomyopathy is the condition: "When Genetic Autopsies Go Awry," *Atlantic,* Oct. 11, 2016.

118 The triggering event for public awareness: Pearson, *Life Project,* 101–3.

118 the Framingham study recruited five thousand local adults: Ibid.; framinghamheartstudy.org.

119 he fed a catheter into an artery in his arm: Nourse, *Body,* 85.

119 build a machine that could oxygenate blood artificially: Le Fanu, *Rise and Fall of Modern Medicine,* 95; National Academy of Sciences, biographical memoir by Harris B. Schumacher Jr., Washington, D.C., 1982.

120 In 1958, a Swedish engineer named Rune Elmqvist: Ashcroft, *Spark of Life,* 152–53.

121 in 2000 he killed himself: *New York Times* obituary, Aug. 21, 2000; "Interview: Dr. Steven E. Nissen," Take One Step, PBS, Aug. 2006, www.pbs.org.

121 To remove a beating heart: *Baylor University Medical Center Proceedings,* Oct. 2017, 476.

122 Frey's sample contained a fungus: Ibid., 247.

122 success rates of 80 percent: Le Fanu, *Rise and Fall of Modern Medicine,* 102.

122 Today some four to five thousand heart transplants: Amidon and Amidon, *Sublime Engine,* 198–99.

123 The young woman's parents argue: *Economist,* April 28, 2018, 56.

123 "Heart disease kills about the same number": Kinch, *Prescription for Change,* 112.

124 By 2000, a million precautionary angioplasties: Welch, *Less Medicine, More Health,* 34–36.

124 "This is really American medicine at its worst": Ibid., 38.

125 A newborn baby contains only about eight ounces: Collis, *Living with a Stranger,* 28.

125 twenty-five thousand miles of blood vessels: Pasternak, *Molecules Within Us,* 58.

125 a single drop of blood: Hill, *Blood,* 14–15.

126 In the United States, plasma sales: *Economist,* May 12, 2018, 12.

126 Hemoglobin has one strange and dangerous quirk: Annals of Medicine, *New Yorker,* Jan. 31, 1970.

126 Each will be shot around your body: Blakelaw and Jennett, *Oxford Companion to the Body,* 85.

127 In severe bleeding, the body: Miller, *Body in Question,* 121–22.

128 They also play important roles in immune response: *Nature,* Sept. 28, 2017, S13.

128 Nearly all Harvey's peers thought him: Zimmer, *Soul Made Flesh,* 74.

128 Harvey couldn't explain how blood circulating: Wootton, *Bad Medicine,* 95–98.

129 Lower transfused about half a pint: "An Account of the Experiment of Transfusion, Practised upon a Man in London," *Proceedings of the Royal Society of London,* Dec. 9, 1667.

132 William Osler, author of *The Principles and Practice of Medicine:* "An Autopsy of Dr. Osler," *New York Review of Books,* May 25, 2000.

132 Although everybody reads and pronounces: Nourse, *Body,* 184.

133 There are some four hundred kinds of antigens: Sanghavi, *Map of the Child,* 64

134 "Blood is a living tissue": Dr. Allan Doctor interview, Oxford, Sept. 18, 2018.

136 For more than fifty years: "The Quest for One of Science's Holy Grails: Artificial Blood," *Stat,* Feb. 27, 2017; "Red Blood Cell Substitutes," *Chemistry World,* Feb. 16, 2018.

137 a $1.6 million saving in costs: "Save Blood, Save Lives," *Nature,* April 2, 2015.

CHAPTER 8: THE CHEMISTRY DEPARTMENT

138 One twelve-year-old boy was left so hungry: Bliss, *Discovery of Insulin,* 37.

139 "wrongly conceived, wrongly conducted": Ibid., 12–13.

140 "The discovery of insulin": "The Pissing Evile," *London Review of Books,* Dec. 1, 1983.

141 Others have suggested an imbalance: "Cause and Effect," *Nature,* May 17, 2012.

141 Between 1980 and 2014, the number of adults: *Nature,* May 26, 2016, 460.

141 That means that insulin levels: "The Edmonton Protocol," *New Yorker,* Feb. 10, 2003.

142 "I love hormones": Interviews with Dr. John Wass, Oxford, March 21 and Sept. 17, 2018.

144 Starling coined the term "hormone": Sengoopta, *Most Secret Quintessence of Life,* 4.

145 History's most famous sufferer: *Journal of Clinical Endocrinology and Metabolism,* Dec. 1, 2006, 4849–53; "The Medical Ordeals of JFK," *Atlantic,* Dec. 2002.

148 Yet in tests where oxytocin: *Nature,* June 25, 2015, 410–12.

148 Perhaps no one has better understood: *Biographical Memoirs of Fellows of the Royal Society,* London, Nov. 1998; *New York Times* obituary, Jan. 19, 1995.

149 In what way exactly testosterone might shorten: Bribiescas, *Men,* 202.

149 there is much greater evidence: *New Scientist,* May 16, 2015, 32.

151 Nonalcoholic fatty liver disease: *Nature,* Nov. 23, 2017, S85; *Annals of Internal Medicine,* Nov. 6, 2018.

154 Each day they process about 190 quarts: Pasternak, *Molecules Within Us,* 60.

155 As we age, the bladder loses elasticity: Nuland, *How We Die,* 55.

155 the urinary world is at least somewhat microbial: *Nature,* Nov. 9, 2017, S40.

156 Probably history's most famous lithotomy: Tomalin, *Samuel Pepys,* 60–65.

157 Pepys for his part marked the anniversary: "Samuel Pepys and His Stones," *Annals of the Royal College of Surgeons* 59 (1977).

CHAPTER 9: IN THE DISSECTING ROOM: THE SKELETON

158 "Feel this," Dr. Ben Ollivere is saying to me: Dr. Ben Ollivere interview, Nottingham, June 23–24, 2017.

161 there was a brief scandal in America: "Yale Students and Dental Professor Took Selfie with Severed Heads," *Guardian,* Feb. 5, 2018.

162 When the great anatomist Andreas Vesalius: Wootton, *Bad Medicine,* 74.

162 William Harvey, in England, was so desperate: Larson, *Severed,* 217.

162 Falloppio and the criminal together: Wootton, *Bad Medicine,* 91.

163 All of his illustrations had to be drawn: *Baylor University Medical Center Proceedings,* Oct. 2009, 342–45.

165 regular exercise helps to stave off Alzheimer's: "Do Our Bones Influence Our Minds?," *New Yorker,* Nov. 1, 2013.

167 It takes one hundred muscles: Collis, *Living with a Stranger,* 56.

167 Studies by NASA have shown: NASA information sheet, "Muscle Atrophy."

168 Sir Charles Bell, the great: *Oxford Dictionary of National Biography,* s.v. "Bell, Sir Charles."

168 What we do have in our thumbs: Roberts, *Incredible Unlikeliness of Being,* 333–35.

170 A good deal of what we know: Francis, *Adventures in Human Being,* 126–27.

170 The average human walks at a pace: "Gait Analysis: Principles and Applications," *American Academy of Orthopaedic Surgeons,* Oct. 1995.

171 ostriches have eliminated this problem: Taylor, *Body by Darwin,* 85.

171 "as early as the eighteenth year": Medawar, *Uniqueness of the Individual,* 109.

171 An estimated 60 percent of adults: Wall, *Pain,* 100–101.

172 surgeons perform over 800,000 joint replacements: "The Coming Revolution in Knee Repair," *Scientific American,* March 2015.

172 Almost no one has heard of Charnley: Le Fanu, *Rise and Fall of Modern Medicine,* 104–8.

173 Three-quarters of men and half of women: Wolpert, *You're Looking Very Well,* 21.

CHAPTER 10: ON THE MOVE: BIPEDALISM AND EXERCISE

175 In 2016, anthropologists at the University of Texas: "Perimortem Fractures in Lucy Suggest Mortality from Fall out of Tall Tree," *Nature*, Sept. 22, 2016.

175 A chimpanzee uses four times: Lieberman, *Story of the Human Body*, 42.

177 Fossil evidence suggests that early hominins: "The Evolution of Marathon Running," *Sports Medicine* 37, no. 4–5 (2007); "Elastic Energy Storage in the Shoulder and the Evolution of High-Speed Throwing in Homo," *Nature*, June 27, 2013.

178 Jeremy Morris, became convinced: Jeremy Morris obituary, *New York Times*, Nov. 7, 2009.

179 Going for regular walks reduces the risk: *New Yorker*, May 20, 2013, 46.

179 Being active for an hour or more: *Scientific American*, Aug. 2013, 71; "Is Exercise Really Medicine? An Evolutionary Perspective," *Current Sports Medicine Reports*, July–Aug. 2015.

179 The ten-thousand-step idea: "Watch Your Step," *Guardian*, Sept. 3, 2018.

179 Only about 20 percent of people: "Is Exercise Really Medicine?"

179 Today the average American walks: Lieberman, *Story of the Human Body*, 217–18.

180 "Some workers have reportedly": *Economist*, Jan. 5, 2019, 50.

180 Modern hunter-gatherers, by contrast: "Is Exercise Really Medicine?"

180 "If you want to understand the human body": Lieberman interview.

180 If everybody else in the world: "Eating Disorder," *Economist*, June 19, 2012.

181 A bodybuilder and a couch potato: "The Fat Advantage," *Nature*, Sept. 15, 2016.

181 the average woman in the United States: *Baylor University Medical Center Proceedings*, Jan. 2016.

181 more than half of today's children: "Interest in Ketogenic Diet Grows for Weight Loss and Type 2 Diabetes," *Journal of the American Medical Association*, Jan. 16, 2018.

181 The current generation of young people: Zuk, *Paleofantasy*, 5.

181 The British are among the tubbiest: *Economist*, March 31, 2018, 30.

182 The global figure for obesity is 13 percent: *Economist*, Jan. 6, 2018, 20.

182 According to one calculation, you must walk: "The Bear's Best Friend," *New York Review of Books*, May 12, 2016.

182 people overestimate the number: "Exercise in Futility," *Atlantic*, April 2016.

182 a worker on a factory floor: Lieberman, *Story of the Human Body*, 217.

182 People who sit a lot: "Are You Sitting Comfortably? Well, Don't," *New Scientist*, June 26, 2013.

182 If you spend an evening: "Our Amazingly Plastic Brains," *Wall Street Journal*, Feb. 6, 2015; "The Futility of the Workout-Sit Cycle," *Atlantic*, Aug. 16, 2016.

183 James Levine, an obesity expert: "Killer Chairs: How Desk Jobs Ruin Your Health," *Scientific American,* Nov. 2014.
183 That alone burned 65 extra calories an hour: *New Scientist,* Aug. 25, 2012, 41.
183 "a pile of rubbish": "The Big Fat Truth," *Nature,* May 23, 2013.

CHAPTER 11: EQUILIBRIUM

185 little creatures have to produce heat: Blumberg, *Body Heat,* 35–38.
185 One area where animals are curiously: West, *Scale,* 197.
186 A typical mammal uses about thirty times: Lane, *Power, Sex, Suicide,* 179.
186 To move more than a very few degrees: Blumberg, *Body Heat,* 206.
187 That experiment largely recalled: Royal Society, "Experiments and Observations in a Heated Room by Charles Blagden, 1774."
187 Curiously, no one knows quite why this happens: Ashcroft, *Life at the Extremes,* 133–34; Blumberg, *Body Heat,* 146–47.
187 An increase of only a degree or so: Davis, *Beautiful Cure,* 113.
188 The idea, incidentally, that we lose most of our heat: "Myth: We Lose Most Heat from Our Heads," *Naked Scientists,* podcast, Oct. 24, 2016.
188 The man who coined the term: *Obituary Notices of Fellows of the Royal Society* 5, no. 15 (Feb. 1947): 407–23; *American National Biography,* s.v. "Cannon, Walter Bradford."
189 paper on the practice of voodoo: "'Voodoo' Death," *American Anthropologist,* April–June 1942.
191 Every day you produce and consume: West, *Scale,* 100.
191 you have only sixty grams: Lane, *Vital Question,* 63.
191 The person who discovered: *Biographical Memoirs,* Royal Society, London.
191 "I was your first wife": *Biochemistry and Biology Molecular Education* 32, no. 1 (2004): 62–66.
192 A child half your height: "Size and Shape," *Natural History,* Jan. 1974.
192 a British airman in World War II: "The Indestructible Alkemade," RAF Museum website, posted Dec. 24, 2014.
194 Consider the case of little Erika Nordby: *Edmonton Sun,* Aug. 28, 2014.
194 Between 1998 and August 2018, almost eight hundred children: Full details can be found at the website noheatstroke.org.
195 The highest permanent settlements in the world: Ashcroft, *Life at the Extremes,* 8.
195 Tenzing Norgay and Raymond Lambert: Ibid., 26.
195 At sea level, about 40 percent of your blood volume: Ibid., 341.
196 Ashcroft notes the case of a pilot: Ibid., 19.
196 In Nazi Germany, healthy prisoners: Annas and Grodin, *Nazi Doctors and the Nuremberg Code,* 25–26.
197 In a typical experiment, Chinese prisoners: Williams and Wallace, *Unit 731,* 42.
197 Some, for unfathomable reasons, were dissected: "Blood and Money," *New York Review of Books,* Feb. 4, 1999.

197 When pregnant women or young children: Lax, *Toxin,* 123.
198 in 1984 a student from Keio University: Williams and Wallace, *Unit 731.*

CHAPTER 12: THE IMMUNE SYSTEM

199 we have some three hundred different types of immune cells: "Ambitious Human Cell Atlas Aims to Catalog Every Type of Cell in the Body," National Public Radio, Aug. 13, 2018.
200 If you are stressed or exhausted: "Department of Defense," *New York Review of Books,* Oct. 8, 1987.
200 Altogether about 5 percent of us suffer: Davis, *Beautiful Cure,* 149.
200 "You could look at it and conclude": Interview with Professor Daniel Davis, University of Manchester, Nov. 30, 2018.
201 "just about the cleverest little cells": Bainbridge, *Visitor Within,* 185.
202 the thymus is a nursery for T cells: Davis, *Compatibility Gene,* 38.
202 "the last person to identify the function": *Lancet,* Oct. 8, 2011, 1290.
203 Faulty inflammation has been linked: "Inflamed," *New Yorker,* Nov. 30, 2015.
204 "the immune system gets so ramped up": Kinch interview.
204 "vivacious, sociable, debonair, brilliant in conversation": "High on Science," *New York Review of Books,* Aug. 16, 1990.
205 "For all the clinical good-will": Medawar, *Uniqueness of the Individual,* 132.
206 Richard Herrick of Marlborough: Le Fanu, *Rise and Fall of Modern Medicine,* 121–23; "A Transplant Makes History," *Harvard Gazette,* Sept. 22, 2011.
207 As of late 2018, 114,000 people: "The Disturbing Reason Behind the Spike in Organ Donations," *Washington Post,* April 17, 2018.
207 People on dialysis live an extra eight years: *Baylor University Medical Center Proceedings,* April 2014.
207 One possible solution would be to use animal transplants: "Genetically Engineering Pigs to Grow Organs for People," *Atlantic,* Aug. 10, 2017.
208 Altogether humans are afflicted by some fifty types: Davis, *Beautiful Cure,* 149.
208 Before 1932, when Burrill Crohn: Blaser, *Missing Microbes,* 177.
208 Lieberman suggests that the overuse: Lieberman, *Story of the Human Body,* 178.
208 autoimmune diseases are grossly sexist: Bainbridge, *X in Sex,* 157; Martin, *Sickening Mind,* 72.
209 The word's first appearance in English: *Oxford English Dictionary.*
209 Roughly 50 percent of people claim: "Skin: Into the Breach," *Nature,* Nov. 23, 2011.
209 one child on an airplane: Pasternak, *Molecules Within Us,* 174.
210 the National Institute of Allergy and Infectious Diseases: "Feed Your Kids Peanuts, Early and Often, New Guidelines Urge," *New York Times,* Jan. 5, 2017.

210 the well-known hygiene hypothesis: "Lifestyle: When Allergies Go West," *Nature,* Nov. 24, 2011; Yong, *I Contain Multitudes,* 122; "Eat Dirt?," *Natural History,* n.d.

CHAPTER 13: DEEP BREATH: THE LUNGS AND BREATHING

212 Every time you breathe, you exhale: *Chemistry World,* Feb. 2018, 66.

213 about 20 percent of all antibiotic prescriptions: *Scientific American,* Feb. 2016, 32.

214 sneeze droplets can travel up to eight meters: "Where Sneezes Go," *Nature,* June 2, 2016; "Why Do We Sneeze?," *Smithsonian,* Dec. 29, 2015.

215 Our lungs can hold about six quarts: "Breathe Deep," *Scientific American,* Aug. 2012.

216 If you are an average-sized adult: West, *Scale,* 152.

216 Before opening his mail: Carter, *Marcel Proust,* 72.

216 Wherever he was in the world: Ibid., 224.

218 asthma remains the fourth leading cause: Jackson, *Asthma,* 159.

218 Japan, for instance, has not seen a great increase: "Lifestyle: When Allergies Go West," *Nature,* Nov. 24, 2011.

218 "You probably think asthma is caused": Interview with Professor Neil Pearce, London School of Hygiene and Tropical Medicine, Nov. 28, 2018.

220 In an asthma attack, the airways narrow: "Asthma: Breathing New Life into Research," *Nature,* Nov. 24, 2011.

220 In what way exactly Western lifestyles: "Lifestyle: When Allergies Go West"; "Asthma and the Westernization 'Package,'" *International Journal of Epidemiology* 31 (2002): 1098–102.

221 A person who smokes cigarettes regularly: "Getting Away with Murder," *New York Review of Books,* July 19, 2007.

223 When Britain's minister of health, Iain Macleod: Wootton, *Bad Medicine,* 263.

223 "No one has established that cigarette smoke": "Getting Away with Murder."

224 the average American adult was smoking: A Reporter at Large, *New Yorker,* Nov. 30, 1963.

224 The number of cigarettes smoked: Smith, *Body,* 329.

224 one of the members of the board: "Cancer: Malignant Maneuvers," *New York Review of Books,* March 6, 2008.

224 As late as 1973, *Nature* ran an editorial: "Get the Placentas," *London Review of Books,* June 2, 2016.

224 The world record for hiccups: *Sioux City Journal,* Jan. 4, 2015.

CHAPTER 14: FOOD, GLORIOUS FOOD

227 Americans today consume about 25 percent: *Baylor University Medical Center Proceedings,* Jan. 2017, 134.

227 The father of caloric measurement: *American National Biography*, s.v. "Atwater, Wilbur Olin"; USDA Agricultural Research Service website; Wesleyan University website.

228 we should eat a lot of meat: McGee, *On Food and Cooking*, 534.

229 You may eat 170 calories' worth of almonds: "Everything You Know About Calories Is Wrong," *Scientific American*, Sept. 2013.

229 "You can't possibly have a large brain": Interview with Professor Daniel Lieberman, London, Oct. 22, 2018.

231 "a figment of the imagination": Gratzer, *Terrors of the Table*, 170.

233 "such a poorly done paper": "Nutrition: Vitamins on Trial," *Nature*, June 25, 2014.

233 Americans can choose from among: "How Did We Get Hooked on Vitamins?," *The Inquiry*, BBC World Service, Dec. 31, 2018.

234 He took up to forty thousand milligrams: "The Dark Side of Linus Pauling's Legacy," Quackwatch, Sept. 14, 2014.

234 Proteins are complicated molecules: Smith, *Body*, 429.

234 Why evolution has wedded us: Challoner, *Cell*, 38.

235 most traditional diets in the world: McGee, *On Food and Cooking*, 534.

235 Virtually all carbohydrates in the diet: Ibid., 803.

236 a 150-gram serving of white rice: *New Scientist*, June 11, 2016, 32.

237 For complex chemical reasons: Lieberman, *Story of the Human Body*, 255.

237 an avocado has five times as much saturated fat: *New Scientist*, Aug. 2, 2014, 35.

237 Not until 2004 did the American Heart Association: Kummerow obituary, *New York Times*, June 1, 2017.

238 The idea has been traced to a 1945 paper: *More or Less*, BBC Radio 4, Jan. 6, 2017.

238 People allowed to drink all the water: Roach, *Grunt*, 133.

238 Drinking too much water: "Can You Drink Too Much Water?," *New York Times*, June 19, 2015; "Strange but True: Drinking Too Much Water Can Kill," *Scientific American*, June 21, 2007.

239 Over a lifetime, we eat about sixty tons of food: Zimmer, *Microcosm*, 56.

239 far more people on Earth suffer from obesity: *Nature*, Feb. 2, 2012, 27.

239 One chocolate chip cookie a week: *New Scientist*, July 18, 2009, 32.

239 The person most responsible: Keys obituary, *Washington Post*, Nov. 2, 2004; Keys obituary, *New York Times*, Nov. 23, 2004; *Journal of Health and Human Behaviour* (Winter 1963): 291–93; *American Journal of Clinical Nutrition* (March 2010).

242 One-fifth of all young people in America: "What Not to Eat," *New York Times*, Jan. 2, 2017; "How Much Harm Can Sugar Do?," *New Yorker*, Sept. 8, 2015.

243 The fruits that Shakespeare ate: Lieberman, *Story of the Human Body*, 265; "Best Before?," *New Scientist*, Oct. 17, 2015.

244 the most popular vegetable in America: *Baylor University Medical Center Proceedings*, April 2011, 158.

244 The average American consumes: "Clearing Up the Confusion About Salt," *New York Times,* Nov. 20, 2017.
245 A meta-analysis at McMaster University: *Chemistry World,* Sept. 2016, 50.
245 "We found that the published literature": *International Journal of Epidemiology,* Feb. 17, 2016.
245 "Well, actually originally it was to impress a girl": Interview with Professor Christopher Gardner, Palo Alto, Calif., Jan. 29, 2018.
247 Roughly 40 percent of people with diabetes: *Nature,* Feb. 2, 2012, 27.
247 "50 percent genetic and 50 percent cheeseburger": *National Geographic,* Feb. 2007, 49.

CHAPTER 15: THE GUTS

248 The surface area of all that tubing: Vogel, *Life's Devices,* 42.
248 Food lingers inside a woman: Blakelaw and Jennett, *Oxford Companion to the Body,* 19.
249 That's why you are constantly told: "Fiber Is Good for You. Now Scientists May Know Why," *New York Times,* Jan. 1, 2018.
249 The rumblings of your gut: Enders, *Gut,* 83.
250 Every year three thousand people: "A Bug in the System," *New Yorker,* Feb. 2, 2015, 30.
250 "but had decided that cooking them": *Food Safety News,* Dec. 27, 2017.
250 According to a USDA study: "Bug in the System," 30.
251 In America, the problems of foodborne illness: Ibid.
252 "People tend to blame the last thing they ate": "What to Blame for Your Stomach Bug? Not Always the Last Thing You Ate," *New York Times,* June 29, 2017.
253 After drifting around for a few years: "Men and Books," *Canadian Medical Association Journal,* June 1959.
254 nearly 400,000 people are hospitalized: "The Global Incidence of Appendicitis: A Systematic Review of Population-Based Studies," *Annals of Surgery,* Aug. 2017.
255 The incidence of acute appendicitis: Blakelaw and Jennett, *Oxford Companion to the Body,* 43.
255 Lipes's bedside manner was not: *New York Times* obituary, April 20, 2005.
256 People flocked to him from all around the world: "Killing Cures," *New York Review of Books,* Aug. 11, 2005.
257 Every gram of feces you produce: Money, *Amoeba in the Room,* 144.
257 Even samples taken from two ends: *Nature,* Aug. 21, 2014, 247.
258 Two strains of *E. coli:* Zimmer, *Microcosm,* 20; Lane, *Power, Sex, Suicide,* 119.
258 *E. coli* wasn't named for him until 1918: *Clinical Infectious Diseases,* Oct. 15, 2007, 1025–29.
258 "The olfactory nerves become paralyzed": Roach, *Gulp,* 253.
259 "many recorded examples of explosion": "Fatal Colonic Explosion During Colonoscopic Polypectomy," *Gastroenterology* 77, no. 6 (1979).

CHAPTER 16: SLEEP

260 In 1989, in an experiment: "Sleep Deprivation in the Rat," *Sleep* 12, no. 1 (1989).

261 People with early signs of hypertension: *Nature*, May 23, 2013, S7.

262 "If sleep does not serve": *Scientific American*, Oct. 2015, 42.

262 Even quite simple creatures like nematodes: *New Scientist*, Feb. 2, 2013, 38–39.

262 Aserinsky's volunteer subject for the first night's test: "The Stubborn Scientist Who Unraveled a Mystery of the Night," *Smithsonian*, Sept. 2003; "Rapid Eye Movement Sleep: Regulation and Function," *Journal of Clinical Sleep Medicine*, June 15, 2013.

263 Sleep is so shallow in these first two stages: Martin, *Counting Sheep*, 98.

264 Typically, a man will be erect: Ibid., 133–39; "Cerebral Hygiene," *London Review of Books*, June 29, 2017.

264 The average person turns over: Martin, *Counting Sheep*, 104.

265 when a dozen airline pilots on long-haul flights: Ibid., 39–40.

266 That may explain why we: Burnett, *Idiot Brain*, 25; Sternberg, *NeuroLogic*, 13–14.

266 One member of an audience shouted: Davis, *Beautiful Cure*, 133.

266 "They struggled to accept that something": Interview with Professor Russell Foster, Brasenose College, Oxford, Oct. 17, 2018.

268 "The pineal is not our soul": Bainbridge, *Beyond the Zonules of Zinn*, 200.

268 When asked to estimate the passage: Shubin, *Universe Within*, 55–67.

269 "Around half of these bestselling drugs": Davis, *Beautiful Cure*, 37.

270 Later start times have been shown: "Let Teenagers Sleep In," *New York Times*, Sept. 20, 2018.

270 Insomnia has been linked to diabetes: "In Search of Forty Winks," *New Yorker*, Feb. 8–15, 2016.

271 women who regularly worked night shifts: "Of Owls, Larks, and Alarm Clocks," *Nature*, March 11, 2009.

271 About 50 percent of people who snore: "Snoring: What to Do When a Punch in the Shoulder Fails," *New York Times*, Dec. 11, 2010.

271 The most extreme and horrifying form: Zeman, *Consciousness*, 46–47; "The Family That Couldn't Sleep," *New York Times*, Sept. 2, 2006.

271 Some authorities think prions may also: *Nature*, April 10, 2014, 181.

272 The condition affects four million people: "The Wild Frontiers of Slumber," *Nature*, March 1, 2018; Zeman, *Consciousness*, 106–9.

273 "I remember when I woke up": *Morning Edition*, National Public Radio, Dec. 27, 2017.

273 Yawning doesn't even correlate reliably: Martin, *Counting Sheep*, 140.

CHAPTER 17: INTO THE NETHER REGIONS

275 "On a Presidential visit to a farm": The story is of course apocryphal.

276 Nettie Stevens deserves to be better known: "Nettie M. Stevens and

the Discovery of Sex Determination by Chromosomes," *Isis*, June 1978; *American National Biography*.

277 It is just an extraordinary coincidence: Bainbridge, *X in Sex*, 66.

277 "literally waited at the foot of the gallows": "The Chromosome Number in Humans: A Brief History," *Nature Reviews Genetics*, Aug. 1, 2006.

277 That number stuck, universally unquestioned: Ridley, *Genome*, 23–24.

278 After countless generations of making: "Vive la Difference," *New York Review of Books*, May 12, 2005.

278 At its current rate of deterioration: "Sorry, Guys: Your Y Chromosome May Be Doomed," *Smithsonian*, Jan. 19, 2018.

279 humans don't actually reproduce: Mukherjee, *Gene*, 357.

279 How many people are unfaithful: "Infidels," *New Yorker*, Dec. 18–25, 2017.

279 In one study, the number of sexual partners: Spiegelhalter, *Sex by Numbers*, 35.

280 Because of funding problems, only 3,432 people: *American Journal of Public Health*, July 1996, 1037–40; "What, How Often, and with Whom?," *London Review of Books*, Aug. 3, 1995.

280 leaving Spiegelhalter to wonder what exactly: Spiegelhalter, *Sex by Numbers*, 2.

281 the median time for sex: Ibid., 218–20.

281 A chimpanzee and a human: "Bonobos Join Chimps as Closest Human Relatives," *Science News*, June 13, 2012.

282 They are more vulnerable to infection: Bribiescas, *Men*, 174–76.

284 "Vaginal secretions [were] the only bodily fluid": Roach, *Bonk*, 12.

285 It is named for Ernst Gräfenberg: *American Journal of Obstetrics and Gynecology*, Aug. 2001, 359.

286 Until the early twentieth century, "clitoris": *Oxford English Dictionary*.

286 The uterus normally weighs two ounces: Cassidy, *Birth*, 80.

287 many mammals get along perfectly well: Bainbridge, *Teenagers*, 254–55.

287 There is also a great deal of uncertainty: "Skin Deep," *New York Review of Books*, Oct. 7, 1999.

288 Authorities seem to be universally agreed: Morris, *Body Watching*, 216; Spiegelhalter, *Sex by Numbers*, 216–17.

CHAPTER 18: IN THE BEGINNING: CONCEPTION AND BIRTH

290 The chances of a successful fertilization: "Not from Venus, Not from Mars," *New York Times*, Feb. 25–26, 2017, international edition.

290 A meta-analysis in the journal: "Yes, Sperm Counts Have Been Steadily Declining," Smithsonian.com, July 26, 2017.

290 "a common class of chemical": "Are Your Sperm in Trouble?," *New York Times*, March 11, 2017.

290 The number of spermatozoa produced: Lents, *Human Errors*, 100.

292 by the age of thirty-five a woman's stock of eggs: "The Divorce of Coitus from Reproduction," *New York Review of Books*, Sept. 25, 2014.

293 Without this, the rate of birth defects: Roberts, *Incredible Unlikeliness of Being*, 344.
294 About 80 percent of mothers-to-be: "What Causes Morning Sickness?," *New York Times*, Aug. 3, 2018.
295 The only truly reliable test: Oakley, *Captured Womb*, 17.
295 Medical students in England weren't required: Epstein, *Get Me Out*, 38.
295 Women were sometimes bled: Oakley, *Captured Womb*, 22.
295 In 1906, an estimated 150,000 American women: Sengoopta, *Most Secret Quintessence of Life*, 16–18.
296 "God knows the number of women": Cassidy, *Birth*, 60.
296 sterilize the air around patients: "The Gruesome, Bloody World of Victorian Surgery," *Atlantic*, Oct. 22, 2017.
296 As late as 1932, one mother in every 238 died: Oakley, *Captured Womb*, 62.
296 It was the rise of penicillin: Cassidy, *Birth*, 61.
297 Yet American women are 70 percent: *Economist*, July 18, 2015, 41.
298 "the least understood organ in the human body": *Scientific American*, Oct. 2017, 38.
300 "Women in labour have pretty much": *Nature*, July 14, 2016, S6.
300 people born by C-section: "The Cesarean-Industrial Complex," *Atlantic*, Sept. 2014.
300 more than 60 percent of Cesareans are done: "Stemming the Global Caesarean Section Epidemic," *Lancet*, Oct. 13, 2018.
301 the rush to clean up babies: Blaser, *Missing Microbes*, 95.
301 *B. infantis*, an important microbe: Yong, *I Contain Multitudes*, 130.
301 by the age of one the average baby: *New Yorker*, Oct. 22, 2012, 33.
301 There is some evidence that a nursing mother: Ben-Barak, *Why Aren't We Dead Yet?*, 68.

CHAPTER 19: NERVES AND PAIN

305 Repeat the experiences, and the patterns: "Show Me Where It Hurts," *Nature*, July 14, 2016.
305 "Pain only emerges when the brain": Interview with Professor Irene Tracey, John Radcliffe Hospital, Oxford, Sept. 18, 2018.
308 The person who first identified nociceptors: *Oxford Dictionary of National Biography*, s.v. "Sherrington, Sir Charles Scott"; *Nature Neuroscience*, June 2010, 429–30.
311 More than half of spinal cord injuries: Annals of Medicine, *New Yorker*, Jan. 25, 2016.
311 Pain, like the nervous system itself: "A Name for Their Pain," *Nature*, July 14, 2016; Foreman, *Nation in Pain*, 22–24.
313 the word is a corruption of the French *demi-craine*: "Headache," *American Journal of Medicine*, Jan. 2018; "Why Migraines Strike," *Scientific American*, Aug. 2008; "A General Feeling of Disorder," *New York Review of Books*, April 23, 2015.

314 "*Donnerwetter*, so it has": Dormandy, *Worst of Evils*, 483.

314 But equally pain is decreased: *Nature Neuroscience*, April 2008, 314.

314 Just having a sympathetic and loving partner: Wolf, *Body Quantum*, vii.

314 In one experiment done by Tracey: *Nature Neuroscience*, April 2008, 314.

314 about 40 percent of adult Americans: Foreman, *Nation in Pain*, 3.

314 Altogether chronic pain affects more people: "The Neuroscience of Pain," *New Yorker*, July 2, 2018.

314 "deaf and blind to other people": Daudet, *In the Land of Pain*, 15.

314 "The drugs we have relieve 50 percent": "Name for Their Pain."

315 Between 1999 and 2014, by one estimate: *Chemistry World*, July 2017, 28; *Economist*, Oct. 28, 2017, 41; "Opioid Nation," *New York Review of Books*, Dec. 6, 2018.

315 opioid fatalities have led to a rise in organ donations: "The Disturbing Reasons Behind the Spike in Organ Donations," *Washington Post*, April 17, 2018.

316 one doctor got good results: "Feel the Burn," *London Review of Books*, Sept. 30, 1999.

316 Even so, 59 percent of those tested: "Honest Fakery," *Nature*, July 14, 2016.

317 Placebos don't shrink tumors: Marchant, *Cure*, 22.

CHAPTER 20: WHEN THINGS GO WRONG: DISEASES

318 In the autumn of 1948, people in the small city: "The Post-viral Syndrome: A Review," *Journal of the Royal College of General Practitioners*, May 1987; "A Disease Epidemic in Iceland Simulating Poliomyelitis," *American Journal of Epidemiology* 2 (1950); "Early Outbreaks of 'Epidemic Neuromyasthenia,'" *Postgraduate Medical Journal*, Nov. 1978; Annals of Medicine, *New Yorker*, Nov. 27, 1965.

319 in 1970, after several years of quiescence: "Epidemic Neuromyasthenia: A Syndrome or a Disease?," *Journal of the American Medical Association*, March 13, 1972.

320 West Nile virus surfaced in New York: Crawford, *Deadly Companions*, 18.

320 Two hundred years later, a very similar illness: "Two Spots and a Bubo," *London Review of Books*, April 21, 2005.

321 Bourbon virus, as it became known: Centers for Disease Control and Prevention, *Emerging Infectious Diseases Journal*, May 2015; "Researchers Reveal That Killer 'Bourbon Virus' Is of the Rare Thogotovirus Genus," *Science Times*, Feb. 22, 2015; "Mysterious Virus That Killed a Farmer in Kansas Is Identified," *New York Times*, Dec. 23, 2014.

321 "Unless doctors are doing laboratory tests": "Deadly Heartland Virus Is Much More Common Than Scientists Thought," National Public Radio, Sept. 16, 2015.

322 Within a few days, 34 were dead: "In Philadelphia 30 Years Ago, an Eruption of Illness and Fear," *New York Times*, Aug. 1, 2006.

322 *Legionella* is widely distributed in soil: "Coping with Legionella," *Public Health*, Nov. 14, 2000.

322 Much the same thing happened: "Early Outbreaks of 'Epidemic Neuromyasthenia.'"
323 Whether or not a disease becomes epidemic: *New Scientist*, May 9, 2015, 30–33.
323 A successful virus is one: "Ebola Wars," *New Yorker*, Oct. 27, 2014.
324 the number of viruses in birds and mammals: "The Next Plague Is Coming. Is America Ready?," *Atlantic*, July–Aug. 2018.
324 "a catastrophe from which we": "Stone Soup," *New Yorker*, July 28, 2014.
325 a shadowy cook and housekeeper: Grove, *Tapeworms, Lice, and Prions*, 334–35; *New Yorker*, Jan. 26, 1935; *American National Biography*, s.v. "Mallon, Mary."
326 The United States has an estimated 5,750 cases each year: CDC figures.
326 The death toll in the twentieth century: "The Awful Diseases on the Way," *New York Review of Books*, June 9, 2016.
326 enough to infect seventeen others: "Bugs Without Borders," *New York Review of Books*, Jan. 16, 2003.
328 In 2014, someone looking through a storage area: U.S. Centers for Disease Control and Prevention, "Media Statement on Newly Discovered Smallpox Specimens," July 8, 2014.
329 Inmates were given pickaxes: "Phrenic Crush," *London Review of Books*, Oct. 2003.
329 she and other inmates were allowed visits: MacDonald, *Plague and I*, 45.
329 Some boroughs of London now have rates: "Killer of the Poor Now Threatens the Wealthy," *Financial Times*, March 24, 2014.
330 The only treatment, even now: *Economist*, April 22, 2017, 54.
330 Bilharz bandaged the pupae of cercariae worms: Kaplan, *What's Eating You?*, ix.
331 a protein called huntingtin: Mukherjee, *Gene*, 280–86.
331 At least forty have been linked to type 2 diabetes: *Nature*, May 17, 2012, S10.
332 "Why a temperate climate": Bainbridge, *Beyond the Zonules of Zinn*, 77–78.
332 Only about two hundred cases of the disorder: Davies, *Life Unfolding*, 197.
332 For 90 percent of rare diseases: *MIT Technology Review*, Nov.–Dec. 2018, 44.
333 "You are most likely going to die": Lieberman, *Story of the Human Body*, 351.
334 only 36 percent less likely to get flu: "The Ghost of Influenza Past and the Hunt for a Universal Vaccine," *Nature*, Aug. 8, 2018.

CHAPTER 21: WHEN THINGS GO VERY WRONG: CANCER

335 Diphtheria, smallpox, and tuberculosis: Bourke, *Fear*, 298–99.
335 "The early history of cancer": Mukherjee, *Emperor of All Maladies*, 44–45.
336 Half of men over sixty: Welch, *Less Medicine, More Health*, 71.

336 A survey of physicians in America: "What to Tell Cancer Patients," *Journal of the American Medical Association* 175, no. 13 (1961).

336 Surveys in Britain at about the same time: Smith, *Body*, 330.

337 "That's why cancers aren't contagious": Interview with Dr. Josef Vormoor, Princess Máxima Center, Utrecht, the Netherlands, Jan. 18–19, 2019.

339 Between birth and the age of forty: Herold, *Stem Cell Wars*, 10.

339 More than half of cases: *Nature*, March 24, 2011, S16.

339 How exactly weight tips the balance: "The Fat Advantage," *Nature*, Sept. 15, 2016; "The Link Between Cancer and Obesity," *Lancet*, Oct. 14, 2017.

339 The first person to notice a connection: *British Journal of Industrial Medicine*, Jan. 1957, 68–70; "Percivall Pott, Chimney Sweeps, and Cancer," *Education in Chemistry*, March 11, 2006.

340 More than eighty thousand chemicals: "Toxicology for the 21st Century," *Nature*, July 8, 2009.

340 Although no one can say to what extent: "Cancer Prevention," *Nature*, March 24, 2011, S22–S23.

340 In the face of opposition: Armstrong, *Gene That Cracked the Cancer Code*, 53, 27–29.

341 Altogether, it has been estimated, pathogens: "The Awful Diseases on the Way," *New York Review of Books*, June 9, 2016.

341 About 10 percent of men: Timmermann, *History of Lung Cancer*, 6–7.

344 There is some evidence that his wife: *Baylor University Medical Center Proceedings*, Jan. 2012.

344 the concept of the radical mastectomy: *American National Biography*, s.v. "Halsted, William Stewart"; "A Very Wide and Deep Dissection," *New York Review of Books*, Sept. 20, 2001; Beckhard and Crane, *Cancer, Cocaine, and Courage*, 111–12.

345 He lost most of his jaw and parts of his skull: Jorgensen, *Strange Glow*, 94.

345 In 1920, four million radium watches: Ibid., 87–88.

346 "he was so badly disfigured": Ibid., 123.

347 Mrs. Lawrence's cancer went into remission: Goodman, McElligott, and Marks, *Useful Bodies*, 81–82.

347 It was subsequently discovered: *American National Biography*, s.v. "Lawrence, John Hundale."

347 From this, it was realized: Armstrong, *Gene That Cracked the Cancer Code*, 53, 253–54; *Nature*, Jan. 12, 2017, 154.

348 The breakthrough moment was in 1968: "Childhood Leukemia Was Practically Untreatable Until Don Pinkel and St. Jude Hospital Found a Cure," *Smithsonian*, July 2016.

349 A significant fraction of childhood cancer deaths: *Nature*, March 30, 2017, 608–9.

349 2.4 million fewer people have died: "We're Making Real Progress Against Cancer. But You May Not Know It if You're Poor," *Vox*, Feb. 2, 2018.

350 no more than 2 to 3 percent of cancer research money: *Nature,* March 24, 2011, S4.

CHAPTER 22: MEDICINE GOOD AND BAD

351 whatever he learned about soil fertility: "The White Plague," *New York Review of Books,* May 26, 1994.

352 Selman Waksman was awarded the Nobel Prize: *Literary Review,* Oct. 2012, 47–48; *Guardian,* Nov. 2, 2002.

353 By one reckoning, life expectancy on Earth: *Economist,* April 29, 2017, 53.

355 "At some point between 1900 and 1912": *Nature,* March 24, 2011, 446.

356 a British epidemiologist named Thomas McKeown: Wootton, *Bad Medicine,* 270–71.

357 McKeown's thesis attracted a good deal: *American Journal of Public Health,* May 2002, 725–29; "White Plague"; Le Fanu, *Rise and Fall of Modern Medicine,* 314–15.

357 males in the East End of Glasgow: "Between Victoria and Vauxhall," *London Review of Books,* June 1, 2017.

358 For every 400 middle-aged Americans: *Economist,* March 25, 2017, 76.

359 Among rich countries, America is at or near: "Why America Is Losing the Health Race," *New Yorker,* June 11, 2014.

359 Even sufferers of cystic fibrosis: "Stunning Gap: Canadians with Cystic Fibrosis Outlive Americans by a Decade," *Stat,* March 13, 2017.

359 One-fifth of all the money: "The US Spends More on Health Care Than Any Other Country," *Washington Post,* Dec. 27, 2016.

359 "Even wealthy Americans are not isolated": "Why America Is Losing the Health Race."

359 A U.S. teenager is twice as likely to be killed: "American Kids Are 70% More Likely to Die Before Adulthood Than Kids in Other Rich Countries," *Vox,* Jan. 8, 2018.

360 A helmeted rider is 70 percent: Insurance Institute for Highway Safety figures.

360 An angiogram, a survey by *The New York Times* found: "The $2.7 Trillion Medical Bill," *New York Times,* June 1, 2013.

360 One commonly accepted yardstick: "Health Spending," OECD Data, data.oecd.org.

362 when 160 gynecologists were asked: Jorgensen, *Strange Glow,* 298.

364 "most doctors take money or gifts": "Drug Companies and Doctors: A Story of Corruption," *New York Review of Books,* Jan. 15, 2009.

364 "they just had better blood-pressure numbers": "When Evidence Says No but Doctors Say Yes," *Atlantic,* Feb. 22, 2017.

365 But when the same drugs were tried on humans: "Frustrated Alzheimer's Researchers Seek Better Lab Mice," *Nature,* Nov. 21, 2018.

366 So for most people there is: "Aspirin to Prevent a First Heart Attack or Stroke," NNT, Jan. 8, 2015, www.thennt.com.

366 low-dose aspirin actually is not effective: National Institute for Health Research press release, July 16, 2018.

CHAPTER 23: THE END

368 more people globally died: *Nature*, Feb. 2, 2012, 27.

368 "Nearly a third of Americans who die": *Economist*, April 29, 2017, 11.

368 In 1940, that probability was reached: "Special Report on Aging," *Economist*, July 8, 2017.

369 if we found a cure for all cancers tomorrow: *Economist*, Aug. 13, 2016, 14.

369 Of nothing is that more true: Hayflick interview, *Nautilus*, Nov. 24, 2016.

369 "For every year of added life": Lieberman, *Story of the Human Body*, 242.

369 In the United States, the elderly constitute: Davis, *Beautiful Cure*, 139.

370 Zhores Medvedev, a Russian biogerontologist: "Rethinking Modern Theories of Ageing and Their Classification," *Anthropological Review* 80, no. 3 (2017).

370 He discovered that cultured human stem cells: "The Disparity Between Human Cell Senescence In Vitro and Lifelong Replication In Vivo," *Nature Biotechnology*, July 1, 2002.

371 A study by geneticists at the University of Utah: University of Utah Genetic Science Learning Center report, "Are Telomeres the Key to Aging and Cancer?"

371 "If all aging was due to telomeres": "You May Have More Control over Aging Than You Think . . . ," *Stat*, Jan. 3, 2017.

372 Most of us would almost certainly: Harman obituary, *New York Times*, Nov. 28, 2014.

372 "It is a massive racket": "Myths That Will Not Die," *Nature*, Dec. 17, 2015; "No Truth to the Fountain of Youth," *Scientific American*, Dec. 29, 2008.

372 "antioxidant supplementation did not lower": "The Free Radical Theory of Aging Revisited," *Antioxidants and Redox Signaling* 19, no. 8 (2013).

373 After the age of forty, the volume of blood: Nuland, *How We Die*, 53.

373 At least two species of whales: *Naked Scientists*, podcast, Feb. 7, 2017.

374 Two principal theories have been advanced: Bainbridge, *Middle Age*, 208–11.

374 It is a myth, incidentally, that menopause: Ibid., 199.

374 A study by the Albert Einstein College of Medicine: *Scientific American*, Sept. 2016, 58.

375 only about one person in ten thousand: "The Patient Talks Back," *New York Review of Books*, Oct. 23, 2008.

375 The Gerontology Research Group: "Keeping Track of the Oldest People in the World," *Smithsonian*, July 8, 2014.

376 Costa Ricans have only about one-fifth: Marchant, *Cure*, 206–11.

378 she might have been suffering: *Literary Review*, Aug. 2016, 35.

378 about 30 percent of elderly people: "Tau Protein—Not Amyloid—May Be Key Driver of Alzheimer's Symptoms," *Science*, May 11, 2016.

379 Virtuous living doesn't eliminate: "Our Amazingly Plastic Brains," *Wall Street Journal,* Feb. 6, 2015.

380 In Britain, dementias cost the National Health Service: *Inside Science,* BBC Radio 4, Dec. 1, 2016.

380 Alzheimer's drugs have a 99.6 percent failure rate: *Chemistry World,* Aug. 2014, 8.

381 Every day, around the world 160,000 people die: World Health Organization statistics.

381 A separate study found evidence: *Journal of Palliative Medicine* 17, no. 3 (2014).

381 Most dying people lose any desire: "What It Feels Like to Die," *Atlantic,* Sept. 9, 2016.

382 Agonal breathing, in which the sufferer: "The Agony of Agonal Respiration: Is the Last Gasp Necessary?," *Journal of Medical Ethics,* June 2002.

382 cancer sufferers receiving palliative care: *Economist,* April 29, 2017, 55.

382 "One review found that": Hatch, *Snowball in a Blizzard,* 7.

382 "A man's corpse looks as though": Nuland, *How We Die,* 122.

383 Some organs function longer than others: "Rotting Reactions," *Chemistry World,* Sept. 2016.

383 decomposition in a sealed coffin: "What's Your Dust Worth?," *London Review of Books,* April 14, 2011.

383 The average grave is visited: *Literary Review,* May 2013, 43.

383 A century ago, only about one person in a hundred: "What's Your Dust Worth?"

BIBLIOGRAPHY

Ackerman, Diane. *A Natural History of the Senses*. London: Chapmans, 1990.

Alberti, Fay Bound. *Matters of the Heart: History, Medicine, and Emotion*. Oxford: Oxford University Press, 2010.

Alcabes, Philip. *Dread: How Fear and Fantasy Have Fueled Epidemics from the Black Death to Avian Flu*. New York: PublicAffairs, 2009.

Al-Khalili, Jim, and Johnjoe McFadden. *Life on the Edge: The Coming Age of Quantum Biology*. London: Bantam Press, 2014.

Allen, John S. *The Lives of the Brain: Human Evolution and the Organ of Mind*. Cambridge, Mass.: Belknap Press, 2009.

Amidon, Stephen, and Thomas Amidon. *The Sublime Engine: A Biography of the Human Heart*. New York: Rodale, 2011.

Andrews, Michael. *The Life That Lives on Man*. London: Faber & Faber, 1976.

Annas, George J., and Michael A. Grodin. *The Nazi Doctors and the Nuremberg Code: Human Rights in Human Experimentation*. Oxford: Oxford University Press, 1992.

Arikha, Noga. *Passions and Tempers: A History of the Humours*. London: Ecco/HarperCollins, 2007.

Armstrong, Sue. *The Gene That Cracked the Cancer Code*. London: Sigma, 2014.

Arney, Kat. *Herding Hemingway's Cats: Understanding How Our Genes Work*. London: Bloomsbury Sigma, 2016.

Ashcroft, Frances. *Life at the Extremes: The Science of Survival*. London: HarperCollins, 2000.

———. *The Spark of Life: Electricity in the Human Body*. London: Allen Lane, 2012.

Ashwell, Ken. *The Brain Book: Development, Function, Disorder, Health*. Buffalo: Firefly Books, 2012.

Bainbridge, David. *Beyond the Zonules of Zinn: A Fantastic Journey Through Your Brain.* Cambridge, Mass.: Harvard University Press, 2008.

——. *Middle Age: A Natural History.* London: Portobello Books, 2012.

——. *Teenagers: A Natural History.* London: Portobello Books, 2009.

——. *A Visitor Within: The Science of Pregnancy.* London: Weidenfeld & Nicolson, 2000.

——. *The X in Sex: How the X Chromosome Controls Our Lives.* Cambridge, Mass.: Harvard University Press, 2003.

Bakalar, Nicholas. *Where the Germs Are: A Scientific Safari.* New York: John Wiley & Sons, 2003.

Ball, Philip. *Bright Earth: The Invention of Colour.* London: Viking, 2001.

——. *H_2O: A Biography of Water.* London: Phoenix Books, 1999.

——. *Stories of the Invisible: A Guided Tour of Molecules.* London: Oxford University Press, 2001.

Barnett, Richard. *Medical London: City of Diseases, City of Cures.* Edited by Mike Jay. London: Strange Attractor Press, 2008.

Bathurst, Bella. *Sound: Stories of Hearing Lost and Found.* London: Profile/Wellcome, 2017.

Beckhard, Arthur J., and William D. Crane. *Cancer, Cocaine, and Courage: The Story of Dr. William Halsted.* New York: Messner, 1960.

Ben-Barak, Idan. *The Invisible Kingdom.* New York: Basic Books, 2009.

——. *Why Aren't We Dead Yet? The Survivor's Guide to the Immune System.* Melbourne: Scribe, 2014.

Bentley, Peter J. *The Undercover Scientist: Investigating the Mishaps of Everyday Life.* London: Random House, 2008.

Berenbaum, May R. *Bugs in the System: Insects and Their Impact on Human Affairs.* Reading, Mass.: Helix Books, 1995.

Birkhead, Tim. *The Most Perfect Thing: Inside (and Outside) a Bird's Egg.* London: Bloomsbury, 2016.

Black, Conrad. *Franklin Delano Roosevelt: Champion of Freedom.* London: Weidenfeld & Nicolson, 2003.

Blakelaw, Colin, and Sheila Jennett, eds. *The Oxford Companion to the Body.* Oxford: Oxford University Press, 2001.

Blaser, Martin. *Missing Microbes: How Killing Bacteria Creates Modern Plagues.* London: Oneworld, 2014.

Bliss, Michael. *The Discovery of Insulin.* Edinburgh: Paul Harris, 1983.

Blodgett, Bonnie. *Remembering Smell: A Memoir of Losing–and Discovering–the Primal Sense.* Boston: Houghton Mifflin Harcourt, 2010.

Blumberg, Mark S. *Body Heat: Temperature and Life on Earth.* Cambridge, Mass.: Harvard University Press, 2002.

Bondeson, Jan. *The Two-Headed Boy, and Other Medical Marvels.* Ithaca, N.Y.: Cornell University Press, 2000.

Bourke, Joanna. *Fear: A Cultural History.* London: Virago, 2005.

Breslaw, Elaine G. *Lotions, Potions, Pills, and Magic: Health Care in Early America.* New York: New York University Press, 2012.

Bribiescas, Richard G. *Men: Evolutionary and Life History.* Cambridge, Mass.: Harvard University Press, 2006.

Brooks, Michael. *At the Edge of Uncertainty: 11 Discoveries Taking Science by Surprise.* London: Profile Books, 2014.

Burnett, Dean. *The Idiot Brain: A Neuroscientist Explains What Your Head Is Really Up To.* London: Guardian Books, 2016.

Campenbot, Robert B. *Animal Electricity: How We Learned That the Body and Brain Are Electric Machines.* Cambridge, Mass.: Harvard University Press, 2016.

Cappello, Mary. *Swallow: Foreign Bodies, Their Ingestion, Inspiration, and the Curious Doctor Who Extracted Them.* New York: New Press, 2011.

Carpenter, Kenneth J. *The History of Scurvy and Vitamin C.* Cambridge, U.K.: Cambridge University Press, 1986.

Carroll, Sean B. *The Serengeti Rules: The Quest to Discover How Life Works and Why It Matters.* Princeton, N.J.: Princeton University Press, 2016.

Carter, William C. *Marcel Proust: A Life.* New Haven, Conn.: Yale University Press, 2000.

Cassidy, Tina. *Birth: A History.* London: Chatto & Windus, 2007.

Challoner, Jack. *The Cell: A Visual Tour of the Building Block of Life.* Lewes: Ivy Press, 2015.

Cobb, Matthew. *The Egg & Sperm Race: The Seventeenth-Century Scientists Who Unravelled the Secrets of Sex, Life, and Growth.* London: Free Press, 2006.

Cole, Simon. *Suspect Identities: A History of Fingerprinting and Criminal Identification.* Cambridge, Mass.: Harvard University Press, 2001.

Collis, John Stewart. *Living with a Stranger: A Discourse on the Human Body.* London: Macdonald & Janes, 1978.

Crawford, Dorothy H. *Deadly Companions: How Microbes Shaped Our History.* Oxford: Oxford University Press, 2007.

——. *The Invisible Enemy: A Natural History of Viruses.* Oxford: Oxford University Press, 2000.

Crawford, Dorothy H., Alan Rickinson, and Ingólfur Johannessen. *Cancer Virus: The Story of Epstein-Barr Virus.* Oxford: Oxford University Press, 2014.

Crick, Francis. *What Mad Pursuit: A Personal View of Scientific Discovery.* London: Weidenfeld & Nicolson, 1989.

Cunningham, Andrew. *The Anatomist Anatomis'd: An Experimental Discipline in Enlightenment Europe.* London: Ashgate, 2010.

Darwin, Charles. *The Expression of the Emotions in Man and Animals.* London: John Murray, 1872.

Daudet, Alphonse. *In the Land of Pain.* London: Jonathan Cape, 2002.

Davies, Jamie A. *Life Unfolding: How the Human Body Creates Itself.* Oxford: Oxford University Press, 2014.

Davis, Daniel M. *The Beautiful Cure: Harnessing Your Body's Natural Defences.* London: Bodley Head, 2018.

——. *The Compatibility Gene.* London: Allen Lane, 2013.

Dehaene, Stanislas. *Consciousness and the Brain: Deciphering How the Brain Codes Our Thoughts.* London: Viking, 2014.

Dittrich, Luke. *Patient H.M.: A Story of Memory, Madness, and Family Secrets.* London: Chatto & Windus, 2016.

Dormandy, Thomas. *The Worst of Evils: The Fight Against Pain.* New Haven, Conn.: Yale University Press, 2006.

Draaisma, Douwe. *Forgetting: Myths, Perils, and Compensations*. New Haven, Conn.: Yale University Press, 2015.

Dunn, Rob. *The Wild Life of Our Bodies: Predators, Parasites, and Partners That Shape Who We Are Today*. New York: HarperCollins, 2011.

Eagleman, David. *The Brain: The Story of You.* Edinburgh: Canongate, 2016.

——. *Incognito: The Secret Lives of the Brain.* New York: Pantheon Books, 2011.

El-Hai, Jack. *The Lobotomist: A Maverick Medical Genius and His Tragic Quest to Rid the World of Mental Illness.* New York: Wiley & Sons, 2005.

Emsley, John. *Nature's Building Blocks: An A–Z Guide to the Elements.* Oxford: Oxford University Press, 2001.

Enders, Giulia. *Gut: The Inside Story of Our Body's Most Under-Rated Organ.* London: Scribe, 2015.

Epstein, Randi Hutter. *Get Me Out: A History of Childbirth from the Garden of Eden to the Sperm Bank.* New York: W. W. Norton, 2010.

Fenn, Elizabeth A. *Pox Americana: The Great Smallpox Epidemic of 1775–82.* Stroud, Gloucestershire: Sutton, 2004.

Finger, Stanley. *Doctor Franklin's Medicine.* Philadelphia: University of Pennsylvania Press, 2006.

Foreman, Judy. *A Nation in Pain: Healing Our Biggest Health Problem.* New York: Oxford University Press, 2014.

Francis, Gavin. *Adventures in Human Being.* London: Profile/Wellcome, 2015.

Froman, Robert. *The Many Human Senses.* London: G. Bell and Sons, 1969.

Garrett, Laurie. *The Coming Plague: Newly Emerging Diseases in a World out of Balance.* New York: Farrar, Straus and Giroux, 1994.

Gawande, Atul. *Better: A Surgeon's Notes on Performance.* London: Profile, 2007.

Gazzaniga, Michael S. *Human: The Science Behind What Makes Us Unique.* New York: Ecco/HarperCollins, 2008.

Gigerenzer, Gerd. *Risk Savvy: How to Make Good Decisions.* London: Allen Lane, 2014.

Gilbert, Avery. *What the Nose Knows: The Science of Scent in Everyday Life.* New York: Crown, 2008.

Glynn, Ian, and Jenifer Glynn. *The Life and Death of Smallpox.* London: Profile Books, 2004.

Goldsmith, Mike. *Discord: The History of Noise.* Oxford: Oxford University Press, 2012.

Goodman, Jordan, Anthony McElligott, and Lara Marks, eds. *Useful Bodies: Humans in the Service of Medical Science in the Twentieth Century.* Baltimore: Johns Hopkins University Press, 2003.

Gould, Stephen Jay. *The Mismeasure of Man.* New York: W. W. Norton, 1981.

Gratzer, Walter. *Terrors of the Table: The Curious History of Nutrition.* Oxford: Oxford University Press, 2005.

Greenfield, Susan. *The Human Brain: A Guided Tour.* London: Weidenfeld & Nicolson, 1997.

Grove, David I. *Tapeworms, Lice, and Prions: A Compendium of Unpleasant Infections*. Oxford: Oxford University Press, 2014.

Hafer, Abby. *The Not-So-Intelligent Designer: Why Evolution Explains the Human Body and Intelligent Design Does Not.* Eugene, Ore.: Cascade Books, 2015.

Hatch, Steven. *Snowball in a Blizzard: The Tricky Problem of Uncertainty in Medicine.* London: Atlantic Books, 2016.

Healy, David. *Pharmageddon.* Berkeley: University of California Press, 2012.

Heller, Joseph, and Speed Vogel. *No Laughing Matter.* London: Jonathan Cape, 1986.

Herbert, Joe. *Testosterone: Sex, Power, and the Will to Win.* Oxford: Oxford University Press, 2015.

Herold, Eve. *Stem Cell Wars: Inside Stories from the Frontlines.* London: Palgrave Macmillan, 2006.

Hill, Lawrence. *Blood: A Biography of the Stuff of Life.* London: Oneworld, 2013.

Hillman, David, and Ulrika Maude. *The Cambridge Companion to the Body in Literature.* Cambridge, U.K.: Cambridge University Press, 2015.

Holmes, Bob. *Flavor: The Science of Our Most Neglected Sense.* New York: W. W. Norton, 2017.

Homei, Aya, and Michael Worboys. *Fungal Disease in Britain and the United States, 1850–2000: Mycoses and Modernity.* Basingstoke: Palgrave Macmillan, 2013.

Ings, Simon. *The Eye: A Natural History.* London: Bloomsbury, 2007.

Inwood, Stephen. *A History of London.* London: Macmillan, 1998.

Jablonski, Nina. *Living Color: The Biological and Social Meaning of Skin Color.* Berkeley: University of California Press, 2012.

——. *Skin: A Natural History.* Berkeley: University of California Press, 2006.

Jackson, Mark. *Asthma: The Biography.* Oxford: Oxford University Press, 2009.

Jones, James H. *Bad Blood: The Tuskegee Syphilis Experiment.* London: Macmillan, 1981.

Jones, Steve. *The Language of the Genes: Biology, History, and the Evolutionary Future.* London: Flamingo, 1994.

——. *No Need for Geniuses: Revolutionary Science in the Age of the Guillotine.* London: Little, Brown, 2016.

Jorgensen, Timothy J. *Strange Glow: The Story of Radiation.* Princeton, N.J.: Princeton University Press, 2016.

Kaplan, Eugene H. *What's Eating You? People and Parasites.* Princeton, N.J.: Princeton University Press, 2010.

Kinch, Michael. *Between Hope and Fear: A History of Vaccines and Human Immunity.* New York: Pegasus Books, 2018.

——. *The End of the Beginning: Cancer, Immunity, and the Future of a Cure.* New York: Pegasus Books, 2019.

——. *A Prescription for Change: The Looming Crisis in Drug Development.* Chapel Hill: University of North Carolina Press, 2016.

Lane, Nick. *Life Ascending: The Ten Great Inventions of Evolution.* London: Profile Books, 2009.

——. *Power, Sex, Suicide: Mitochondria and the Meaning of Life.* Oxford: Oxford University Press, 2005.

Larson, Frances, *Severed: A History of Heads Lost and Heads Found*. London: Granta, 2014.

Lax, Alistair J. *Toxin: The Cunning of Bacterial Poisons*. Oxford: Oxford University Press, 2005.

Lax, Eric. *The Mould in Dr. Florey's Coat*. London: Little, Brown, 2004.

Leavitt, Judith Walzer. *Typhoid Mary: Captive to the Public's Health*. Boston: Beacon Press, 1995.

Le Fanu, James. *The Rise and Fall of Modern Medicine*. London: Abacus, 1999.

——. *Why Us? How Science Discovered the Mystery of Ourselves*. London: Harper Press, 2009.

Lents, Nathan H. *Human Errors: A Panorama of Our Glitches from Pointless Bones to Broken Genes*. Boston: Houghton Mifflin Harcourt, 2018.

Lieberman, Daniel E. *The Evolution of the Human Head*. Cambridge, Mass.: Belknap Press, 2011.

——. *The Story of the Human Body: Evolution, Health, and Disease*. New York: Pantheon Books, 2013.

Linden, David J. *Touch: The Science of Hand, Heart, and Mind*. London: Viking, 2015.

Lutz, Tom. *Crying: The Natural and Cultural History of Tears*. New York: W. W. Norton, 1999.

MacDonald, Betty. *The Plague and I*. London: Hammond, 1948.

Macinnis, Peter. *The Killer Beans of Calabar, and Other Stories*. Sydney: Allen & Unwin, 2004.

Macpherson, Gordon. *Black's Medical Dictionary*. 39th ed. London: A. & C. Black, 1999.

Maddox, John. *What Remains to Be Discovered: Mapping the Secrets of the Universe, the Origins of Life, and the Future of the Human Race*. London: Macmillan, 1998.

Marchant, Jo. *Cure: A Journey into the Science of Mind over Body*. Edinburgh: Canongate, 2016.

Martin, Paul. *Counting Sheep: The Science and Pleasures of Sleep and Dreams*. London: HarperCollins, 2002.

——. *The Sickening Mind: Brain, Behaviour, Immunity, and Disease*. London: HarperCollins, 1997.

McGee, Harold. *On Food and Cooking: The Science and Lore of the Kitchen*. London: Unwin Hyman, 1986.

McNeill, Daniel. *The Face*. London: Hamish Hamilton, 1999.

Medawar, Jean. *A Very Decided Preference: Life with Peter Medawar*. Oxford: Oxford University Press, 1990.

Medawar, P. B. *The Uniqueness of the Individual*. New York: Dover, 1981.

Money, Nicholas P. *The Amoeba in the Room: Lives of the Microbes*. Oxford: Oxford University Press, 2014.

Montagu, Ashley. *The Elephant Man: A Study in Human Dignity*. London: Allison & Busby, 1972.

Morris, Desmond. *Bodywatching: A Field Guide to the Human Species*. London: Jonathan Cape, 1985.

Morris, Thomas. *The Heart of the Matter: A History of the Heart in Eleven Operations*. London: Bodley Head, 2017.

Mouritsen, Ole G., and Klavs Styrbaek. *Umami: Unlocking the Secrets of the Fifth Taste*. New York: Columbia University Press, 2014.

Mukherjee, Siddhartha. *The Emperor of All Maladies: A Biography of Cancer*. London: Fourth Estate, 2011.

——. *The Gene: An Intimate History*. London: Bodley Head, 2016.

Newman, Lucile F., ed. *Hunger in History: Food Shortage, Poverty, and Deprivation*. Oxford: Basil Blackwell, 1990.

Nourse, Alan E. *The Body*. Amsterdam: Time-Life International, 1965.

Nuland, Sherwin B. *How We Die*. London: Chatto & Windus, 1994.

Oakley, Ann. *The Captured Womb: A History of the Medical Care of Pregnant Women*. Oxford: Blackwell, 1984.

O'Hare, Mick, ed. *Does Anything Eat Wasps? And 101 Other Questions*. London: Profile Books, 2005.

O'Malley, Charles D., and J. B. de C. M. Saunders. *Leonardo da Vinci on the Human Body*. New York: Henry Schuman, 1952.

O'Sullivan, Suzanne. *Brainstorm: Detective Stories from the World of Neurology*. London: Chatto & Windus, 2018.

Owen, Adrian. *Into the Grey Zone: A Neuroscientist Explores the Border Between Life and Death*. London: Faber & Faber, 2017.

Pasternak, Charles A. *The Molecules Within Us: Our Body in Health and Disease*. New York: Plenum, 2001.

Pearson, Helen. *The Life Project: The Extraordinary Story of Our Ordinary Lives*. London: Allen Lane, 2016.

Perrett, David. *In Your Face: The New Science of Human Attraction*. London: Palgrave Macmillan, 2010.

Perutz, Max. *I Wish I'd Made You Angry Earlier: Essays on Science, Scientists, and Humanity*. Cold Spring Harbor, N.Y.: Cold Spring Harbor Laboratory Press, 1998.

Peto, James, ed. *The Heart*. New Haven, Conn.: Yale University Press, 2007.

Platoni, Kara. *We Have the Technology: How Biohackers, Foodies, Physicians, and Scientists Are Transforming Human Perception One Sense at a Time*. New York: Basic Books, 2015.

Pollack, Robert. *Signs of Life: The Language and Meanings of DNA*. London: Viking, 1994.

Postgate, John. *The Outer Reaches of Life*. Cambridge, U.K.: Cambridge University Press, 1991.

Prescott, John. *Taste Matters: Why We Like the Foods We Do*. London: Reaktion Books, 2012.

Richardson, Sarah. *Sex Itself: The Search for Male and Female in the Human Genome*. Chicago: University of Chicago Press, 2013.

Ridley, Matt. *Genome: The Autobiography of a Species in 23 Chapters*. London: Fourth Estate, 1999.

Rinzler, Carol Ann. *Leonardo's Foot: How 10 Toes, 52 Bones, and 66 Muscles Shaped the World*. New York: Bellevue Literary Press, 2013.

Roach, Mary. *Bonk: The Curious Coupling of Science and Sex*. New York: W. W. Norton, 2008.

——. *Grunt: The Curious Science of Humans at War*. New York: W. W. Norton, 2016.

——. *Gulp: Adventures in the Alimentary Canal.* New York: W. W. Norton, 2013.
Roberts, Alice. *The Incredible Unlikeliness of Being: Evolution and the Making of Us.* London: Heron Books, 2014.
Roberts, Callum. *The Ocean of Life.* London: Allen Lane, 2012.
Roberts, Charlotte, and Keith Manchester. *The Archaeology of Disease.* 3rd ed. Stroud, Gloucestershire: History Press, 2010.
Roossinck, Marilyn J. *Virus: An Illustrated Guide to 101 Incredible Microbes.* Brighton: Ivy Press, 2016.
Roueché, Berton. *Curiosities of Medicine: An Assembly of Medical Diversions, 1552–1962.* London: Victor Gollancz, 1963.
Rutherford, Adam. *A Brief History of Everyone Who Ever Lived: The Stories in Our Genes.* London: Weidenfeld & Nicolson, 2016.
——. *Creation: The Origin of Life.* London: Viking, 2013.
Sanghavi, Darshak. *A Map of the Child: A Pediatrician's Tour of the Body.* New York: Henry Holt, 2003.
Scerri, Eric. *A Tale of Seven Elements.* Oxford: Oxford University Press, 2013.
Selinus, Olle. *Essentials of Medical Geology: Impacts of the Natural Environment on Public Health.* Amsterdam: Elsevier, 2005.
Sengoopta, Chandak. *The Most Secret Quintessence of Life: Sex, Glands, and Hormones, 1850–1950.* Chicago: University of Chicago Press, 2006.
Shepherd, Gordon M. *Neurogastronomy: How the Brain Creates Flavor and Why It Matters.* New York: Columbia University Press, 2012.
Shorter, Edward. *Bedside Manners: The Troubled History of Doctors and Patients.* London: Viking, 1986.
Shubin, Neil. *The Universe Within: A Scientific Adventure.* London: Allen Lane, 2012.
——. *Your Inner Fish: A Journey into the 3.5-Billion-Year History of the Human Body.* London: Allen Lane, 2008.
Sinnatamby, Chummy S. *Last's Anatomy: Regional and Applied.* London: Elsevier, 2006.
Skloot, Rebecca. *The Immortal Life of Henrietta Lacks.* London: Macmillan, 2010.
Smith, Anthony. *The Body.* London: Allen and Unwin, 1968.
Spence, Charles. *Gastrophysics: The New Science of Eating.* London: Viking, 2017.
Spiegelhalter, David. *Sex by Numbers: The Statistics of Sexual Behaviour.* London: Profile/Wellcome, 2015.
Stark, Peter. *Last Breath: Cautionary Tales from the Limits of Human Endurance.* New York: Ballantine Books, 2001.
Starr, Douglas. *Blood: An Epic History of Medicine and Commerce.* London: Little, Brown, 1999.
Sternberg, Eliezer J. *NeuroLogic: The Brain's Hidden Rationale Behind Our Irrational Behavior.* New York: Pantheon Books, 2015.
Stossel, Scott. *My Age of Anxiety: Fear, Hope, Dread, and the Search for Peace of Mind.* London: William Heinemann, 2014.
Tallis, Raymond. *The Kingdom of Infinite Space: A Fantastical Journey Around Your Head.* London: Atlantic Books, 2008.
Taylor, Jeremy. *Body by Darwin: How Evolution Shapes Our Health and Transforms Medicine.* Chicago: University of Chicago Press, 2015.

Thwaites, J. G. *Modern Medical Discoveries*. London: Routledge & Kegan Paul, 1958.

Timmermann, Carsten. *A History of Lung Cancer: The Recalcitrant Disease*. London: Palgrave/Macmillan, 2014.

Tomalin, Claire. *Samuel Pepys: The Unequalled Self*. London: Viking, 2002.

Trumble, Angus. *The Finger: A Handbook*. London: Yale University Press, 2010.

Tucker, Holly. *Blood Work: A Tale of Medicine and Murder in the Scientific Revolution*. New York: W. W. Norton, 2011.

Ungar, Peter S. *Evolution's Bite: A Story of Teeth, Diet, and Human Origins*. Princeton, N.J.: Princeton University Press, 2017.

Vaughan, Adrian. *Isambard Kingdom Brunel: Engineering Knight-Errant*. London: John Murray, 1991.

Vogel, Steven. *Life's Devices: The Physical World of Animals and Plants*. Princeton, N.J.: Princeton University Press, 1988.

Wall, Patrick. *Pain: The Science of Suffering*. London: Weidenfeld & Nicolson, 1999.

Welch, Gilbert. *Less Medicine, More Health: Seven Assumptions That Drive Too Much Medical Care*. Boston: Beacon Press, 2015.

West, Geoffrey. *Scale: The Universal Laws of Life and Death in Organisms, Cities, and Companies*. London: Weidenfeld & Nicolson, 2017.

Wexler, Alice. *The Woman Who Walked into the Sea: Huntington's and the Making of a Genetic Disease*. New Haven, Conn.: Yale University Press, 2008.

Williams, Peter, and David Wallace. *Unit 731: The Japanese Army's Secret of Secrets*. London: Hodder & Stoughton, 1989.

Winston, Robert. *The Human Mind: And How to Make the Most of It*. London: Bantam Press, 2003.

Wolf, Fred Alan. *The Body Quantum: The New Physics of Body, Mind, and Health*. New York: Macmillan, 1986.

Wolpert, Lewis. *You're Looking Very Well: The Surprising Nature of Getting Old*. London: Faber and Faber, 2011.

Wootton, David. *Bad Medicine: Doctors Doing Harm Since Hippocrates*. Oxford: Oxford University Press, 2006.

Wrangham, Richard. *Catching Fire: How Cooking Made Us Human*. London: Profile Books, 2009.

Yong, Ed. *I Contain Multitudes*. London: Bodley Head, 2016.

Zeman, Adam. *Consciousness: A User's Guide*. New Haven, Conn.: Yale University Press, 2002.

———. *A Portrait of the Brain*. New Haven, Conn.: Yale University Press, 2008.

Zimmer, Carl. *Microcosm:* E. coli *and the New Science of Life*. New York: Pantheon Books, 2008.

———. *A Planet of Viruses*. Chicago: University of Chicago Press, 2011.

———. *Soul Made Flesh: The Discovery of the Brain—and How It Changed the World*. London: William Heinemann, 2004.

Zuk, Marlene. *Paleofantasy: What Evolution Really Tells Us About Sex, Diet, and How We Live*. New York: W. W. Norton, 2013.

———. *Riddled with Life: Friendly Worms, Ladybug Sex, and the Parasites That Make Us Who We Are*. Orlando, Fla.: Harvest/Harcourt, 2007.

INDEX

ILLUSTRATION CREDITS

Page 1 Royal Collection Trust © Her Majesty Queen Elizabeth II, 2019 / Bridgeman Images

Page 2 top: © Photo Researchers / Mary Evans Picture Library
bottom left: Wolf Suschitzky / The LIFE Images Collection / Getty Images
bottom right: Granger / Bridgeman Images

Page 3 top: Bettmann / Getty Images
bottom left: Wellcome Collection
bottom right: Mütter Museum of the College of Physicians of Philadelphia

Page 4 top: Granger / Bridgeman Images
bottom left: Nationaal Archief / Collectie Spaarnestad / ANP / Bridgeman Images
bottom right: Louis Washkansky, Popperfoto / Getty Images

Page 5 top: Wellcome Collection
bottom left: Minneapolis Public Library Collection, Audio-Visual Department, Abraham Lincoln Presidential Library & Museum
bottom right: Keystone-France / Getty Images

Page 6 top: Wellcome Collection
bottom left: Bridgeman Images
bottom right: © SZ Photo / Scherl / Bridgeman Images

Page 7 top: Hulton Archive / Getty Images
bottom: Wellcome Images

Page 8 top left: © Ken Welsh / Bridgeman Images

top right: Wellcome Collection
bottom: © King's College London / Mary Evans Picture Library

Page 9 top left: Wellcome Collection
top right: Bettmann / Getty Images
bottom: Bettmann / Getty Images

Page 10 top: Topham Picturepoint © 1999
bottom: Wallace Kirkland / Getty Images

Page 11 top: Granger / Bridgeman Images
bottom: Keystone-France / Getty Images

Page 12 top left: Heritage Image Partnership Ltd /Alamy Stock Photo
top right: INTERFOTO / Alamy Stock Photo
bottom: Ernst Gräfenberg, Museum of Contraception and Abortion, Vienna

Page 13 top: Neil Harding / Getty Images
center: Dr Yorgos Nikas / Science Photo Library/Getty Images
bottom: Science History Images / Alamy Stock Photo

Page 14 top: Wellcome Collection
center: Keystone-France / Getty Images
bottom: Science History Images / Alamy Stock Photo

Page 15 top: Wellcome Collection
bottom: Hulton-Deutsch Collection / Getty Images

Page 16 bottom left: Getty Images
bottom right: Science History Images / Alamy Stock Photo

ABOUT THE AUTHOR

Bill Bryson's bestselling books include *A Walk in the Woods, Notes from a Small Island, I'm a Stranger Here Myself, In a Sunburned Country, A Short History of Nearly Everything* (which earned him the 2004 Aventis Prize and the European Union's Descartes Prize), *The Life and Times of the Thunderbolt Kid, At Home,* and *One Summer.* He is a former chancellor of Durham University and is an honorary fellow of the Royal Society of London, the Royal Society of Chemistry, and the Kavli Institute of Particle Physics at the University of California at Santa Barbara. He lives in England with his wife.